LHC Physics

Scottish Graduate Series

LHC Physics

Edited by

T. Binoth • C. Buttar

P. J. Clark • E.W.N. Glover

CRC Press
Taylor & Francis Group
Boca Raton London New York

CRC Press is an imprint of the
Taylor & Francis Group, an **informa** business

A TAYLOR & FRANCIS BOOK

CRC Press
Taylor & Francis Group
6000 Broken Sound Parkway NW, Suite 300
Boca Raton, FL 33487-2742

First issued in paperback 2019

© 2012 by The Scottish Universities Summer School in Physics and/or the individual contributors
CRC Press is an imprint of Taylor & Francis Group, an Informa business

No claim to original U.S. Government works

ISBN-13: 978-1-4398-3770-2 (hbk)
ISBN-13: 978-0-367-38146-2 (pbk)

Library of Congress Cataloging-in-Publication Data

LHC physics / editors, T. Binoth ... [et al.].
  p. cm. -- (Scottish graduate series)
  Includes bibliographical references and index.
  ISBN 978-1-4398-3770-2 (hardback)
  1. Large Hadron Collider (France and Switzerland) 2. Particles (Nuclear physics) I. Binoth, T.

QC787.P73L44 2012
539.7'36--dc23   2012000319

Visit the Taylor & Francis Web site at
http://www.taylorandfrancis.com

and the CRC Press Web site at
http://www.crcpress.com

SUSSP Proceedings

SUSSP Proceedings (continued)

Lecturers

Keith Ellis	Fermi National Accelerator Laboratory, Batavia, Illinois, USA.
Sven Heinemeyer	Instituto de Fisica de Cantabria (CSIC), Santander, Spain.
Torbjorn Sjostrand	Lund University, Lund, Sweden.
Gino Isidori	Laboratori Nazionali, Frascati, Italy.
John Ellis	CERN, Geneva, Switzerland.
Gustaaf Brooijmans	Columbia University, New York, USA.
Raimond Snellings	NIKHEF, Amsterdam, the Netherlands.
Gunther Dissertori	Eidgenössiche Technische Hochschule Zürich, Zürich, Switzerland.
Albert De Roeck	CERN, Geneva, Switzerland and Antwerp University, Antwerp, Belgium.
Philippe Lebrun	CERN, Geneva, Switzerland.
Glen Cowan	Royal Holloway, University of London, UK.
Philippe Charpentier	CERN, Geneva, Switzerland.

The students of the school (alphabetical order)

Henso ABREU
Gauthier ALEXANDRE
Steven BEALE
Catherine BERNACIAK
Maria Ilaria BESANA
Maria BORGIA
Cristina BOTTA
Bruno BOYER
Ayse CAGIL
Maria CEPEDA
Greig COWAN
Dirk DAMMANN
Valerio DAO
Michel DE CIAN
Frederico DE GUIO
Alice DECHAMBRE
Francesco DETTTORI
Giovanni DIANA
Alexandru DOBRIN
Caterina DOGLIONI
Alessia D'ORAZIO
Giacomo DOVIER
Jula DRAEGER
Eliane EPPLE
Erica FANCHINI

Gemma FARDELL
Marco FELICIANGELI
Sylvain FICHET
Conor FITZPATRICK
Alistair GEMMELL
Paola GIOVANNINI
Marc GRABALOSA
Philippe GROS
Stefan GUINDON
Anna HENRICHS
Elizabeth HINES
Florian HIRSCH
Laura JEANTY
Subrata KHAN
Vasiliki KOUSKOURA
Josh KUNKLE
Vicente LACUESTA
Eve LE MENEDEU
Tatjana LENZ
Grazia LUPARELLO
Pawel MALECKI
Anja MARUNOVIC
Ioannis NOMIDIS
Cristina OROPEZA BARRERA
Alexander PENSON

Andreas PETRIDIS
Michele PINAMONTI
Sahill PODDAR
Eugenia PUCCIO
Albert PUIG
Thomas PUNZ
Michael RAMMENSEE
Ann-Karin SANCHEZ
Javier SANTAOLALLA
Manuel SCHILLER
Evelyn SCHMIDT
Genevieve STEELE
Benjamin STIEGER
Kathrin STOERIG
Kristaq SUXO
Hajrah TABASSAM
Daniele TROCINO
Dustin URBANIEC
Naomi VAN DER KOLK
Liv WIIK
Evan WULF
Howard Wells WULSIN
Ross YOUNG
Carolin ZENDLER
Giovanni ZEVI DELLA PORTA

Organising Committee

Professor A. Doyle	University of Glasgow	*Co-Director*
Professor N. Glover	IPPP, University of Durham	*Co-Director*
Dr V. Martin	University of Edinburgh	*Secretary*
Prof. C.D. Froggatt	University of Glasgow	*Treasurer*
Dr S. Eisenhardt	University of Edinburgh	*Secretary and Treasurer (deputising)*
Dr T. Binoth	University of Edinburgh	*Co-Editor*
Dr P. Clark	University of Edinburgh	*Co-Editor*
Dr C. Buttar	University of Glasgow	*Co-Editor*
Prof. P. Clarke	University of Edinburgh	*Co-Steward*
Dr C. Parkes	University of Glasgow	*Co-Steward*
Dr S. Scott	University of Glasgow	*Administration*

Preface

The Standard Model of particle physics was established 40 years ago and describes in fantastic detail the wealth of experimental data from high-energy particle colliders. However, a key ingredient of this theory, the Higgs boson, has so far remained elusive. The Large Hadron Collider (LHC), constructed over the last decade at CERN in Geneva, is the highest-energy collider ever built. It is designed to open up the new territory of TeV-scale (Terascale) physics, where ground-breaking discoveries are expected. With first collisions expected in late 2009, the summer of 2009 was an opportune moment to study LHC phenomenology with an emphasis both on the first years of data taking at the LHC, and on the experimental and theoretical tools needed to exploit that potential.

The lectures and discussion classes included an introduction to the theoretical and phenomenological framework of hadron collisions, and current theoretical models of frontier physics, as well as overviews of the main detector components, the initial calibration procedures and physics samples, and early LHC results. Explicit examples of physics analyses were drawn from the current Tevatron experience to help inform these exchanges. The school covered all these at a pedagogical level, starting with a basic introduction to the Standard Model and its most likely extensions. Theoretical training was supplemented by courses on the detector capabilities and search strategies. In summary, the aim of the school was to equip young particle physicists with the basic tools to extract the maximum benefit from the various LHC experiments.

Following the pattern many recent successful schools, we held the school in St. Andrews in August 2009, using the facilities of the Chemistry Department and the John Burnet Hall of Residence. This location is ideal for a school of this size (72 students and 20 lecturers and other staff) and character. The 72 participants came from the UK and Europe (Belgium, Croatia, France, Germany, Greece, Italy, the Netherlands, Poland, Spain, Sweden, Switzerland and Turkey), as well as the USA and India. There was a good gender balance (for a physics summer school at least) with 32 of the 72 being female.

There was ample opportunity for discussion and exposition, with participants (both students and lecturers) able to discuss the contents of the lectures and related issues throughout the school. The informal and relaxed atmosphere of St. Andrews and the John Burnet Hall contributed enormously to fostering this interaction. The lecturers all contributed hugely to the success of the school, based on their acknowledged world prominence and international impact in their research area, proven ability to deliver interesting and lively lectures and commitment to contribute meaningfully to discussions with the students.

These proceedings provide a unique insight into what was discussed in the 32 lectures and five evening discussion classes. They incorporate the theoretical foundations, anticipated experimental signatures and analysis tools that will be needed by all those interested in LHC physics. They provide a thorough introduction to the physics of the LHC for those inspired by the scale of the endeavour but, perhaps, overawed by its scale and complexity. The LHC provides a rich seam of physics, building upon the wealth of data from the Tevatron. These proceedings are a useful guide to the relevant areas of the Standard Model, which will be challenged in the coming years by the data from the LHC, and where we may find vital evidence as to where to look beyond it. We hope that you will find all that you want to know about the LHC, and more, in this book.

It is becoming increasingly important that we take our message to the public. Consequently, on Wednesday 26 August, we held a schools outreach event in the afternoon, with a lecture for the general public in the evening. Altogether over 150 students attended the afternoon event. Hands-on displays and experiments were provided by the Particle Physics for Scottish Schools project (PP4SS) which is supported by the School of Physics and Astronomy, the University of Edinburgh and by STFC. The exhibition was staffed by 10 volunteer research students from the School and experienced demonstrators familiar with the PP4SS setups. The audience was encouraged to try out the displays and experiments and was naturally drawn into discussions about the topics on display and also others often further afield. The main lecture (repeated to an audience of over 100 in the evening) was given by Professor Brian Cox (Manchester University) and titled "CERN's Big Bang Machine: The Large Hadron Collider." As one might expect, it was an outstanding event and thoroughly enjoyed by all who attended.

Summer Schools are not just about science; they are about dialogue, discussion, meeting people and forming lifelong friendships. The School succeeded in this secondary aim, aided by a full social programme. The SUSSP has a tradition of hard work accompanied by the opportunity to sample Scottish culture, through trips to local castles (Dunottar, Glamis and Falkland Palace), a memorable whisky tasting led by David Wishart, a traditional ceilidh, a visit to Edinburgh, Haggis, pipers and, of course, putting on the "Himalayas" mini golf course. For many, the highlight was Peter Higgs attending the School Banquet, and giving many the opportunity of talking and being photographed with him. The stand-out feature at the banquet was the after-dinner speech by Alan Walker, who gave a very humorous and entertaining account of the history and traditions of the SUSSP.

The feedback on the school was enormously positive. 100% of students found the school either "extremely useful" or "very useful." 100% of the students either "strongly agreed" or "agreed" with the statement that the school had better equipped them with the tools to extract the maximum benefit from the LHC. We thank our co-organisers Thomas Binoth, Phil Clark, Peter Clarke, Stephan Eisenhardt, Victoria Martin, Leanne O'Donnell, Craig Buttar, Colin Froggatt and Suzanne Scott, who all ensured that the School was a success.

We gratefully acknowledge the support of the UK Science and Technology Facilities Council (STFC), the European Science Foundation (ESF), the Scottish Universities Physics Alliance (SUPA), the Institute for Particle Physics Phenomenology in Durham, the Scottish Universities Summer Schools in Physics (particularly Ken Bowler and Alan Walker), the Institute of Physics, the Physics and Astronomy Departments of the Universities of Edinburgh and Glasgow and Chemistry Department of St. Andrews without which the school would not have been possible.

Shortly after the School, on 3 January 2010, we were all saddened to hear of the premature and unexpected death of Thomas Binoth in an avalanche in the Diemtigtal Valley, south of Bern, Switzerland. Thomas was instrumental in setting up and creating a lively environment at the summer school as well as being a co-organiser and proceedings editor. Thomas' stimulating questions at the end of many lectures reminded us that we are all, ultimately, students seeking answers to profound questions. We remember Thomas not just as an outstanding inquiring physicist but also as a friend. These proceedings are dedicated, with love and admiration, to Thomas.

Tony Doyle and Nigel Glover
Co-Directors

Left: Alan Walker, Peter Higgs and one of our "students," Sven Heinemeyer at the School Banquet.
Right: Thomas Binoth, Nigel Glover and Tony Doyle putting on the "Himalayas."

Thomas Binoth

This volume is dedicated to the memory of our dear friend and colleague, Thomas Binoth, who died in an avalanche on 3 January 2010, while skiing in the Diemtigtal Valley, 25 miles south of Bern, Switzerland.

Thomas at the top of a mountain in the Ecrins National Parc in the French Alps, Summer 2008. (Photograph by Gudrun Heinrich.)

Contents

Section I: Theoretical Foundations

Perturbative QCD and the Parton Model

R. Keith Ellis

Fermilab, Batavia, IL, USA

1 Definition and Properties of QCD

1.1 Motivation for QCD

The discovery that the strong interactions can be described by a non-Abelian gauge theory based on the colour group SU(3) is one of the triumphs of twentieth century physics. The current focus of research in collider physics regards QCD as an established tool used to predict reaction rates and distributions, so I will be quite brief on the motivation for QCD.

The original motivation for the colour degree of freedom came from the properties of the lightest baryons. As an example we can consider the pion-nucleon resonance, the $\Delta^{++}(1232)$, which has a measured isospin (spin-parity) of $I(J^P) = \frac{3}{2}(\frac{3}{2}^+)$. The quark model views this resonance as made of three up-quarks in a state of zero orbital angular momentum,

$$\Delta^{++}\left(J_z = \frac{3}{2}\right) = |u \uparrow u \uparrow u \uparrow>, \qquad (1)$$

which is therefore symmetric under the isospin, spin and spatial degrees of freedom in apparent violation of Fermi statistics. This difficulty can be removed by introducing a colour degree of freedom. Each quark q_a occurs in three colours $a = r, g, b$ and the observed baryons are in a completely anti-symmetric state with respect to colour. The introduction of the colour degree of freedom, however, raises the question why the spectrum shows no sign of the states which would correspond to excitations of this degree of freedom. Before the advent of QCD and the property of confinement, the colour degree of freedom had to be supplemented with a dictum that only colour singlets survive as observed physical states.

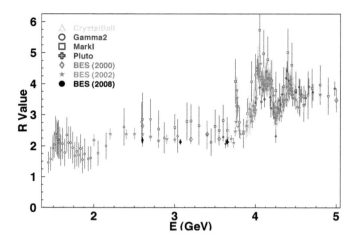

Figure 1. *A compilation of results on R taken from Ref. [1] showing approximate agreement with Eqs. (3,4)*

An additional motivation for the colour degree of freedom came from the total hadronic cross section in e^+e^- annihilation. Consider the ratio R of the e^+e^- total hadronic cross section to the cross section for the production of a pair of point-like, charge-one objects such as muons,

$$R_{e^+e^-} = \frac{\sigma(e^+e^- \to \text{hadrons})}{\sigma(e^+e^- \to \mu^+\mu^-)}. \tag{2}$$

The virtual photon excites all electrically charged constituent-anticonstituent pairs from the vacuum. At low energy the virtual photon excites only the u, d and s quarks, each of which occurs in $N = 3$ colours.

$$R = N\sum_i Q_i^2 = 3\left[\left(\frac{2}{3}\right)^2 + \left(-\frac{1}{3}\right)^2 + \left(-\frac{1}{3}\right)^2\right] = 2. \tag{3}$$

For centre-of-mass energies $E_{\text{cm}} \geq 4$ GeV, one is above the threshold for the production of pairs of c quarks, and so

$$R = 3\left[2 \times \left(\frac{2}{3}\right)^2 + 2 \times \left(-\frac{1}{3}\right)^2\right] = \frac{10}{3}. \tag{4}$$

The data on R are in reasonable agreement with the prediction of the three-colour model as shown in Fig. 1. The agreement is further improved once perturbative strong interactions are included.

Given that there are three colours of quarks and if we assume that we want to describe the interactions between quarks by a non-Abelian gauge

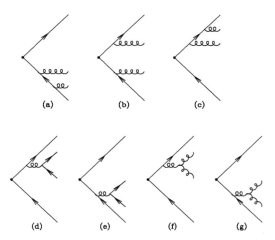

Figure 2. *Feynman diagrams for $e^+e^- \to 4$ jet events. Only the final state of the reaction is shown.*

theory, what gauge groups are acceptable? We shall consider the three groups, $U(3), SU(3)$ and $O(3)$.

The group $U(3)$ would lead to nine gluons, since a unitary 3×3 matrix has nine independent parameters (because it has nine complex parameters, subject to nine real constraints). The ninth gluon $(r\bar{r} + g\bar{g} + b\bar{b})/\sqrt{3}$ would be invariant under U(3) transformations. It would therefore be a colour singlet; it would propagate without confinement leading to a long-range strong force, in contrast with observation. The group $O(3)$ with only three gluons can also be excluded. Since $O(3)$ has real representations, the representation q and $\bar{q}$ transform in an equivalent way under $O(3)$ rotations. The group $O(3)$ would therefore predict that if $q\bar{q}$ states form, qq states should also be bound, leading to mesons of fractional charge.

Finally, four jet events are generated in e^+e^- annihilation due to radiation from a basic quark-antiquark pair as shown in Fig. 2, where the curly lines represent gluons which we shall talk more about in the next section. Let us define N as the number of colours and N_G as the number of gluons. The angular structure of 4-jet events is controlled by the weights of the different diagrams in Fig. 2, which in turn are sensitive to the ratio of N and N_G. For example, the Delphi collaboration [2] obtained the result $N_C/N_G = 0.38 \pm 0.1$, to be compared with the $SU(3)$ $(3/8 = 0.375)$ and $U(3)$ $(3/9 = 0.3333)$.

The acceptance that QCD was the correct theory of strong interactions was not immediate. Once QCD came on the scene it took time before the new features of the theory, like asymptotic freedom, confinement and factorisation, were understood enough that one could calculate physical quantities and gain confidence in the correctness of the theory.

1.2 Lagrangian of QCD

The Feynman rules for perturbative QCD follow from the Lagrangian

$$\mathcal{L} = -\frac{1}{4} F^A_{\alpha\beta} F^{\alpha\beta}_A + \sum_{\text{flavours}} \bar{q}_a (i\not{D} - m)_{ab} q_b + \mathcal{L}_{\text{gauge-fixing}} + \mathcal{L}_{\text{ghost}}. \qquad (5)$$

$F^A_{\alpha\beta}$ is the field strength tensor for the spin-1 gluon field $\mathcal{A}^A_\alpha$,

$$F^A_{\alpha\beta} = \partial_\alpha \mathcal{A}^A_\beta - \partial_\beta \mathcal{A}^A_\alpha - g f^{ABC} \mathcal{A}^B_\alpha \mathcal{A}^C_\beta. \qquad (6)$$

The upper-case indices A, B, C run over the eight colour degrees of freedom of the gluon field. The third "non-Abelian" term in Eq. (6) distinguishes QCD from QED, giving rise to triplet and quartic gluon self-interactions and ultimately to the property of asymptotic freedom. The numbers f^{ABC} $(A, B, C = 1, ..., 8)$ are the structure constants of the SU(3) colour group. The quark fields q_a $(a = 1, 2, 3)$ are in triplet colour representation and D is the covariant derivative, which has the two following forms, depending on whether it acts on a triplet or an octet field:

$$(D_\alpha)_{ab} = \partial_\alpha \delta_{ab} + ig \left(t^C \mathcal{A}^C_\alpha \right)_{ab}, \qquad (7)$$

$$(D_\alpha)_{AB} = \partial_\alpha \delta_{AB} + ig (T^C \mathcal{A}^C_\alpha)_{AB}, \qquad (8)$$

where t and T are matrices in the fundamental (triplet) and adjoint (octet) representations of SU(3), respectively:

$$[t^A, t^B] = i f^{ABC} t^C, \quad [T^A, T^B] = i f^{ABC} T^C, \qquad (9)$$

where $(T^A)_{BC} = -i f^{ABC}$. We use the metric $g^{\alpha\beta} = \text{diag}(1, -1, -1, -1)$ and set $\hbar = c = 1$. $\not{D}$ is symbolic notation for $\gamma^\alpha D_\alpha$. The normalisation of the t matrices is chosen so that

$$\text{Tr}\, t^A t^B = T_R\, \delta^{AB}, \quad T_R = \frac{1}{2}, \qquad (10)$$

and the colour matrices obey the relations

$$\sum_A t^A_{ab} t^A_{bc} = C_F\, \delta_{ac}, \quad C_F = \frac{N^2 - 1}{2N},$$

$$\sum_A t^A_{ab} t^A_{cd} = \frac{1}{2} \left[\delta_{ad} \delta_{cb} - \frac{1}{N} \delta_{ab} \delta_{cd} \right],$$

$$\text{Tr}\, T^C T^D = \sum_{A,B} f^{ABC} f^{ABD} = C_A\, \delta^{CD}, \quad C_A = N. \qquad (11)$$

Thus $C_F = \frac{4}{3}$ and $C_A = 3$ for SU(3). The QCD coupling strength is defined (in analogy with the fine structure constant of electrodynamics) as $\alpha_S \equiv g^2/4\pi$.

1.3 Gauge Invariance

The QCD Lagrangian is invariant under local gauge transformations. That is, one can redefine the quark fields independently at every point in space-time,

$$q_a(x) \to q_a'(x) = \exp(it \cdot \theta(x))_{ab} q_b(x) \equiv \Omega(x)_{ab} q_b(x), \tag{12}$$

without changing physical content of the theory. The covariant derivative is so called because it transforms under gauge transformations in the same way as the quark field itself,

$$D_\alpha q(x) \to D_\alpha' q'(x) \equiv \Omega(x) D_\alpha q(x), \tag{13}$$

where we omit the colour labels of quark fields from now on. We can use Eq. (13) to derive the transformation of the gluon field $\mathcal{A}$ under local gauge transformations,

$$\begin{aligned}
D_\alpha' q'(x) &= \big(\partial_\alpha + igt \cdot \mathcal{A}_\alpha'\big)\Omega(x)q(x), \\
&\equiv (\partial_\alpha \Omega(x))q(x) + \Omega(x)\partial_\alpha q(x) + igt \cdot \mathcal{A}_\alpha' \Omega(x)q(x),
\end{aligned} \tag{14}$$

where $t \cdot \mathcal{A}_\alpha \equiv \sum_A t^A \mathcal{A}_\alpha^A$. Hence

$$t \cdot \mathcal{A}_\alpha' = \Omega(x)t \cdot \mathcal{A}_\alpha \Omega^{-1}(x) + \frac{i}{g}\big(\partial_\alpha \Omega(x)\big)\Omega^{-1}(x). \tag{15}$$

The transformation property of gluon field strength $F_{\alpha\beta}$ in Eq. (6) is

$$t \cdot F_{\alpha\beta}(x) \to t \cdot F_{\alpha\beta}'(x) = \Omega(x)F_{\alpha\beta}(x)\Omega^{-1}(x). \tag{16}$$

This should be contrasted with the gauge-invariance of the QED field strength. The QCD field strength is not gauge invariant because of self-interactions of the gluons. The carriers of the colour force are themselves coloured, unlike the electrically neutral photon. The first term in Eq. (5) can be written as a trace in colour space,

$$-\frac{1}{4}F_{\alpha\beta}^A F_A^{\alpha\beta} \equiv -\frac{1}{2}\mathrm{Tr}\, t \cdot F_{\alpha\beta}\, t \cdot F^{\alpha\beta}, \tag{17}$$

which is manifestly invariant under the transformation of Eq. (16). Note also that there is no gauge-invariant way of including a gluon mass. A term such as

$$m^2 \mathcal{A}^\alpha \mathcal{A}_\alpha \tag{18}$$

is not gauge invariant. This is similar to the QED result for mass of the photon. On the other hand the quark mass term in Eq. (6) is gauge invariant.

1.4   Feynman Rules

We can use the free piece of QCD Lagrangian, Eq. (5), to obtain the quark and gluon propagators. For example the inverse gluon propagator is determined by the action

$$-S = -i \int d^4x \, \mathcal{L}. \tag{19}$$

Introducing the Fourier transform of the gluon field,

$$\mathcal{A}_\mu^A(p) = \frac{1}{(2\pi)^4} \int d^4x \, \mathcal{A}_\mu^A(x) \exp(-ip \cdot x), \tag{20}$$

we find that the piece of the action quadratic in the gluon field can be written as

$$-S = \frac{1}{2}(2\pi)^4 \int d^4p \, \mathcal{A}_A^\alpha(p) \, \Gamma^{(2)}_{\{AB,\,\alpha\beta\}}(p) \, \mathcal{A}_B^\beta(p). \tag{21}$$

The gluon propagator, the inverse of $\Gamma^{(2)}_{\{AB,\,\alpha\beta\}}$, is impossible to define without the gauge-fixing term. The choice

$$\mathcal{L}_{\text{gauge-fixing}} = -\frac{1}{2\,\lambda}\left(\partial^\alpha \mathcal{A}_\alpha^A\right)^2 \tag{22}$$

defines *covariant gauges* with gauge parameter λ. The inverse gluon propagator is then

$$\Gamma^{(2)}_{\{AB,\,\alpha\beta\}}(p) = i\delta_{AB}\left[p^2 g_{\alpha\beta} - (1 - \frac{1}{\lambda})p_\alpha p_\beta\right]. \tag{23}$$

Without the gauge-fixing term the function in Eq. (23) would have no inverse. This is a consequence of the fact that we have too many degrees of freedom, related by gauge transformations, all of which are physically equivalent. After introducing the gauge-fixing parameter the resulting propagator is given in Table 1. The $i\varepsilon$ prescription for the pole of the propagator is determined by causality, as in QED.

All physical results will be independent of the gauge parameter λ because the physical content of the theory is unchanged by the method by which we remove the superfluous degrees of freedom. The choices $\lambda = 1$ or $\lambda = 0$ are referred to as the Feynman and Landau gauges. For convenience, we usually use the Feynman gauge because it simplifies the form of the propagator. In non-Abelian theories like QCD, the covariant gauge-fixing term must be supplemented by a *ghost term* which we do not discuss here. The ghost field, shown by dotted lines in Table 1, cancels unphysical degrees of freedom of gluon which would otherwise propagate in covariant gauges. Note that the propagators are determined from $-S$, and the interactions from S. Thus we have the quark propagator in momentum space obtained by setting $\partial^\alpha = -ip^\alpha$ for an incoming field.

$$\delta^{AB}\left[-g^{\alpha\beta}+(1-\lambda)\frac{p^\alpha p^\beta}{p^2+i\epsilon}\right]\frac{i}{p^2+i\epsilon}$$

$$\delta^{AB}\frac{i}{(p^2+i\epsilon)}$$

$$\delta^{ab}\frac{i}{(\not{p}-m+i\epsilon)_{ji}}$$

$$-g\ f^{ABC}[(p-q)^\gamma g^{\alpha\beta}+(q-r)^\alpha g^{\beta\gamma}+(r-p)^\beta g^{\gamma\alpha}]$$
$$\text{(all momenta incoming)}$$

$$-ig^2\ f^{XAC}f^{XBD}\left[g^{\alpha\beta}g^{\gamma\delta}-g^{\alpha\delta}g^{\beta\gamma}\right]$$
$$-ig^2\ f^{XAD}f^{XBC}\left[g^{\alpha\beta}g^{\gamma\delta}-g^{\alpha\gamma}g^{\beta\delta}\right]$$
$$-ig^2\ f^{XAB}f^{XCD}\left[g^{\alpha\gamma}g^{\beta\delta}-g^{\alpha\delta}g^{\beta\gamma}\right]$$

$$g\ f^{ABC}q^\alpha$$

$$-ig\ (t^A)_{cb}\ (\gamma^\alpha)_{ji}$$

Table 1. *Feynman rules for QCD.*

1.5 Asymptotic Freedom

1.5.1 Running Coupling

In order to introduce the concept of the running coupling, consider a dimensionless physical observable R which depends on a single large energy scale, Q. The ratio of the hadronic cross section to the muon pair production cross section in e^+e^- annihilation is an example of such a variable. We assume that $Q \gg m$ where m is any mass, so that we can set $m \to 0$ (assuming this limit exists). Dimensional analysis then suggests that R should be independent of Q.

This simple scale invariance is, however, not true in quantum field theory. The calculation of R as a perturbation series in the coupling (which in QCD is

$\alpha_S = g^2/4\pi$) requires renormalisation to remove ultraviolet divergences. This introduces a second mass scale μ, which is the scale at which subtractions to remove divergences are performed. Therefore R depends also on the ratio Q/μ and is not constant. The renormalised coupling α_S also depends on μ.

The choice of the scale μ is completely arbitrary. Therefore, if we hold the bare coupling fixed, the ratio R cannot depend on μ. Since R is dimensionless, any dependence on μ can only enter through Q^2/μ^2 or the renormalised coupling α_S. Hence

$$\mu^2 \frac{d}{d\mu^2} R\left(\frac{Q^2}{\mu^2}, \alpha_S\right) \equiv \left[\mu^2 \frac{\partial}{\partial\mu^2} + \mu^2 \frac{\partial\alpha_S}{\partial\mu^2} \frac{\partial}{\partial\alpha_S}\right] R = 0. \qquad (24)$$

Introducing

$$\tau = \ln\left(\frac{Q^2}{\mu^2}\right), \quad \beta(\alpha_S) = \mu^2 \frac{\partial\alpha_S}{\partial\mu^2}, \qquad (25)$$

Eq. (24) can be written as

$$\left[-\frac{\partial}{\partial\tau} + \beta(\alpha_S)\frac{\partial}{\partial\alpha_S}\right] R = 0. \qquad (26)$$

This renormalisation group equation is solved using the running coupling $\alpha_S(Q)$ which is defined implicitly in terms of the β-function.

$$\tau = \int_{\alpha_S}^{\alpha_S(Q)} \frac{dx}{\beta(x)}, \quad \alpha_S(\mu) \equiv \alpha_S. \qquad (27)$$

Using Eq. (27) we can derive a pair of auxiliary results

$$\frac{\partial\alpha_S(Q)}{\partial\tau} = \beta(\alpha_S(Q)), \quad \frac{\partial\alpha_S(Q)}{\partial\alpha_S} = \frac{\beta(\alpha_S(Q))}{\beta(\alpha_S)}. \qquad (28)$$

Hence the solution to Eq. (24) is,

$$R\left(\frac{Q^2}{\mu^2}, \alpha_S\right) = R(1, \alpha_S(Q)). \qquad (29)$$

Thus all scale dependence in R comes from running of $\alpha_S(Q)$. In the following subsection we shall see QCD is asymptotically free, so that $\alpha_S(Q) \to 0$ as $Q \to \infty$. Thus for large Q we can safely use perturbation theory. Then knowledge of $R(1, \alpha_S)$ to fixed order allows us to predict the variation of R with Q.

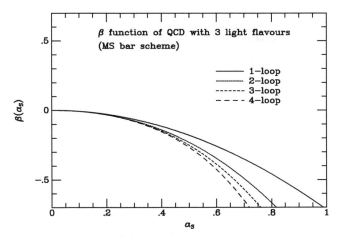

Figure 3. *The β-function of QCD at various loop orders.*

1.5.2 Beta Function

As indicated by Eq. (28) the running of the QCD coupling α_S is determined by the β function, which has the expansion

$$\beta(\alpha_S) \;=\; -b\alpha_S^2(1 + b'\alpha_S) + \mathcal{O}(\alpha_S^4), \tag{30}$$

with

$$b = \frac{(11C_A - 2N_f)}{12\pi}, \qquad b' = \frac{(17C_A^2 - 5C_A N_f - 3C_F N_f)}{2\pi(11C_A - 2N_f)}, \tag{31}$$

and where N_f is the number of "active" light flavours. Terms up to $\mathcal{O}(\alpha_S^5)$ are known [3–7]. The behaviour of the β-function calculated at various loop orders is shown in Fig. 3.

If both $\alpha_S(\mu^2)$ and $\alpha_S(Q^2)$ are in the perturbative region it makes sense to truncate the series on the right-hand side and solve the resulting differential equation for $\alpha_S(Q^2)$.

For example, neglecting the b' and higher coefficients in Eq. (30) gives the solution

$$\alpha_S(Q^2) = \frac{\alpha_S(\mu^2)}{1 + \alpha_S(\mu^2)bt}, \quad t = \ln\frac{Q^2}{\mu^2}. \tag{32}$$

This gives the relation between $\alpha_S(Q^2)$ and $\alpha_S(\mu^2)$, if both are in the perturbative region. Evidently as t becomes very large, the running coupling $\alpha_S(Q^2)$ decreases to zero. This is the property of asymptotic freedom. Indeed at large Q^2 we may write

$$\alpha_S(Q^2) = \frac{1}{b\ln(Q^2/\Lambda^2)}. \tag{33}$$

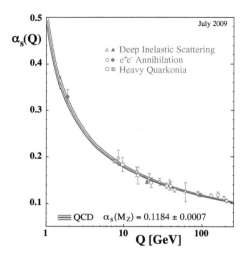

Figure 4. *The running of the strong coupling as a function of the energy Q from Ref. [8].*

If $d\alpha_S/d\tau = -b\alpha_S^2(1 + b'\alpha_S)$ and we change the definition of α_S such that

$$\alpha_S \to \bar{\alpha}_S(1 + c\bar{\alpha}_S) \tag{34}$$

it follows that $d\bar{\alpha}_S/d\tau = -b\bar{\alpha}_S^2(1 + b'\bar{\alpha}_S) + O(\bar{\alpha}_S^4)$. The first two coefficients b, b' are thus invariant under scheme change. This is reassuring since the behaviour of the beta function for small coupling leads to the property of asymptotic freedom; it would be unacceptable if this property were to depend on a particular definition of the QCD coupling.

1.6   Measurements of the Running Coupling α_S

For a recent review of measurements of α_S see Ref. [8]. As shown in Fig. 4 the evidence that α_S falls off with Q is conclusive. The current world average as shown in Fig. 4 is

$$\alpha_S(M_Z) = 0.1184 \pm 0.0007. \tag{35}$$

The current error on the determination of α_S is less than 1%, so that uncertainty in the value of α_S is not a major source of error in the determination of hadronic cross sections.

1.7   Infrared Divergences

Despite the fact that the QCD coupling is small at high energy, not all physical quantities can be calculated using perturbation in the strong coupling α_S.

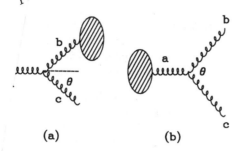

(a)   (b)

racelike branching: gluon splitting on incoming line.
iching: gluon splitting on outgoing line

juence of the fact that soft or collinear gluon emission gives
rared divergences in perturbation theory. Light quarks ($m_q \ll$
*to divergences in the limit $m_q \to 0$, which are known as mass
ies.

*spacelike branching, as illustrated in Fig. 5(a), we have ($p_a^2 = p_c^2 = 0$),

$$p_b^2 = (p_a - p_c)^2 = -2E_a E_c (1 - \cos\theta) \leq 0. \tag{36}$$

We see that the propagator factor $1/p_b^2$ diverges both as $E_c \to 0$, which is a
soft singularity and as $\theta \to 0$ which is a collinear or mass singularity. If a and
b are quarks, with velocities v_a, v_b, the inverse propagator factor is

$$p_b^2 - m_q^2 = -2E_a E_c (1 - v_a \cos\theta) \leq 0, \tag{37}$$

where v_a is the velocity of the quark a rescaled by the energy E_a. Hence as
$E_c \to 0$ the soft divergence remains; in addition the collinear enhancement
becomes a divergence as $v_a \to 1$, i.e., when the quark mass is negligible. If the
emitted parton c is a quark, a numerator factor coming from the quark-gluon
vertex cancels the potential divergence as $E_c \to 0$.

For the case of timelike branching of a gluon shown in Fig. 5(b) we have

$$p_a^2 = (p_b + p_c)^2 = 2E_b E_c (1 - \cos\theta) \geq 0. \tag{38}$$

This expression diverges when either emitted gluon is soft (E_b or $E_c \to 0$)
or when the opening angle $\theta \to 0$. If b and/or c are quarks, there is a
collinear/mass singularity in the $m_q \to 0$ limit, but soft quark divergences
are cancelled by a numerator factor.

Infrared divergences indicate dependence on long-distance aspects of QCD
not correctly described by perturbation theory. We can still use perturbation
theory to perform calculations, provided we limit ourselves to two classes of
observables:

- Infrared safe quantities, i.e., those insensitive to soft or collinear branching. Infrared divergences in a perturbation theory calculation either

cancel between real and virtual contributions or are
matic factors. Such quantities are determined primari
distance physics; long-distance effects give power correc by kine-
by inverse powers of a large momentum scale.

- Factorisable quantities, i.e., those in which infrared sens
  absorbed into an overall non-perturbative factor, to be de
  perimentally.

In the next subsection we consider infrared safe definitions of jets. Fa
quantities will be considered in Section 2.

1.8 Infrared Safety and Jet Algorithms

Jets are recognized by the eye when looking at an event display from a h
energy collider detector and it is easy to imagine that such a jet is the resul
the fragmentation of a single hard parton. A jet algorithm is the way in whi
we formalize this concept so that both experiment and theory can produce
rates for jet cross sections [9]. A jet definition should be simple to implement
both in an experimental analysis and in a theoretical calculation and should
yield finite cross sections at any order of perturbation theory. In view of the
discussion given above a jet measure can only give finite cross sections if it
is insensitive to soft and collinear emission. Thus for any jet measure F we
have the following two requirements.

$$F_{\{is\}}^{(n+1)}(p_A, p_B; p_1, \ldots p_i, \ldots p_{n+1})$$
$$\overset{p_i \to 0}{\to} F_{\{is\}}^{(n)}(p_A, p_B; p_1, \ldots p_{n+1}), \tag{39}$$

$$F_{\{is\}}^{(n+1)}(p_A, p_B; p_1, \ldots p_i, p_j, \ldots p_{n+1})$$
$$\overset{p_i \| p_j}{\to} F_{\{is\}}^{(n)}(p_A, p_B; p_1, \ldots, p_i + p_j, \ldots p_{n+1}). \tag{40}$$

In Eq. (40) p_A and p_B are the momenta of the incoming hadrons and p_i is an
arbitrary final state momentum, which, when the momentum becomes very
small, leaves the jet measure unchanged.

1.8.1 Sequential Recombination Jet Algorithms

The k_T algorithm in e^+e^- collision is defined by:

- For each pair of particles i, j work out the separation

$$y_{ij} = \frac{2\min(E_i^2, E_j^2)(1 - \cos\theta_{ij})}{Q^2}, \tag{41}$$

where E_i and E_j are the energies of particles i, j and θ_{ij} is the angle
between them. In the collinear limit, this expression reduces to k_T^2/Q^2

where k_T is the transverse momentum of the softer parton with respect to the harder.

- Find the minimum y_{min} of all the y_{ij}.

- If y_{min} is smaller than a jet resolution threshold y_{cut} combine i and j into a single pseudo particle.

- Iterate this procedure from step one until all pseudoparticles have a separation greater than y_{cut}.

This algorithm is infrared and collinear safe because any soft or collinear particle will be the first to be clustered into a pseudoparticle. The end result will thus be the same whether the resultant pseudo particle branches or not.

2   Parton Branching and the DGLAP Equation

In this lecture we will consider a second class of observables which are not infrared finite, and thus are not completely calculable in perturbation theory. However, by examining the terms which give rise to logarithms in perturbation theory, we can predict the change in these quantities as the energy scale is increased. These logarithms are due to parton branching. But before treating branching in detail we shall derive the solution of the Dirac equation for free massless fermions, using the Weyl representation for the gamma matrices. This solution will be used in the calculation of the branching probabilities.

2.1   Solution of the Dirac Equation in the Weyl Basis

We choose an explicit representation for the gamma matrices. The most commonly used representation for the 4×4 gamma matrices is the Bjorken and Drell representation [10] which is given by

$$\gamma^0 = \begin{pmatrix} 1 & 0 \\ 0 & -1 \end{pmatrix}, \quad \gamma^i = \begin{pmatrix} 0 & \sigma^i \\ -\sigma^i & 0 \end{pmatrix}, \quad \gamma_5 = \begin{pmatrix} 0 & 1 \\ 1 & 0 \end{pmatrix}, \quad (42)$$

but any representation of the gamma matrices which satisfies the anticommutation relation

$$\gamma^\mu \gamma^\nu + \gamma^\nu \gamma^\mu = 2g^{\mu\nu}, \quad (43)$$

is acceptable. The Weyl representation is more suitable at high energy

$$\gamma^0 = \begin{pmatrix} 0 & 1 \\ 1 & 0 \end{pmatrix}, \quad \gamma^i = \begin{pmatrix} 0 & -\sigma^i \\ \sigma^i & 0 \end{pmatrix}, \quad \gamma_5 = \begin{pmatrix} 1 & 0 \\ 0 & -1 \end{pmatrix}, \quad (44)$$

because it leads to a slightly simpler form for the solution of the Dirac equation for massless fermions. The 2×2 Pauli matrices σ^i are given by

$$\sigma^1 = \begin{pmatrix} 0 & 1 \\ 1 & 0 \end{pmatrix}, \quad \sigma^2 = \begin{pmatrix} 0 & -i \\ i & 0 \end{pmatrix}, \quad \sigma^3 = \begin{pmatrix} 1 & 0 \\ 0 & -1 \end{pmatrix}. \quad (45)$$

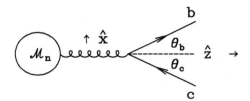

Figure 6. *Definition of kinematics for timelike parton branching.*

and $\mathbf{1}$ is the 2×2 identity matrix. In the Weyl representation upper and lower components correspond to the two different helicity states because, as can be seen from Eq. (44), the chirality projector $\frac{1}{2}(1 \pm \gamma_5)$ projects out the upper or lower components. In order to derive an explicit solution for the massless Dirac equation $\not{p}u(p) = 0$ it is useful to write out an explicit expression for $\not{p}$ in the Weyl representation. Since $\not{p} = \gamma^0 p^0 - \gamma^1 p^1 - \gamma^2 p^2 - \gamma^3 p^3$ we find

$$\not{p} = \begin{pmatrix} 0 & 0 & p^+ & p^1 - ip^2 \\ 0 & 0 & p^1 + ip^2 & p^- \\ p^- & -p^1 + ip^2 & 0 & 0 \\ -p^1 - ip^2 & p^+ & 0 & 0 \end{pmatrix}, \qquad (46)$$

where $p^\pm = p^0 \pm p^3$. The massless spinor solutions of the Dirac equation are

$$u_+(p) = \begin{bmatrix} \sqrt{p^+} \\ \sqrt{p^-}e^{i\varphi_p} \\ 0 \\ 0 \end{bmatrix}, \quad u_-(p) = \begin{bmatrix} 0 \\ 0 \\ \sqrt{p^-}e^{-i\varphi_p} \\ -\sqrt{p^+} \end{bmatrix}, \qquad (47)$$

in a normalisation in which $u_\pm^\dagger u_\pm = 2p^0$ and where

$$e^{\pm i\varphi_p} \equiv \frac{p^1 \pm ip^2}{\sqrt{(p^1)^2 + (p^2)^2}} = \frac{p^1 \pm ip^2}{\sqrt{p^+ p^-}}. \qquad (48)$$

In this representation the Dirac conjugate spinors are

$$\bar{u}_+(p) \equiv u_+^\dagger(p)\gamma^0 = [0, 0, \sqrt{p^+}, \sqrt{p^-}e^{-i\varphi_p}], \qquad (49)$$

$$\bar{u}_-(p) = \left[\sqrt{p^-}e^{i\varphi_p}, -\sqrt{p^+}, 0, 0\right]. \qquad (50)$$

2.2 Branching Probabilities

Consider a gluon travelling in the positive z direction which splits into a quark-antiquark pair as shown in Fig. 6. The kinematics and notation for the branching of parton a into $b + c$ are $p = (E, p_x, p_y, p_z)$,

$$p_a = \left(E_a + \frac{p_a^2}{4E_a}, 0, 0, E_a - \frac{p_a^2}{4E_a}\right),$$

$$p_b = (E_b, +E_b \sin\theta_b, 0, +E_b \cos\theta_b),$$
$$p_c = (E_c, -E_c \sin\theta_c, 0, +E_c \cos\theta_c). \tag{51}$$

We assume that $p_b^2, p_c^2 \ll p_a^2 \equiv t$. If we take a to be an outgoing parton, this process is referred to as timelike branching since $t > 0$. The opening angle is $\theta = \theta_b + \theta_c$. Defining the energy fraction as

$$z = E_b/E_a = 1 - E_c/E_a, \tag{52}$$

we have for small angles $t = 2E_bE_c(1 - \cos\theta) = z(1 - z)E_a^2\theta^2$. Further using transverse momentum conservation we have that $z\theta_b = (1 - z)\theta_c$ so that

$$\theta = \frac{1}{E_a}\sqrt{\frac{t}{z(1-z)}} = \frac{\theta_b}{1-z} = \frac{\theta_c}{z}. \tag{53}$$

Taking the small angle approximation, Eq. (51) becomes

$$p_a = \left(E_a + \frac{p_a^2}{4E_a}, 0, 0, E_a - \frac{p_a^2}{4E_a}\right),$$
$$p_b \sim (E_b, +E_b\theta_b, 0, +E_b),$$
$$p_c \sim (E_c, -E_c\theta_c, 0, +E_c). \tag{54}$$

Thus, for example, from Eq. (49),

$$u_+^\dagger(p_b) = \sqrt{2E_b}\left[1, \frac{\theta_b}{2}, 0, 0\right], \quad u_-^\dagger(p_b) = \sqrt{2E_b}\left[0, 0, \frac{\theta_b}{2}, -1\right], \tag{55}$$

and correspondingly from Eq. (47) we have

$$u_+(p_c) \equiv v_-(p_c) = \sqrt{2E_c}\begin{bmatrix} 1 \\ -\frac{\theta_c}{2} \\ 0 \\ 0 \end{bmatrix}, \quad u_-(p_c) \equiv v_+(p_c) = \sqrt{2E_c}\begin{bmatrix} 0 \\ 0 \\ \frac{\theta_c}{2} \\ -1 \end{bmatrix}. \tag{56}$$

Hence for polarization vectors $\varepsilon_{in} = (0, 1, 0, 0), \varepsilon_{out} = (0, 0, 1, 0)$,

$$-g\,\bar{u}_+^b \not{\varepsilon}_{in} v_-^c = g\sqrt{E_bE_c}(\theta_b - \theta_c) = g\sqrt{z(1-z)}(1-2z)E_a\theta, \tag{57}$$

$$-g\,\bar{u}_+^b \not{\varepsilon}_{out} v_-^c = ig\sqrt{E_bE_c}(\theta_b + \theta_c) = ig\sqrt{z(1-z)}E_a\theta. \tag{58}$$

The matrix element relation for the branching is

$$|\mathcal{M}_{n+1}|^2 \sim \frac{g^2}{t}T_RF(z; \varepsilon_a, \lambda_b, \lambda_c)|\mathcal{M}_n|^2 \tag{59}$$

where the colour factor is now $\mathrm{Tr}(t^At^A)/8 = T_R = 1/2$. The non-vanishing functions $F(z; \varepsilon_a, \lambda_b, \lambda_c)$ for quark and antiquark helicities λ_b and λ_c are given

ε_a	λ_b	λ_c	$F(z; \varepsilon_a, \lambda_b, \lambda_c)$
in	$\pm$	$\mp$	$(1-2z)^2$
out	$\pm$	$\mp$	1

Table 2. *Results for the splitting amplitude $q\bar{q} \leftarrow g$.*

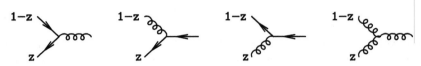

Figure 7. *The four branching probabilities.*

in Table 2. Summing over the polarizations we get for the gluon to $q\bar{q}$ branching,

$$2\Big[(1-2z)^2 + 1\Big] = 4(z^2 + (1-z)^2). \tag{60}$$

In general we find that

$$\hat{P}_{ba}(z) = \int \frac{d\phi}{2\pi} \, C \, F, \tag{61}$$

where $\hat{P}_{ba}(z)$ is the appropriate splitting function and C is a colour factor. Thus in the collinear approximation we can represent the cross section for the emission of a single extra parton as

$$d\sigma_{n+1} = d\sigma_n \, \frac{dt}{t} \, dz \, \frac{\alpha_S}{2\pi} \, \hat{P}_{ba}(z). \tag{62}$$

There are four possible parton branchings as shown in Fig. 7. Including all the color factors we find the results for the unregulated branching probabilities:

$$
\begin{aligned}
\hat{P}_{qg}(z) &= T_R \left[z^2 + (1-z)^2 \right], \\
\hat{P}_{qq}(z) &= C_F \left[\frac{1+z^2}{(1-z)} \right], \\
\hat{P}_{gq}(z) &= C_F \left[\frac{1+(1-z)^2}{z} \right], \\
\hat{P}_{gg}(z) &= C_A \left[\frac{z}{(1-z)} + \frac{1-z}{z} + z(1-z) \right].
\end{aligned}
\tag{63}
$$

2.3   DGLAP Equation and Its Solution

Consider the enhancement of higher-order contributions due to multiple small-angle parton emission, for example in deep inelastic scattering (DIS). In this

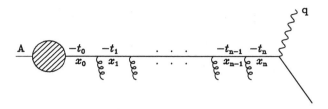

Figure 8. *Initial state branching in deep inelastic lepton scattering.*

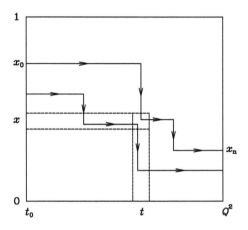

Figure 9. *The parton shower viewed as a path in x- and t-space.*

experiment the incoming quark from the target hadron, initially with low virtual mass-squared $-t_0$ and carrying a fraction x_0 of hadron's momentum, moves to larger virtual masses and lower momentum fractions by successive small-angle emissions, as shown in Fig. 8, and is finally struck by a photon of virtual mass-squared $q^2 = -Q^2$.

The cross section will depend on Q^2 and on the momentum fraction distribution of partons seen by the virtual photon at this scale, $f(x, Q^2)$. To derive the evolution equation for the Q^2-dependence of $f(x, Q^2)$, we first introduce a pictorial representation of evolution in Fig. 9.

We represent a sequence of branchings by a path in (t, x)-space. Each branching is a step downwards in x, at a value of t equal to (minus) the virtual mass-squared after the branching. At $t = t_0$, paths have a distribution of starting points $f(x_0, t_0)$ characteristic of a target hadron at that scale. The distribution $f(x, t)$ of partons at scale t is just the x-distribution of paths at that scale.

Consider the change in the parton distribution $f(x, t)$ when t is increased to $t + \delta t$. This is the number of paths arriving in element $(\delta t, \delta x)$ minus the number leaving that element, divided by δx. The number arriving is the branching probability times the parton density integrated over all higher momenta $x' = x/z$,

$$
\begin{aligned}
\delta f_{\text{in}}(x,t) &= \frac{\delta t}{t} \int_x^1 dx'\, dz \frac{\alpha_S}{2\pi} \hat{P}(z) f(x',t)\, \delta(x - zx') \\
&= \frac{\delta t}{t} \int_0^1 \frac{dz}{z} \frac{\alpha_S}{2\pi} \hat{P}(z) f(x/z,t).
\end{aligned} \tag{64}
$$

For the number leaving the element, one must integrate over lower momenta $x' = zx$:

$$
\begin{aligned}
\delta f_{\text{out}}(x,t) &= \frac{\delta t}{t} f(x,t) \int_0^x dx'\, dz \frac{\alpha_S}{2\pi} \hat{P}(z)\, \delta(x' - zx) \\
&= \frac{\delta t}{t} f(x,t) \int_0^1 dz \frac{\alpha_S}{2\pi} \hat{P}(z).
\end{aligned} \tag{65}
$$

Therefore the change in population of the element is

$$
\begin{aligned}
\delta f(x,t) &= \delta f_{\text{in}} - \delta f_{\text{out}} \\
&= \frac{\delta t}{t} \int_0^1 dz \frac{\alpha_S}{2\pi} \hat{P}(z) \left[\frac{1}{z} f(x/z,t) - f(x,t) \right].
\end{aligned} \tag{66}
$$

We introduce the plus-prescription with the definition

$$
\int_0^1 dx\, f(x)\, g(x)_+ = \int_0^1 dx\, [f(x) - f(1)]\, g(x). \tag{67}
$$

Using this we can define the regularized splitting function

$$
P(z) = \hat{P}(z)_+. \tag{68}
$$

The plus-prescription, like the Dirac-delta function, is only defined under the integral sign. The plus-prescription includes some of the effects of virtual diagrams which up to now we have not considered.

2.4   DGLAP

Taking the limit of an infinitesimal element in Eq. (66) we obtain the Dokshitzer–Gribov–Lipatov–Altarelli–Parisi (DGLAP) evolution equation [11]:

$$
t \frac{\partial}{\partial t} f(x,t) = \int_x^1 \frac{dz}{z} \frac{\alpha_S}{2\pi} P(z) f\left(\frac{x}{z}, t \right). \tag{69}
$$

Here $f(x,t)$ represents the parton momentum fraction distribution inside the incoming hadron, probed at scale t. For the case of timelike branching, it represents instead the hadron momentum fraction distribution produced by an outgoing parton. For timelike branching the boundary conditions and the direction of evolution are different, but the evolution equation remains the same.

For several different types of partons, one must take into account different processes by which partons of type i can enter or leave the element $(\delta t, \delta x)$. This leads to coupled DGLAP evolution equations of the form

$$t\frac{\partial}{\partial t}f_i(x,t) = \sum_j \int_x^1 \frac{dz}{z}\frac{\alpha_S}{2\pi}P_{ij}(z)f_j\left(\frac{x}{z},t\right). \tag{70}$$

A quark ($i = q$) can enter the element via either $q \to qg$ or $g \to q\bar{q}$, but can only leave via $q \to qg$. Thus the plus-prescription applies only to the $q \to qg$ part, giving

$$
\begin{aligned}
P_{qq}(z) &= \hat{P}_{qq}(z)_+ = C_F\left(\frac{1+z^2}{1-z}\right)_+, \\
P_{qg}(z) &= \hat{P}_{qg}(z) = T_R\left[z^2 + (1-z)^2\right].
\end{aligned}
\tag{71}
$$

A gluon can arrive either from $g \to gg$ (2 contributions) or from $q \to qg$ (or $\bar{q} \to \bar{q}g$). Thus the number arriving is

$$
\begin{aligned}
\delta f_{g,\text{in}} &= \frac{\delta t}{t}\int_0^1 dz\frac{\alpha_S}{2\pi}\left\{\hat{P}_{gg}(z)\left[\frac{f_g(x/z,t)}{z} + \frac{f_g(x/(1-z),t)}{1-z}\right]\right. \\
&\qquad \left. + \frac{\hat{P}_{qq}(z)}{1-z}\left[f_q\left(\frac{x}{1-z},t\right) + f_{\bar{q}}\left(\frac{x}{1-z},t\right)\right]\right\} \\
&= \frac{\delta t}{t}\int_0^1 \frac{dz}{z}\frac{\alpha_S}{2\pi}\left\{2\hat{P}_{gg}(z)f_g\left(\frac{x}{z},t\right)\right. \\
&\qquad \left. + \hat{P}_{qq}(1-z)\left[f_q\left(\frac{x}{z},t\right) + f_{\bar{q}}\left(\frac{x}{z},t\right)\right]\right\}.
\end{aligned}
\tag{72}
\tag{73}
$$

A gluon can leave by splitting into either gg or $q\bar{q}$, so that

$$\delta f_{g,\text{out}} = \frac{\delta t}{t}f_g(x,t)\int_0^1 dz\frac{\alpha_S}{2\pi}\left[\hat{P}_{gg}(z) + N_f\hat{P}_{qg}(z)\,dz\right]. \tag{74}$$

After some manipulation we find

$$
\begin{aligned}
P_{gg}(z) &= 2C_A\left[\left(\frac{z}{1-z} + \frac{1}{2}z(1-z)\right)_+ + \frac{1-z}{z} + \frac{1}{2}z(1-z)\right] \\
&\quad - \frac{2}{3}N_f T_R\,\delta(1-z), \\
P_{gq}(z) &= P_{g\bar{q}}(z) = \hat{P}_{qq}(1-z) = C_F\frac{1+(1-z)^2}{z}.
\end{aligned}
\tag{75}
$$

Using the definition of the plus-prescription, we can check that

$$\left(\frac{z}{1-z} + \frac{1}{2}z(1-z)\right)_+ = \frac{z}{(1-z)_+} + \frac{1}{2}z(1-z) + \frac{11}{12}\delta(1-z), \quad (76)$$

$$\left(\frac{1+z^2}{1-z}\right)_+ = \frac{1+z^2}{(1-z)_+} + \frac{3}{2}\delta(1-z), \quad (77)$$

so P_{qq} and P_{gg} can be written in more common forms,

$$P_{qq}(z) = C_F\left[\frac{1+z^2}{(1-z)_+} + \frac{3}{2}\delta(1-z)\right],$$

$$P_{gg}(z) = 2C_A\left[\frac{z}{(1-z)_+} + \frac{1-z}{z} + z(1-z)\right] + 2\pi b\,\delta(1-z), \quad (78)$$

where b is given in Eq. (31)

2.5   Solution of the DGLAP Equation

The structure of the DGLAP equation is,

$$t\frac{\partial}{\partial t}f(x,t) = \int_x^1 \frac{dz}{z}\frac{\alpha_S}{2\pi}P(z)f(x/z,t). \quad (79)$$

Given $f_i(x,t)$ at some scale $t = t_0$, the structure of the DGLAP equation means we can compute its form at any other scale. One strategy for doing this is to take moments (Mellin transforms) with respect to x:

$$\tilde{f}_i(N,t) = \int_0^1 dx\, x^{N-1}\, f_i(x,t). \quad (80)$$

The inverse Mellin transform is defined by

$$f_i(x,t) = \frac{1}{2\pi i}\int_C dN\, x^{-N}\, \tilde{f}_i(N,t), \quad (81)$$

where the contour C is parallel to the imaginary axis and to the right of all singularities of the integrand. After Mellin transformation, the convolution in the DGLAP equation becomes simply a product:

$$t\frac{\partial}{\partial t}\tilde{f}_i(x,t) = \sum_j \gamma_{ij}(N,\alpha_S)\tilde{f}_j(N,t). \quad (82)$$

From the above expressions, Eqs. (71,78), for $P_{ij}(z)$ we find

$$\gamma_{qq}^{(0)}(N) = C_F\left[-\frac{1}{2} + \frac{1}{N(N+1)} - 2\sum_{k=2}^N \frac{1}{k}\right],$$

$$\gamma_{qg}^{(0)}(N) = T_R\left[\frac{(2+N+N^2)}{N(N+1)(N+2)}\right],$$

$$\gamma_{gg}^{(0)}(N) = 2C_A\left[-\frac{1}{12}+\frac{1}{N(N-1)}+\frac{1}{(N+1)(N+2)}-\sum_{k=2}^{N}\frac{1}{k}\right]$$

$$-\frac{2}{3}N_f T_R,$$

$$\gamma_{gq}^{(0)}(N) = C_F\left[\frac{(2+N+N^2)}{N(N^2-1)}\right]. \tag{83}$$

In deriving this equation we have used

$$\int_0^1 dz\ z^{N-1}\frac{1}{(1-z)_+} \equiv \int_0^1 dz\ \frac{z^{N-1}-1}{(1-z)} = -\sum_{k=1}^{N-1}\frac{1}{k}, \tag{84}$$

which is easily derived by using the identity

$$\sum_{k=1}^{N-1} z^{k-1} = \frac{1-z^{N-1}}{1-z}. \tag{85}$$

Consider a combination of parton distributions which is a flavour non-singlet, e.g., $f_V = f_{q_i} - f_{\bar{q}_i}$ or $f_{q_i} - f_{q_j}$. Then the mixing with the flavour-singlet gluons drops out and the solution for fixed α_S is

$$\tilde{f}_V(N,t) = \tilde{f}_V(N,t_0)\left(\frac{t}{t_0}\right)^{\gamma_{qq}(N,\alpha_S)}. \tag{86}$$

We see that the dimensionless function f_V, instead of being a scale-independent function of x as expected from dimensional analysis, has scaling violation: its moments vary like powers of the scale t (hence the name anomalous dimensions). For running coupling $\alpha_S(t)$, the scaling violation is power-behaved in $\ln t$ rather than t. Using the leading-order formula, Eq. (33), $\alpha_S(t) = 1/b\ln(t/\Lambda^2)$, we find

$$\tilde{f}_V(N,t) = \tilde{f}_V(N,t_0)\left(\frac{\alpha_S(t_0)}{\alpha_S(t)}\right)^{d_{qq}(N)}, \tag{87}$$

where $d_{qq}(N) = \gamma_{qq}^{(0)}(N)/2\pi b$. Now $d_{qq}(1) = 0$ and $d_{qq}(N) < 0$ for $N \geq 2$. Thus as t increases f_V *decreases* at large x and *increases* at small x. Physically, this is due to increase in the phase space for gluon emission by quarks as t increases, leading to loss of momentum. This qualitative behaviour is clearly visible in the HERA data shown in Fig. 10.

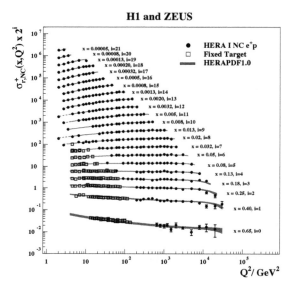

Figure 10. *Measurements of the generalized structure function from HERA [12]. For clarity, the points have been displaced by an x dependent shift.*

2.6   Flavour Singlet Combination

For the flavour-singlet parton densities, we can define the combination

$$\Sigma = \sum_i \left(q_i + \bar{q}_i\right). \tag{88}$$

Then we obtain

$$t\frac{\partial \Sigma}{\partial t} = \frac{\alpha_S(t)}{2\pi}\left[P_{qq} \otimes \Sigma + 2N_f P_{qg} \otimes g\right], \tag{89}$$

$$t\frac{\partial g}{\partial t} = \frac{\alpha_S(t)}{2\pi}\left[P_{gq} \otimes \Sigma + P_{gg} \otimes g\right]. \tag{90}$$

Thus the flavour-singlet quark distribution Σ mixes with the gluon distribution g. These equations are most easily solved by direct numerical integration in x-space starting with input distributions obtained from data.

3   Processes with Two Incoming Hadrons

Since the LHC is a proton–proton collider we are primarily interested in processes with two incoming hadrons. According to the QCD improved parton

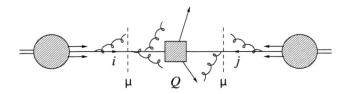

Figure 11. *QCD parton model for hadron–hadron collisions.*

model illustrated in Fig. 11, the form of the cross section is

$$\sigma(s) = \sum_{i,j} \int dx_1 dx_2 f_i(x_1, \mu^2) f_j(x_2, \mu^2) \hat{\sigma}_{ij} \left(\hat{s}, \alpha_S(\mu^2), \frac{Q^2}{\mu^2} \right),$$

$$(91)$$

where the incoming hadron momenta are P_1, P_2 ($s = (P_1 + P_2)^2$), μ^2 is the factorisation scale, $\hat{\sigma}_{ij}$ is the subprocess cross section for parton types i, j and the f_i are the distributions of the longitudinal momentum of the partons of type i inside the incoming hadrons. Notice that the factorisation scale is in principle arbitrary: it separates the interactions that are considered part of the hard scattering from the interactions which belong to the evolution of the initial state (the parton shower). The μ-dependence of the parton distributions f_i is controlled by the DGLAP equation. Unlike e^+e^- or ep collisions, we may also have interactions between spectator partons, leading to a soft underlying event and/or multiple hard scattering.

Eq. (91) is predictive because it allows us to use parton distributions measured in deeply inelastic lepton–nucleon scattering (DIS) and apply them to make predictions for hadron–hadron collisions. This requires the property of factorisation to hold, i.e., that the parton distributions in hadron H_2 are unaffected (before the moment of impact) by the strongly interacting particle (hadron H_1) that is hurtling towards it. Such a property is not *a priori* evident, because the constituents of H_1 give rise to long-range colour fields.

3.1 Factorisation of the Cross Section

Why does the factorisation property hold and when should it fail? For a heuristic argument let us consider the simplest hard process involving two hadrons,

$$H_1(P_1) + H_2(P_2) \to V + X, \qquad (92)$$

where V is a colour singlet vector boson.

Do the partons in hadron H_1, through the influence of their colour fields, change the distribution of partons in hadron H_2 before the vector boson is produced? Gluons which are emitted long before the collision are potentially troublesome.

A simple model [13] from classical electrodynamics can give some insight. The vector potential due to an electromagnetic current density J is given by

$$A^\mu(t, \vec{x}) = \int dt' d\vec{x}' \, \frac{J^\mu(t', \vec{x}')}{|\vec{x} - \vec{x}'|} \, \delta(t' + |\vec{x} - \vec{x}'| - t), \tag{93}$$

where the delta function provides the retarded behaviour required by causality. Consider a particle (a constituent of H_1) with charge e travelling in the positive z direction with constant velocity β. The non-zero components of the current density due to this particle are

$$\begin{aligned} J^t(t', \vec{x}') &= e\delta(\vec{x}' - \vec{r}(t')), \\ J^z(t', \vec{x}') &= e\beta\delta(\vec{x}' - \vec{r}(t')), \quad \vec{r}(t') = \beta t' \hat{z}, \end{aligned} \tag{94}$$

where $\hat{z}$ is a unit vector in the z direction. At an observation point (the supposed position of hadron H_2) described by coordinates x, y and z, the vector potential can be obtained by integration using the current density given in Eq. (94). After integrating over x' we obtain

$$A^z(t, \vec{x}) = \beta A^t(t, \vec{x}) = e\beta \int dt' \frac{1}{t - t'} \delta(t' - t + \sqrt{x^2 + y^2 + (z - \beta t')^2}). \tag{95}$$

The relevant retarded solution to the quadratic constraint imposed by the δ-function is

$$(1 - \beta^2)t' = t - \beta z - \sqrt{[(x^2 + y^2)(1 - \beta^2) + (z - \beta t)^2]}. \tag{96}$$

The Jacobian for the transformation of the δ-function evaluated in Eq. (96) is

$$\begin{aligned} J &= 1 - \frac{\beta(z - \beta t')}{\sqrt{[x^2 + y^2 + (z - \beta t')^2]}} \\ &= \frac{t - t' - \beta(z - \beta t')}{t - t'} \\ &= \frac{\sqrt{[(x^2 + y^2)(1 - \beta^2) + (z - \beta t)^2]}}{t - t'}. \end{aligned} \tag{97}$$

Hence the result for the four components of the vector potential is

$$\begin{aligned} A^t(t, \vec{x}) &= \frac{e\gamma}{\sqrt{[x^2 + y^2 + \gamma^2(z - \beta t)^2]}}, \\ A^x(t, \vec{x}) &= 0, \\ A^y(t, \vec{x}) &= 0, \\ A^z(t, \vec{x}) &= \frac{e\gamma\beta}{\sqrt{[x^2 + y^2 + \gamma^2(z - \beta t)^2]}}, \end{aligned} \tag{98}$$

where $\gamma^2 = 1/(1 - \beta^2)$. By supposition the target hadron H_2 is at rest near the origin, so that $\gamma \approx s/(2m^2)$.

Note that for large γ and fixed non-zero $(z - \beta t)$ some components of the potential tend to a constant independent of γ, suggesting that there will be non-zero fields which are not in coincidence with the arrival of the particle, even at high energy, i.e., even in the limit of large γ.

However, at large γ the potential is a pure gauge piece, $A^\mu = \partial^\mu \chi$, where χ is a scalar function. Therefore at high energy the expression for the vector potential A, Eq. (98), contains large fields which ultimately have no effect. For example, the electric field along the z direction is

$$E^z(t, \vec{x}) = F^{tz} \equiv \frac{\partial A^z}{\partial t} + \frac{\partial A^t}{\partial z} = \frac{e\gamma(z - \beta t)}{[x^2 + y^2 + \gamma^2(z - \beta t)^2]^{\frac{3}{2}}}. \quad (99)$$

The leading terms in γ cancel and the field strengths are of order $1/\gamma^2$ and hence of order m^4/s^2. The model suggests the force experienced by a charge in the hadron H_2, at any fixed time before the arrival of the quark, decreases as m^4/s^2. This force can thus be dropped at high energy and factorisation will continue to hold up to terms of order $1/s^2$. However, this cancellation is not immediate in the quantized theory which is formulated in terms of the vector potential, A.

3.2 Parton Luminosity

Using Eq. (91) we may come up with a rough-and-ready way to estimate cross sections. Let us define the parton luminosity $\frac{dL_{ij}}{d\tau}$

$$\tau \frac{dL_{ij}}{d\tau} = \frac{1}{1 + \delta_{ij}} \int_0^1 dx_1 dx_2 \Big[\big(x_1 f_i(x_1, \mu^2) \, x_2 f_j(x_2, \mu^2)\big) + \big(1 \leftrightarrow 2\big) \Big] \delta(\tau - x_1 x_2). \quad (100)$$

We may write any hadronic cross section, for which Eq. (91) holds, in terms of the parton luminosity

$$\sigma(s) = \sum_{\{ij\}} \int_{\tau_0}^1 \frac{d\tau}{\tau} \left[\frac{1}{s} \frac{dL_{ij}}{d\tau} \right] \left[\hat{s}\hat{\sigma}_{ij} \right], \quad (101)$$

where $\hat{s} = \tau s$. In Eq. (101) the first object in square brackets has the dimensions of a cross section, whereas the second expression in square brackets $[\hat{s}\hat{\sigma}]$ is dimensionless and in a crude approximation can be considered to be given by coupling constants alone. τ_0 is the minimum energy squared of the hard interaction, rescaled by the centre-of-mass energy squared. The logarithmic integral in Eq. (101) will contribute a factor of a few and can be neglected when performing order of magnitude estimates. Plots of the quark-antiquark and gluon-gluon luminosities are given in Fig. 12. As an example of the utility of these plots let us consider the production of a 200 GeV object $(\sqrt{\hat{s}} = 0.2 \text{ TeV})$ produced by gluon fusion at $\sqrt{s} = 7, 14 \text{ TeV}$. The resultant luminosities are 4×10^5 and 1.5×10^6 pb, respectively, so that a Higgs boson

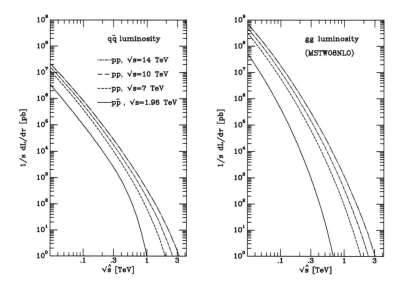

Figure 12. *Parton luminosity plots for $q\bar{q}$ and gg at various energies.*

with this mass is produced four times more frequently at a 14 TeV collider than at a 7 TeV collider operating at the same hadronic luminosity. Further examples of parton luminosity plots may also be found in Ref. [15], where ratio plots are also given.

3.3   Lepton Pair Production

The mechanism for lepton pair production by the materialization of a virtual photon is shown in Fig. 13. This is the simplest process involving two incoming hadrons, since at the Born level there are no coloured particles in the final state. Q is the four-momentum of the virtual photon and also the combined four-momentum of the lepton antilepton pair. In the centre-of-mass frame of the two hadrons, the momenta of the incoming partons are

$$p_1 = \frac{\sqrt{s}}{2}(x_1, 0, 0, x_1), \quad p_2 = \frac{\sqrt{s}}{2}(x_2, 0, 0, -x_2). \tag{102}$$

The square of the $q\bar{q}$ collision energy $\hat{s}$ is related to the overall hadron–hadron collision energy by $\hat{s} = (p_1 + p_2)^2 = x_1 x_2 s$.

The parton model cross section for this process is:

$$
\begin{aligned}
\frac{d\sigma}{dQ^2} &= \int_0^1 dx_1 dx_2 \sum_q \{f_q(x_1)f_{\bar{q}}(x_2) + (q \leftrightarrow \bar{q})\} \frac{d\hat{\sigma}}{dQ^2}(q\bar{q} \to l^+l^-) \\
&= \frac{\sigma_0}{Ns} \int_0^1 \frac{dx_1}{x_1} \frac{dx_2}{x_2} \, \delta(1-z) \left[\sum_q Q_q^2 \{f_q(x_1)f_{\bar{q}}(x_2) + (q \leftrightarrow \bar{q})\} \right].
\end{aligned}
$$

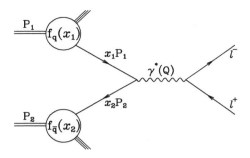

Figure 13. *Lepton pair production in the Drell–Yan model.*

Because the colour neutral photon can only be produced by the annihilation of a quark and antiquark of the same colour, there is a colour average factor of $1/N$ where $N = 3$.

$$\frac{d\hat{\sigma}}{dQ^2} = \frac{\sigma_0}{N} Q_q^2 \, \delta(\hat{s} - Q^2), \qquad \sigma_0 = \frac{4\pi\alpha^2}{3Q^2}, \quad \text{cf. } e^+e^- \text{ annihilation.} \quad (103)$$

For later convenience we have introduced the variable $z = \frac{Q^2}{\hat{s}} = \frac{Q^2}{x_1 x_2 s}$. The sum here is over quarks only and the $\bar{q}q$ contributions are indicated explicitly. This mechanism can also be applied to W and Z production, and with a simple extension also to the production of vector-boson pairs, and has come to be known as the Drell–Yan process.

3.3.1 Radiative Corrections to the Drell–Yan Process

The factorisation property of perturbative QCD allows one to systematically improve the estimate of cross sections by calculating higher-order terms in perturbation theory. The radiative corrections to the Drell–Yan process, shown in Fig. 14, have been calculated in Ref. [14]. The real matrix element contains divergences as $t \to 0$ or $u \to 0$ or both. We will control the divergences by continuing the dimensionality of space-time, $d = 4 - 2\epsilon$.

The contribution of the real diagrams in Fig. 14(b) to the invariant matrix element squared (in d dimensions) is

$$
\begin{aligned}
|M|^2 \;\sim\; & g^2 (\mu^2)^\epsilon C_F \left[(1 - \epsilon) \left(\frac{u}{t} + \frac{t}{u} \right) + \frac{2Q^2 s}{ut} - 2\epsilon \right] \\
=\; & g^2 (\mu^2)^\epsilon C_F \left[\left(\frac{1 + z^2}{1 - z} - \epsilon(1 - z) \right) \left(\frac{-s}{t} + \frac{-s}{u} \right) - 2 \right], \quad (104)
\end{aligned}
$$

where $z = Q^2/s$, $s + t + u = Q^2$.

We will ignore for simplicity the diagrams with incoming gluons shown in Fig. 14(c). The two-body phase space for the production of a virtual photon

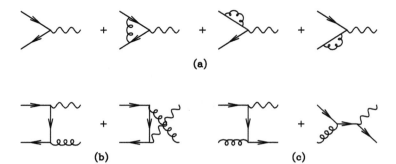

Figure 14. *Radiative corrections to the Drell–Yan process.*

and a gluon of energy E is given in d dimensions as

$$PS = \frac{c_\Gamma}{4\pi} \int_0^\infty dE E^{1-2\epsilon} \int_{-1}^1 d\cos\theta (1 - \cos^2\theta)^{-\epsilon} \delta(s - Q^2 - 2\sqrt{s}E), \quad (105)$$

with $c_\Gamma = (4\pi)^\epsilon / \Gamma(1-\epsilon)$. Performing the integration over the gluon energy E and making the change of variables $y = \frac{1}{2}(1 + \cos\theta)$ the phase space becomes

$$PS = \frac{c_\Gamma}{8\pi} \Big(\frac{1}{Q^2}\Big)^\epsilon z^\epsilon (1-z)^{1-2\epsilon} \int_0^1 dy (y(1-y))^{-\epsilon}. \quad (106)$$

In terms of Q^2, y and z, the invariants s, t and u are given by

$$s = \frac{Q^2}{z}, \quad t = -\frac{Q^2}{z}(1-z)(1-y), \quad u = -\frac{Q^2}{z}(1-z)y. \quad (107)$$

Note that the real diagrams contain collinear singularities, $u \to 0, t \to 0$ and soft singularities, $z \to 1$. The coefficient of the divergence in Eq. (104) is the unregulated branching probability $\hat{P}_{qq}(z)$. Performing the phase space integration and using identities such as

$$z^\epsilon (1-z)^{-1-\epsilon} = -\frac{1}{\epsilon}\delta(1-z) + \frac{1}{(1-z)_+} - \epsilon \left[\frac{\ln(1-z)}{1-z}\right]_+ + \epsilon \frac{\ln z}{1-z} + O(\epsilon^2) \quad (108)$$

the total contribution of the real diagrams is found to be

$$
\begin{aligned}
\sigma_R = {}& \frac{\alpha_S}{2\pi} C_F \left(\frac{\mu^2}{Q^2}\right)^\epsilon c_\Gamma \left[\left(\frac{2}{\epsilon^2} + \frac{3}{\epsilon} - \frac{\pi^2}{3}\right)\delta(1-z) - \frac{2}{\epsilon} P_{qq}(z) \right. \\
& \left. + \; 4(1+z^2)\left[\frac{\ln(1-z)}{1-z}\right]_+ - 2\frac{1+z^2}{(1-z)}\ln z \right],
\end{aligned}
\quad (109)
$$

with $P_{qq}(z)$ given by Eq. (78).

The contribution of the Born plus virtual diagrams to the $\gamma^* q\bar{q}$ vertex shown in Fig. 14(a) is

$$\Gamma^\mu = \bar{v}(-p_2)\gamma^\mu u(p_1) \left\{ 1 + \frac{\alpha_S}{4\pi} c_\Gamma \left(\frac{\mu^2}{-Q^2} \right)^\epsilon \left[-\frac{2}{\epsilon^2} - \frac{3}{\epsilon} - 8 \right] \right\} + O(\epsilon), \quad (110)$$

where $\sqrt{Q^2}$ is the off-shellness of the virtual photon. As a consequence the result for the Born + virtual cross section is

$$\sigma_V = \delta(1-z) \left[1 + \frac{\alpha_S}{2\pi} C_F \left(\frac{\mu^2}{Q^2} \right)^\epsilon c_\Gamma \left(-\frac{2}{\epsilon^2} - \frac{3}{\epsilon} - 8 + \pi^2 \right) \right]. \quad (111)$$

Adding up the real and virtual contributions we get

$$\begin{aligned} \sigma_{R+V} &= \frac{\alpha_S}{2\pi} C_F \left(\frac{\mu^2}{Q^2} \right)^\epsilon c_\Gamma \left[\left(\frac{2\pi^2}{3} - 8 \right) \delta(1-z) - \frac{2}{\epsilon} P_{qq}(z) \right. \\ &+ \left. 4(1+z^2) \left[\frac{\ln(1-z)}{1-z} \right]_+ - 2\frac{1+z^2}{(1-z)} \ln z \right]. \end{aligned} \quad (112)$$

The divergences, proportional to the branching probability, are universal. We will factorise them into the parton distributions. We perform the mass factorisation by subtracting the counterterm,

$$2\frac{\alpha_S}{2\pi} C_F \left[\frac{-c_\Gamma}{\epsilon} P_{qq}(z) \right]. \quad (113)$$

The final result for the short distance cross section is

$$\begin{aligned} \hat{\sigma} &= \delta(1-z) + \frac{\alpha_S}{2\pi} C_F \left[\left(\frac{2\pi^2}{3} - 8 \right) \delta(1-z) + 4(1+z^2) \left[\frac{\ln(1-z)}{1-z} \right]_+ \right. \\ &- \left. 2\frac{1+z^2}{(1-z)} \ln z + 2 P_{qq}(z) \ln \frac{Q^2}{\mu^2} \right]. \end{aligned} \quad (114)$$

We can calculate a similar correction for the incoming gluon diagrams of Fig. 14(c). The magnitude of the $O(\alpha_S)$ correction depends on the lepton pair mass and on the overall collision energy. For W and Z production in $p\bar{p}$ collisions at 2 TeV the $O(\alpha_S)$ correction increases the lowest-order cross section by about $20 - 30\%$. This correction is needed to achieve agreement with the experimental data.

3.4   Heavy Quark Production, Leading Order

The leading-order processes for the production of a heavy quark Q of mass m in hadron–hadron collisions are shown in Fig. 15

$$(a) \qquad q(p_1) + \bar{q}(p_2) \to Q(p_3) + \overline{Q}(p_4)$$

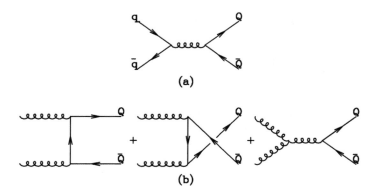

(a)

(b)

Figure 15. *Lowest-order contributions to heavy quark production.*

$$(b) \qquad g(p_1) + g(p_2) \to Q(p_3) + \overline{Q}(p_4), \tag{115}$$

where the four-momenta of the partons are given in brackets. The resultant lowest-order matrix elements for the pair production of heavy quarks are given in Table 3.

Process	$\overline{\sum}|\mathcal{M}|^2/g^4$
$q\,\overline{q} \to Q\,\overline{Q}$	$\frac{4}{9}\left(\tau_1^2 + \tau_2^2 + \frac{\rho}{2}\right)$
$g\,g \to Q\,\overline{Q}$	$\left(\frac{1}{6\tau_1\tau_2} - \frac{3}{8}\right)\left(\tau_1^2 + \tau_2^2 + \rho - \frac{\rho^2}{4\tau_1\tau_2}\right)$

Table 3. *Lowest-order matrix elements for the pair production of heavy quarks.*

The symbol $\overline{\sum}$ indicates that the matrix elements squared have been averaged (summed) over initial (final) colours and spins. We have introduced the following notation for the ratios of scalar products:

$$\tau_1 = \frac{2p_1 \cdot p_3}{\hat{s}}, \quad \tau_2 = \frac{2p_2 \cdot p_3}{\hat{s}}, \quad \rho = \frac{4m^2}{\hat{s}}, \quad \hat{s} = (p_1 + p_2)^2. \tag{116}$$

The short-distance cross section is obtained from the invariant matrix element in the usual way:

$$d\hat{\sigma}_{ij} = \frac{1}{2\hat{s}} \frac{d^3p_3}{(2\pi)^3 2E_3} \frac{d^3p_4}{(2\pi)^3 2E_4} (2\pi)^4 \delta^4(p_1 + p_2 - p_3 - p_4) \overline{\sum}|\mathcal{M}_{ij}|^2. \tag{117}$$

The first factor is the flux factor for massless incoming particles. The other terms come from the phase space for $2 \to 2$ scattering.

In terms of the rapidity $y = \frac{1}{2}\ln((E + p_z)/(E - p_z))$ and transverse momentum, p_T, the relativistically invariant phase space volume element of the final-state heavy quarks is

$$\frac{d^3p}{E} = dy \, d^2p_T. \tag{118}$$

The result for the invariant cross section may be written as

$$\frac{d\sigma}{dy_3 dy_4 d^2p_T} = \frac{1}{16\pi^2 \hat{s}^2} \sum_{ij} x_1 f_i(x_1, \mu^2) \, x_2 f_j(x_2, \mu^2) \overline{\sum} |\mathcal{M}_{ij}|^2. \tag{119}$$

x_1 and x_2 are fixed if we know the transverse momenta and rapidity of the outgoing heavy quarks. In the centre-of-mass system of the incoming hadrons we may write

$$
\begin{aligned}
p_1 &= \tfrac{1}{2}\sqrt{s}(x_1, 0, 0, x_1), \\
p_2 &= \tfrac{1}{2}\sqrt{s}(x_2, 0, 0, -x_2), \\
p_3 &= (m_T \cosh y_3, p_T, 0, m_T \sinh y_3), \\
p_4 &= (m_T \cosh y_4, -p_T, 0, m_T \sinh y_4).
\end{aligned} \tag{120}
$$

Applying energy and momentum conservation, we obtain,

$$x_1 = \frac{m_T}{\sqrt{s}}\left(e^{y_3} + e^{y_4}\right), x_2 = \frac{m_T}{\sqrt{s}}\left(e^{-y_3} + e^{-y_4}\right), \hat{s} = 2m_T^2(1 + \cosh \Delta y). \tag{121}$$

The quantity $m_T = \sqrt{(m^2 + p_T^2)}$ is the transverse mass of the heavy quarks and $\Delta y = y_3 - y_4$ is the rapidity difference between them. In these variables the leading-order cross section is

$$
\begin{aligned}
\frac{d\sigma}{dy_3 dy_4 d^2 p_T} &= \frac{1}{64\pi^2 m_T^4 (1 + \cosh(\Delta y))^2} \\
&\quad \times \sum_{ij} x_1 f_i(x_1, \mu^2) \, x_2 f_j(x_2, \mu^2) \overline{\sum} |\mathcal{M}_{ij}|^2.
\end{aligned} \tag{122}
$$

Expressed in terms of m, m_T and Δy, the matrix elements for the two processes are

$$\overline{\sum} |\mathcal{M}_{q\bar{q}}|^2 = \frac{4g^4}{9}\left(\frac{1}{1 + \cosh(\Delta y)}\right)\left(\cosh(\Delta y) + \frac{m^2}{m_T^2}\right), \tag{123}$$

$$\overline{\sum} |\mathcal{M}_{gg}|^2 = \frac{g^4}{24}\left(\frac{8\cosh(\Delta y) - 1}{1 + \cosh(\Delta y)}\right)\left(\cosh(\Delta y) + 2\frac{m^2}{m_T^2} - 2\frac{m^4}{m_T^4}\right). \tag{124}$$

As the rapidity separation Δy between the two heavy quarks becomes large

$$\overline{\sum} |\mathcal{M}_{q\bar{q}}|^2 \sim \text{constant}, \quad \overline{\sum} |\mathcal{M}_{gg}|^2 \sim \exp \Delta y. \tag{125}$$

The cross section is damped at large Δy and heavy quarks produced by $q\bar{q}$ annihilation are more closely correlated in rapidity than those produced by gg fusion.

3.5   Applicability of Perturbation Theory

Consider the propagators in the diagrams in Fig. 15.

$$
\begin{aligned}
(p_1 + p_2)^2 &= & 2p_1.p_2 &= 2m_T^2(1 + \cosh \Delta y), \\
(p_1 - p_3)^2 - m^2 &= & -2p_1.p_3 &= -m_T^2(1 + e^{-\Delta y}), \\
(p_2 - p_3)^2 - m^2 &= & -2p_2.p_3 &= -m_T^2(1 + e^{\Delta y}).
\end{aligned}
\tag{126}
$$

Note that the propagators are all off-shell by a quantity of at least of order m^2.

Thus for a sufficiently heavy quark we expect the methods of perturbation theory to be applicable. It is the mass m (which by supposition is very much larger than the scale of the strong interactions Λ) which provides the large scale in heavy quark production. We expect corrections of order Λ/m. This does not address the issue of whether the charm or bottom mass is large enough to be adequately described by perturbation theory.

3.6   Scale Dependence

The next-to-leading order calculation of heavy quark production is by now well known [16, 17]. In this section I wish to illustrate the dependence on the unphysical parameter, μ. The physical predictions should be invariant under changes of μ at the appropriate order in perturbation theory. If we have performed a calculation to $O(\alpha_S^3)$, variations of the scale μ will lead to corrections of $O(\alpha_S^4)$, $\mu^2 \frac{d}{d\mu^2}\sigma = O(\alpha_S^4)$. The short distance cross section can be written as

$$
\hat{\sigma}_{ij}(s, m^2, \mu^2) = \frac{\alpha_S^2(\mu^2)}{m^2} c_{ij}\left(\rho, \frac{\mu^2}{m^2}\right).
\tag{127}
$$

Equation (127) completely describes the short-distance cross section for the production of a heavy quark of mass m in terms of the functions c_{ij}, where the indices i and j specify the types of the incoming partons. The variable μ is the renormalisation and factorisation scale. The dimensionless functions c_{ij} have the following perturbative expansion:

$$
c_{ij}\left(\rho, \frac{\mu^2}{m^2}\right) = c_{ij}^{(0)}(\rho) + 4\pi\alpha_S(\mu^2)\left[c_{ij}^{(1)}(\rho) + \bar{c}_{ij}^{(1)}(\rho)\ln\left(\frac{\mu^2}{m^2}\right)\right] + O(\alpha_S^2),
\tag{128}
$$

where ρ is defined in Eq. (116).

The term $\bar{c}^{(1)}$, which controls the μ dependence of the higher-order perturbative contributions, is fixed in terms of the lower-order result $c^{(0)}$:

$$
\begin{aligned}
\bar{c}_{ij}^{(1)}(\rho) &= \frac{1}{8\pi^2}\left[4\pi b c_{ij}^{(0)}(\rho) - \int_\rho^1 dz_1 \sum_k c_{kj}^{(0)}\left(\frac{\rho}{z_1}\right) P_{ki}^{(0)}(z_1)\right. \\
&\qquad \left. - \int_\rho^1 dz_2 \sum_k c_{ik}^{(0)}\left(\frac{\rho}{z_2}\right) P_{kj}^{(0)}(z_2)\right].
\end{aligned}
\tag{129}
$$

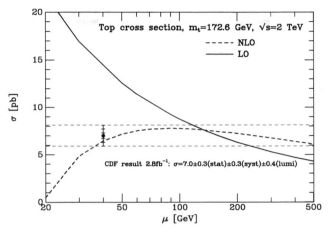

Figure 16. *The cross section for the production of $t\bar{t}$ pairs at $\sqrt{s} = 1.96$ TeV.*

In obtaining this result we have used the renormalisation group equation for the running coupling

$$\mu^2 \frac{d}{d\mu^2} \alpha_S(\mu^2) = -b\alpha_S^2 + \dots \qquad (130)$$

and the lowest-order form of the DGLAP equation

$$\mu^2 \frac{d}{d\mu^2} f_i(x, \mu^2) = \frac{\alpha_S(\mu^2)}{2\pi} \sum_k \int_x^1 \frac{dz}{z} P_{ik}^{(0)}(z) f_k \left(\frac{x}{z}, \mu^2\right) + \dots . \qquad (131)$$

This illustrates an important point which is a general feature of renormalisation group improved perturbation series in QCD. The coefficient of the perturbative correction depends on the choice made for the scale μ, but the scale dependence changes the result in such a way that the physical result is independent of that choice. Thus the scale dependence is formally small because it is of higher order in α_S. This does not assure us that the scale dependence is actually *numerically* small for all series. A pronounced dependence on the scale μ is a signal of an untrustworthy perturbation series.

The inclusion of the higher-order terms leads to a stabilization of the top cross section. This is illustrated in Fig. 3.6 which shows that the top cross section calculated in NLO is less dependent on the scale μ than the LO result.

3.7 Epilogue

We do not know in detail what will be observed at the LHC. But we can be sure of one thing. The discovery of new physics at the LHC will require a detailed understanding of QCD, since most signals for new physics will have doppelgangers generated by the standard model. For a more extensive description of QCD I refer you to Ref. [18]. I wish you good hunting!

Acknowledgements

I thank the organizers of the school for their hospitality in St. Andrews. I would also like to thank Nigel Glover and Ciaran Williams for comments on the manuscript.

References

[1] M. Ablikim *et al.* [BES Collaboration], Phys. Lett. B **677** (2009) 239 [arXiv:0903.0900 [hep-ex]].

[2] P. Abreu *et al.* [DELPHI Collaboration], Phys. Lett. B **414** (1997) 401.

[3] W. E. Caswell, Phys. Rev. Lett. **33** (1974) 244.

[4] D. R. T. Jones, Nucl. Phys. B **75** (1974) 531.

[5] O. V. Tarasov, A. A. Vladimirov and A. Y. Zharkov, Phys. Lett. B **93** (1980) 429.

[6] S. A. Larin and J. A. M. Vermaseren, Phys. Lett. B **303** (1993) 334 [arXiv:hep-ph/9302208].

[7] T. van Ritbergen, J. A. M. Vermaseren and S. A. Larin, Phys. Lett. B **400** (1997) 379 [arXiv:hep-ph/9701390].

[8] S. Bethke, Eur. Phys. J. **C64** (2009) 689. [arXiv:0908.1135 [hep-ph]].

[9] G. P. Salam, Eur. Phys. J. **C67** (2010) 637. [arXiv:0906.1833 [hep-ph]].

[10] J. D. Bjorken and S. D. Drell, McGraw-Hill (1964). ISBN 0-07-005493-2.

[11] G. Altarelli and G. Parisi, Nucl. Phys. B **126** (1977) 298.

[12] F. D. Aaron *et al.* [H1 Collaboration and ZEUS Collaboration], JHEP **1001** (2010) 109 [arXiv:0911.0884 [hep-ex]].

[13] R. Basu, A. J. Ramalho and G. Sterman, Nucl. Phys. B **244** (1984) 221.

[14] G. Altarelli, R. K. Ellis and G. Martinelli, Nucl. Phys. B **157** (1979) 461.

[15] C. Quigg, arXiv:0908.3660 [hep-ph].

[16] P. Nason, S. Dawson and R. K. Ellis, Nucl. Phys. B **303** (1988) 607.

[17] M. Czakon and A. Mitov, Nucl. Phys. B **824** (2010) 111 [arXiv:0811.4119 [hep-ph]].

[18] R. K. Ellis, W. J. Stirling and B. R. Webber, Camb. Monogr. Part. Phys. Nucl. Phys. Cosmol. **8** (1996) 1.

Higgs and Electroweak Physics

Sven Heinemeyer

Instituto de Física de Cantabria (CSIC), Santander, Spain

1   Introduction

A major goal of the particle physics program at the high-energy frontier, recently pursued at the Fermilab Tevatron collider and now taken up by the CERN Large Hadron Collider (LHC), is to unravel the nature of electroweak symmetry breaking. While the existence of the massive electroweak gauge bosons $(W^\pm, Z)$, together with the successful description of their behavior by non-abelian gauge theory, requires some form of electroweak symmetry breaking to be present in nature, the underlying dynamics is not known yet. An appealing theoretical suggestion for such dynamics is the Higgs mechanism [1], which implies the existence of one or more Higgs bosons (depending on the specific model considered). Therefore, the search for Higgs bosons is a major cornerstone in the physics programs of past, present and future high-energy colliders.

Many theoretical models employing the Higgs mechanism in order to account for electroweak symmetry breaking have been studied in the literature, of which the most popular ones are the Standard Model (SM) [2] and the Minimal Supersymmetric Standard Model (MSSM) [3]. Within the SM, the Higgs boson is the last undiscovered particle, whereas the MSSM has a richer Higgs sector, containing three neutral and two charged Higgs bosons. Among alternative theoretical models beyond the SM and the MSSM, the most prominent are the Two Higgs Doublet Model (THDM) [4], non-minimal supersymmetric extensions of the SM (e.g., extensions of the MSSM by an extra singlet superfield [5]), little Higgs models [6] and models with more than three spatial dimensions [7].

We will discuss the Higgs boson sector in the SM and the MSSM. This includes their connection to electroweak precision physics and the searches for the SM and supersymmetric (SUSY) Higgs bosons at the LHC. While the LHC will discover the SM Higgs boson and, in the case that the MSSM is realized in nature, almost certainly also one or more SUSY Higgs bosons, a "cleaner" experimental environment, such as at the ILC, will be needed to measure all the Higgs boson characteristics [8, 9].

2 The SM and the Higgs

2.1 Higgs: Why and How?

We start by looking at one of the most simple Lagrangians, that of QED:

$$\mathcal{L}_{\text{QED}} = -\frac{1}{4}F_{\mu\nu}F^{\mu\nu} + \bar{\psi}(i\gamma^\mu D_\mu - m)\psi. \tag{1}$$

Here D_μ denotes the covariant derivative

$$D_\mu = \partial_\mu + i\,e\,A_\mu. \tag{2}$$

ψ is the electron spinor, and A_μ is the photon vector field. The QED Lagrangian is invariant under the local $U(1)$ gauge symmetry,

$$\psi \quad \rightarrow \quad e^{-i\alpha(x)}\psi, \tag{3}$$

$$A_\mu \quad \rightarrow \quad A_\mu + \frac{1}{e}\partial_\mu\alpha(x). \tag{4}$$

Introducing a mass term for the photon,

$$\mathcal{L}_{\text{photon mass}} = \frac{1}{2}m_A^2 A_\mu A^\mu, \tag{5}$$

however, is not gauge-invariant. Applying Eq. (4) yields

$$\frac{1}{2}m_A^2 A_\mu A^\mu \rightarrow \frac{1}{2}m_A^2 \left[A_\mu A^\mu + \frac{2}{e}A^\mu \partial_\mu\alpha + \frac{1}{e^2}\partial_\mu\alpha\,\partial^\mu\alpha \right]. \tag{6}$$

A way out is the Higgs mechanism [1]. The simplest implementation uses one elementary complex scalar Higgs field Φ that has a vacuum expectation value v (vev) that is constant in space and time. The Lagrangian of the new Higgs field reads

$$\mathcal{L}_\Phi = \mathcal{L}_{\Phi,\text{kin}} + \mathcal{L}_{\Phi,\text{pot}} \tag{7}$$

with

$$\mathcal{L}_{\Phi,\text{kin}} \quad = \quad (D_\mu\Phi)^* (D^\mu\Phi), \tag{8}$$

$$-\mathcal{L}_{\Phi,\text{pot}} \quad = \quad V(\Phi) = \mu^2|\Phi|^2 + \lambda|\Phi|^4. \tag{9}$$

Here λ has to be chosen positive to have a potential bounded from below. μ^2 can be either positive or negative, where we will see that $\mu^2 < 0$ yields the desired vev, as will be shown below. The complex scalar field Φ can be parameterised by two real scalar fields ϕ and η,

$$\Phi(x) = \frac{1}{\sqrt{2}}\phi(x)e^{i\eta(x)}, \tag{10}$$

yielding

$$V(\phi) = \frac{\mu^2}{2}\phi^2 + \frac{\lambda}{4}\phi^4. \tag{11}$$

Minimizing the potential one finds

$$\frac{dV}{d\phi}\Big|_{\phi=\phi_0} = \mu^2\phi_0 + \lambda\phi_0^3 \overset{!}{=} 0. \tag{12}$$

Only for $\mu^2 < 0$ this yields the desired non-trivial solution

$$\phi_0 = \sqrt{\frac{-\mu^2}{\lambda}} \left(= \langle\phi\rangle =: v\right). \tag{13}$$

The picture simplifies more by going to the "unitary gauge," $\alpha(x) = -\eta(x)/v$, which yields a real-valued Φ everywhere. The kinetic term now reads

$$(D_\mu\Phi)^* (D^\mu\Phi) \to \frac{1}{2}(\partial_\mu\phi)^2 + \frac{1}{2}e^2q^2\phi^2 A_\mu A^\mu, \tag{14}$$

where q is the charge of the Higgs field, which can now be expanded around its vev,

$$\phi(x) = v + H(x). \tag{15}$$

The remaining degree of freedom, $H(x)$, is a real scalar boson, the Higgs boson. The Higgs boson mass and self-interactions are obtained by inserting Eq. (15) into the Lagrangian (neglecting a constant term),

$$-\mathcal{L}_{\text{Higgs}} = \frac{1}{2}m_H^2 H^2 + \frac{\kappa}{3!}H^3 + \frac{\xi}{4!}H^4, \tag{16}$$

with

$$m_H^2 = 2\lambda v^2, \quad \kappa = 3\frac{m_H^2}{v}, \quad \xi = 3\frac{m_H^2}{v^2}. \tag{17}$$

Similarly, Eq. (15) can be inserted in Eq. (14), yielding (neglecting the kinetic term for ϕ),

$$\mathcal{L}_{\text{Higgs–photon}} = \frac{1}{2}m_A^2 A_\mu A^\mu + e^2q^2 v H A_\mu A^\mu + \frac{1}{2}e^2q^2 H^2 A_\mu A^\mu \tag{18}$$

where the second and third terms describe the interaction between the photon and one or two Higgs bosons, respectively, and the first term is the photon mass,

$$m_A^2 = e^2q^2v^2. \tag{19}$$

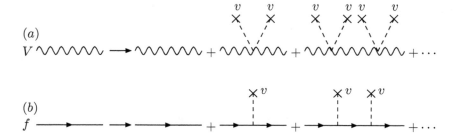

Figure 1. *Generation of a gauge boson mass (a) and a fermion mass (b) via the interaction with the vev of the Higgs field.*

Another important feature can be observed: the coupling of the photon to the Higgs is proportional to its own mass squared.

Similarly a gauge-invariant Lagrangian can be defined to give mass to the chiral fermion $\psi = (\psi_L, \psi_R)^T$,

$$\mathcal{L}_{\text{fermion mass}} = y_\psi \psi_L^\dagger \, \Phi \, \psi_R + \text{c.c.}, \tag{20}$$

where y_ψ denotes the dimensionless Yukawa coupling. Inserting $\Phi(x) = (v + H(x))/\sqrt{2}$ one finds

$$\mathcal{L}_{\text{fermion mass}} = m_\psi \psi_L^\dagger \psi_R + \frac{m_\psi}{v} H \, \psi_L^\dagger \psi_R + \text{c.c.}, \tag{21}$$

with

$$m_\psi = y_\psi \frac{v}{\sqrt{2}}. \tag{22}$$

Again the important feature can be observed: by construction the coupling of the fermion to the Higgs boson is proportional to its own mass m_ψ.

The "creation" of a mass term can be viewed from a different angle. The interaction of the gauge field or the fermion field with the scalar background field, i.e., the vev, shifts the masses of these fields from zero to non-zero values. This is shown graphically in Fig. 1 for the gauge boson (a) and the fermion (b) field.

The shift in the propagators reads (with p being the external momentum and $g = eq$ in Eq. (19)):

$$(a) \quad \frac{1}{p^2} \rightarrow \frac{1}{p^2} + \sum_{k=1}^{\infty} \frac{1}{p^2} \left[\left(\frac{gv}{2}\right) \frac{1}{p^2} \right]^k = \frac{1}{p^2 - m_V^2} \text{ with } m_V^2 = g^2 \frac{v^2}{4}, \tag{23}$$

$$(b) \quad \frac{1}{\not{p}} \rightarrow \frac{1}{\not{p}} + \sum_{k=1}^{\infty} \frac{1}{\not{p}} \left[\left(\frac{y_\psi v}{2}\right) \frac{1}{\not{p}} \right]^k = \frac{1}{\not{p} - m_\psi} \text{ with } m_\psi = y_\psi \frac{v}{\sqrt{2}}. \tag{24}$$

2.2 SM Higgs Theory

We now turn to the electroweak sector of the SM, which is described by the gauge symmetry $SU(2)_L \times U(1)_Y$. The bosonic part of the Lagrangian is given by

$$\mathcal{L}_{\text{bos}} = -\frac{1}{4}B_{\mu\nu}B^{\mu\nu} - \frac{1}{4}W^a_{\mu\nu}W^{\mu\nu}_a + |D_\mu\Phi|^2 - V(\Phi), \tag{25}$$

$$V(\Phi) = \mu^2|\Phi|^2 + \lambda|\Phi|^4. \tag{26}$$

Φ is a complex scalar doublet with charges $(2, 1)$ under the SM gauge groups,

$$\Phi = \begin{pmatrix} \phi^+ \\ \phi^0 \end{pmatrix}, \tag{27}$$

and the electric charge is given by $Q = T^3 + \frac{1}{2}Y$, where T^3 is the third component of the weak isospin. We furthermore have

$$D_\mu = \partial_\mu + ig\frac{\tau^a}{2}W_{\mu a} + ig'\frac{Y}{2}B_\mu, \tag{28}$$

$$B_{\mu\nu} = \partial_\mu B_\nu - \partial_\nu B_\mu, \tag{29}$$

$$W^a_{\mu\nu} = \partial_\mu W^a_\nu - \partial_\nu W^a_\mu - gf^{abc}W_{\mu b}W_{\nu c}. \tag{30}$$

Here, g and g' are the $SU(2)_L$ and $U(1)_Y$ gauge couplings, respectively, τ^a are the Pauli matrices, and f^{abc} are the $SU(2)$ structure constants.

Choosing $\mu^2 < 0$, the minimum of the Higgs potential is found at

$$\langle\Phi\rangle = \frac{1}{\sqrt{2}}\begin{pmatrix} 0 \\ v \end{pmatrix} \quad \text{with} \quad v := \sqrt{\frac{-\mu^2}{\lambda}}. \tag{31}$$

$\Phi(x)$ can now be expressed through the vev, the Higgs boson and three Goldstone bosons $\phi_{1,2,3}$,

$$\Phi(x) = \frac{1}{\sqrt{2}}\begin{pmatrix} \phi_1(x) + i\phi_2(x) \\ v + H(x) + i\phi_3(x) \end{pmatrix}. \tag{32}$$

Diagonalising the mass matrices of the gauge bosons, one finds that the three massless Goldstone bosons are absorbed as longitudinal components of the three massive gauge bosons, $W^\pm_\mu, Z_\mu$, while the photon A_μ remains massless,

$$W^\pm_\mu = \frac{1}{\sqrt{2}}\left(W^1_\mu \mp iW^2_\mu\right), \tag{33}$$

$$Z_\mu = c_{\text{w}}W^3_\mu - s_{\text{w}}B_\mu, \tag{34}$$

$$A_\mu = s_{\text{w}}W^3_\mu + c_{\text{w}}B_\mu. \tag{35}$$

Here we have introduced the weak mixing angle $\theta_W = \arctan(g'/g)$, and $s_{\mathrm{w}} := \sin\theta_W$, $c_{\mathrm{w}} := \cos\theta_W$. The Higgs-gauge boson interaction Lagrangian reads

$$\mathcal{L}_{\text{Higgs-gauge}} = \left[M_W^2 W_\mu^+ W^{-\mu} + \frac{1}{2} M_Z^2 Z_\mu Z^\mu \right] \left(1 + \frac{H}{v} \right)^2 \\ - \frac{1}{2} M_H^2 H^2 - \frac{\kappa}{3!} H^3 - \frac{\xi}{4!} H^4, \tag{36}$$

with

$$M_W = \frac{1}{2} gv, \quad M_Z = \frac{1}{2} \sqrt{g^2 + g'^2}\, v, \tag{37}$$

$$(M_H^{\text{SM}} :=)\ M_H = \sqrt{2\lambda}\, v, \quad \kappa = 3\frac{M_H^2}{v}, \quad \xi = 3\frac{M_H^2}{v^2}. \tag{38}$$

From the measurement of the gauge boson masses and couplings one finds $v \approx 246$ GeV. Furthermore the two massive gauge boson masses are related via

$$\frac{M_W}{M_Z} = \frac{g}{\sqrt{g^2 + g'^2}} = c_{\mathrm{w}}. \tag{39}$$

We now turn to the fermion masses, where we take the top- and bottom-quark masses as a representative example. The Higgs–fermion interaction Lagrangian reads

$$\mathcal{L}_{\text{Higgs-fermion}} = y_b Q_L^\dagger \Phi\, b_R + y_t Q_L^\dagger \Phi_c\, t_R + \text{h.c.} \tag{40}$$

$Q_L = (t_L, b_L)^T$ is the left-handed $SU(2)_L$ doublet. Going to the "unitary gauge" the Higgs field can be expressed as

$$\Phi(x) = \frac{1}{\sqrt{2}} \begin{pmatrix} 0 \\ v + H(x) \end{pmatrix}, \tag{41}$$

and it is obvious that this doublet can give masses only to the bottom(-type) fermion(s). A way out is the definition of

$$\Phi_c = i\sigma^2 \Phi^* = \frac{1}{\sqrt{2}} \begin{pmatrix} v + H(x) \\ 0 \end{pmatrix}, \tag{42}$$

which is employed to generate the top(-type) mass(es) in Eq. (40). Inserting Eqs. (41) and (42) into Eq. (40) yields

$$\mathcal{L}_{\text{Higgs-fermion}} = m_b \bar{b} b \left(1 + \frac{H}{v} \right) + m_t \bar{t} t \left(1 + \frac{H}{v} \right), \tag{43}$$

where we have used $\bar{\psi}\psi = \psi_L^\dagger \psi_R + \psi_R^\dagger \psi_L$ and $m_b = y_b v/\sqrt{2}$, $m_t = y_t v/\sqrt{2}$.

Figure 2. *Diagrams contributing to the evolution of the Higgs self-interaction* λ *at the tree level (left) and at the one-loop level (middle and right).*

The mass of the SM Higgs boson, M_H^{SM}, is the last remaining free parameter in the model. However, it is possible to derive bounds on M_H^{SM} derived from theoretical considerations [10–12] and from experimental precision data. Here we review the first approach, while the latter one is followed in Section 2.4. Evaluating loop diagrams as shown in the middle and right of Fig. 2 yields the renormalisation group equation (RGE) for λ,

$$\frac{d\lambda}{dt} = \frac{3}{8\pi^2}\left[\lambda^2 + \lambda y_t^2 - y_t^4 + \frac{1}{16}\left(2g^4 + (g^2 + g'^2)^2\right)\right], \qquad (44)$$

with $t = \log(Q^2/v^2)$, where Q is the energy scale.
For large $M_H^2 \propto \lambda$, Eq. (44) reduces to

$$\frac{d\lambda}{dt} = \frac{3}{8\pi^2}\lambda^2 \qquad (45)$$

$$\Rightarrow \quad \lambda(Q^2) = \frac{\lambda(v^2)}{1 - \frac{3\lambda(v^2)}{8\pi^2}\log\left(\frac{Q^2}{v^2}\right)}. \qquad (46)$$

For $\frac{3\lambda(v^2)}{8\pi^2}\log\left(\frac{Q^2}{v^2}\right) = 1$ one finds that λ diverges (it runs into the "Landau pole"). Requiring $\lambda(\Lambda) < \infty$ yields an upper bound on M_H^2 depending up to which scale Λ the Landau pole should be avoided,

$$\lambda(\Lambda) < \infty \quad \Rightarrow \quad M_H^2 \leq \frac{8\pi^2 v^2}{3\log\left(\frac{\Lambda^2}{v^2}\right)}. \qquad (47)$$

For small $M_H^2 \propto \lambda$, Eq. (44) becomes

$$\frac{d\lambda}{dt} = \frac{3}{8\pi^2}\left[-y_t^4 + \frac{1}{16}\left(2g^4 + (g^2 + g'^2)^2\right)\right] \qquad (48)$$

$$\Rightarrow \quad \lambda(Q^2) = \frac{3\lambda(v^2)}{8\pi^2}\left[-y_t^4 + \frac{1}{16}\left(2g^4 + (g^2 + g'^2)^2\right)\right]\log\left(\frac{Q^2}{v^2}\right). \quad (49)$$

Demanding $V(v) < V(0)$, corresponding to $\lambda(\Lambda) > 0$, one finds a lower bound

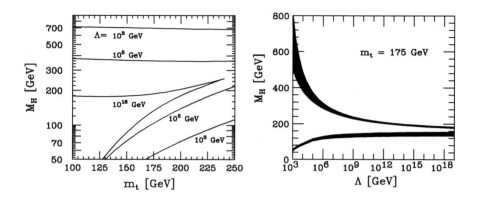

Figure 3. *Bounds on the mass of the Higgs boson in the SM. Λ denotes the energy scale up to which the model is valid [10–12].*

on M_H^2 depending on Λ,

$$\lambda(\Lambda) > 0 \;\Rightarrow\; M_H^2 \;>\; \frac{v^2}{4\pi^2}\left[-y_t^4 + \frac{1}{16}\left(2g^4 + (g^2 + g'^2)^2\right)\right]\log\left(\frac{\Lambda^2}{v^2}\right). \quad (50)$$

The combination of the upper bound in Eq. (47) and the lower bound in Eq. (50) on M_H is shown in Fig. 3. Requiring the validity of the SM up to the GUT scale yields a limit on the SM Higgs boson mass of 130 GeV $\lesssim M_H^{\mathrm{SM}} \lesssim$ 180 GeV.

2.3   SM Higgs Boson Searches at the LHC

An SM-like Higgs boson can be produced in many channels at the LHC as shown in Fig. 4 (taken from Ref. [13], where also the relevant original references can be found). The corresponding discovery potential for an SM-like Higgs boson of ATLAS is shown in Fig. 5 [14], where similar results have been obtained for CMS [15]. With $10\,\mathrm{fb}^{-1}$ a $5\,\sigma$ discovery is expected for $M_H^{\mathrm{SM}} \gtrsim 130$ GeV. For lower masses a higher integrated luminosity will be needed; see also Ref. [9] for a recent overview. The largest production cross section is reached by $gg \to H$, which, however, will be visible only in the decay to SM gauge bosons. A precise mass measurement of $\delta M_h^{\mathrm{exp}} \approx 200$ MeV can be provided by the decays $H \to \gamma\gamma$ at lower Higgs masses and by $H \to ZZ^{(*)} \to 4\ell$ at higher masses. This guarantees the detection of the new state and a precise mass measurement over the relevant parameter space within the SM.

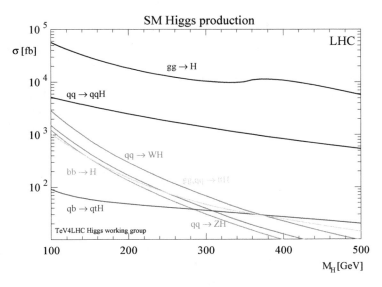

Figure 4. *The various production cross sections for an SM-like Higgs boson at the 14 TeV LHC are shown as a function of M_H (taken from Ref. [13], where also the relevant references can be found).*

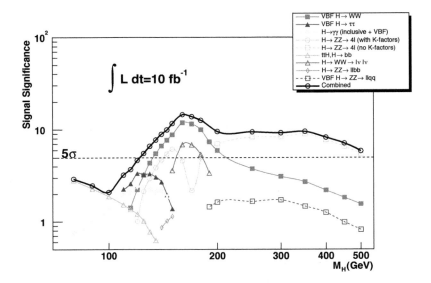

Figure 5. *Significance of a Higgs signal, measured at ATLAS with $10\,fb^{-1}$ of data at 14 TeV [14]. Similar results have been obtained for CMS [15].*

2.4 Electroweak Precision Observables

Within the SM the electroweak precision observables (EWPO) have been used to constrain the last unknown parameter of the model, the Higgs–boson mass M_H^{SM}. Originally the EWPO comprise over a thousand measurements of "realistic observables" (with partially correlated uncertainties) such as cross sections, asymmetries, branching ratios, etc. This huge set is reduced to 17 so-called "pseudo-observables" by the LEP [16] and Tevatron [17] Electroweak working groups. The "pseudo-observables" (again called EWPO in the following) comprise the W boson mass M_W, the width of the W boson, Γ_W, as well as various Z pole observables: the effective weak mixing angle, $\sin^2\theta_{\text{eff}}$, Z decay widths to SM fermions, $\Gamma(Z \to f\bar{f})$, the invisible and total width, Γ_{inv} and Γ_Z, forward-backward and left-right asymmetries, A_{FB}^f and A_{LR}^f, and the total hadronic cross section, σ_{had}^0. The Z pole results including their combination are final [18]. Experimental progress from the Tevatron comes from measurements of M_W and m_t. (Also the error combination for M_W and Γ_W from the four LEP experiments has not been finalised yet due to not-yet-final analyses on the color-reconnection effects.)

The EWPO that give the strongest constraints on M_H^{SM} are M_W, A_{FB}^b and A_{LR}^e. The value of $\sin^2\theta_{\text{eff}}$ is extracted from a combination of various A_{FB}^f and A_{LR}^f, where A_{FB}^b and A_{LR}^e give the dominant contribution.

The one-loop contributions to M_W can be decomposed as follows [19]

$$M_W^2 \left(1 - \frac{M_W^2}{M_Z^2}\right) = \frac{\pi\alpha}{\sqrt{2}G_F}\left(1 + \Delta r\right), \tag{51}$$

$$\Delta r_{1-\text{loop}} = \Delta\alpha - \frac{c_{\text{w}}^2}{s_{\text{w}}^2}\Delta\rho + \Delta r_{\text{rem}}(M_H^{\text{SM}}). \tag{52}$$

The first term, $\Delta\alpha$, contains large logarithmic contributions as $\log(M_Z/m_f)$ and contributes $\sim 6\%$. The second term contains the ρ parameter [20], being $\Delta\rho \sim m_t^2$. This term amounts to $\sim 3.3\%$. The quantity $\Delta\rho$,

$$\Delta\rho = \frac{\Sigma^Z(0)}{M_Z^2} - \frac{\Sigma^W(0)}{M_W^2}, \tag{53}$$

parameterises the leading universal corrections to the electroweak precision observables induced by the mass splitting between fields in an isospin doublet. $\Sigma^{Z,W}(0)$ denote the transverse parts of the unrenormalised Z and W boson self-energies at zero momentum transfer, respectively. The final term in Eq. (52) is $\Delta r_{\text{rem}} \sim \log(M_H^{\text{SM}}/M_W)$, with a size of $\sim 1\%$ correction, yields the constraints on M_H^{SM}. The fact that the leading correction involving M_H^{SM} is logarithmic also applies to the other EWPO. Starting from two-loop order, terms $\sim (M_H^{\text{SM}}/M_W)^2$ also appear. The SM prediction of M_W as a function of m_t for the range $M_H^{\text{SM}} = 114$ GeV ... 1000 GeV is shown as the dark shaded band in Fig. 6 [16]. The upper edge with $M_H^{\text{SM}} = 114$ GeV corresponds to the

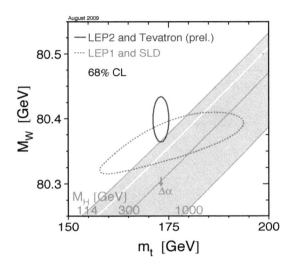

Figure 6. *Prediction for M_W in the SM as a function of m_t for the range $M_H = 114$ GeV . . . 1000 GeV [16]. The prediction is compared with the present experimental results for M_W and m_t as well as with the indirect constraints obtained from EWPO.*

lower limit on M_H^{SM} obtained at LEP [21]. The prediction is compared with the direct experimental result [22, 23],

$$M_W^{\mathrm{exp}} = 80.399 \pm 0.023 \text{ GeV,} \qquad (54)$$
$$m_t^{\mathrm{exp}} = 173.1 \pm 1.3 \text{ GeV,} \qquad (55)$$

shown as the dotted ellipse and with the indirect results for M_W and m_t as obtained from EWPO (solid ellipse). The direct and indirect determinations have significant overlap, representing a non-trivial success for the SM. However, it should be noted that the experimental value of M_W is somewhat higher than the region allowed by the LEP Higgs bounds: $M_H^{\mathrm{SM}} \approx 60$ GeV is preferred as a central value by the measurement of M_W and m_t.

The effective weak mixing angle is evaluated from various asymmetries and other EWPO as shown in Fig. 7 [24]. The average determination yields $\sin^2 \theta_{\mathrm{eff}} = 0.23153 \pm 0.00016$ with a $\chi^2/\mathrm{d.o.f}$ of 11.8/5, corresponding to a probability of 3.7% [24]. The large χ^2 is driven by the two single most-precise measurements, A_{LR}^e by SLD and A_{FB}^b by LEP, where the earlier (latter) one prefers a value of $M_H^{\mathrm{SM}} \sim 32(437)$ GeV [25]. The two measurements differ by more than $3\,\sigma$. The averaged value of $\sin^2 \theta_{\mathrm{eff}}$, as shown in Fig. 7, prefers $M_H^{\mathrm{SM}} \sim 110$ GeV [25].

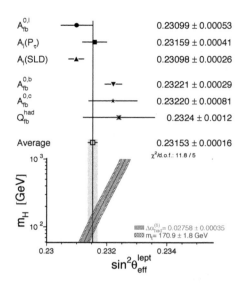

Figure 7. *Prediction for* $\sin^2 \theta_{\text{eff}}$ *in the SM as a function of* M_H^{SM} *for* $m_t = 170.9 \pm 1.8$ GeV *and* $\Delta \alpha_{\text{had}}^5 = 0.02758 \pm 0.00035$ *[24]. The prediction is compared with the present experimental results for* $\sin^2 \theta_{\text{eff}}$ *as averaged over several individual measurements.*

The indirect M_H^{SM} determination for several individual EWPO is given in Fig. 8. Shown are the central values of M_H^{SM} and the one σ errors [16]. The dark-shaded vertical band indicates the combination of the various single measurements in the $1\,\sigma$ range. The vertical line shows the lower LEP bound for M_H^{SM} [21]. It can be seen that M_W, A_{LR}^e and A_{FB}^b give the most precise indirect M_H^{SM} determination, where only the latter one pulls the preferred M_H^{SM} value up, yielding a averaged value of [16]

$$M_H^{\text{SM}} = 87^{+35}_{-26} \text{ GeV}, \tag{56}$$

still compatible with the direct LEP bound of [21],

$$M_H^{\text{SM}} \geq 114.4 \text{ GeV at 95\% C.L.} \tag{57}$$

Thus, the measurement of A_{FB}^b prevents the SM from being incompatible with the direct bound and the indirect constraints on M_H^{SM}.

In the left plot of Fig. 9 [16] we show the result for the global fit to M_H^{SM} including all EWPO, but not including the direct search bounds from LEP and the Tevatron. $\Delta \chi^2$ is shown as a function of M_H^{SM}, yielding Eq. (56) as best fit with an upper limit of 157 GeV at 95% C.L. The theory (intrinsic) uncertainty in the SM calculations (as evaluated with TOPAZ0 [26] and ZFITTER [27]) are represented by the thickness of the shaded band around the parabola. The width of the parabola itself, on the other hand, is determined by the experimental precision of the measurements of the EWPO and the input parameters. The result changes somewhat if the direct bounds on M_H^{SM} from LEP [21] and the Tevatron [28] are taken into account as shown in the right plot of Fig. 9. The upper limit reduced to $M_H^{\text{SM}} \lesssim 150$ GeV at the 95% C.L. [29]. In this analysis a Tevatron exclusion of 160 GeV $< M_H^{\text{SM}} < 170$ GeV [28] was assumed. The most recent limit is slightly smaller,

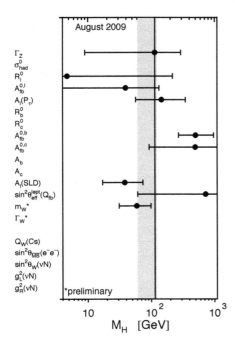

Figure 8. *Indirect constraints on* $M_H^{\rm SM}$ *from various EWPO. Shown are the central values and the one* σ *errors [16]. The dark shaded vertical band indicates the combination of the various single measurements in the* 1σ *range. The vertical line shows the lower bound of* $M_H^{\rm SM} \geq 114.4$ GeV *obtained at LEP [21].*

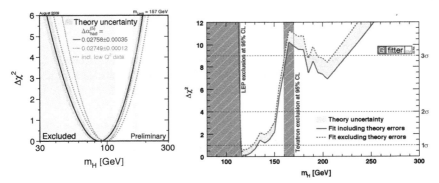

Figure 9. $\Delta\chi^2$ *curve derived from all EWPO measured at LEP, SLD, CDF and D0, as a function of* $M_H^{\rm SM}$, *assuming the SM to be the correct theory of nature. Left: The direct bounds on* $M_H^{\rm SM}$ *are not included [16]. Right: The bounds from LEP [21] and the Tevatron [30] are included [29].*

163 GeV $< M_H^{\rm SM} <$ 166 GeV [30], however the picture is expected to vary only very little with this shift.

The current and anticipated future experimental uncertainties for $\sin^2\theta_{\rm eff}$, M_W and m_t are summarized in Table 1. Also shown is the relative precision of the indirect determination of M_H^{SM} [24]. Each column represents the combined results of all detectors and channels at a given collider, taking into account correlated systematic uncertainties; see Refs. [31–34] for details. The indirect M_H^{SM} determination has to be compared with the (possible) direct measurement at the LHC [14,15] and the ILC [35],

$$\delta M_H^{\mathrm{SM,exp,LHC}} \approx 200 \text{ MeV}, \tag{58}$$

$$\delta M_H^{\mathrm{SM,exp,ILC}} \approx 50 \text{ MeV}. \tag{59}$$

	Now	Tevatron	LHC	ILC	ILC with GigaZ
$\delta \sin^2 \theta_{\mathrm{eff}} (\times 10^5)$	16	—	14–20	—	1.3
δM_W [MeV]	23	20	15	10	7
δm_t [GeV]	1.3	1.0	1.0	0.2	0.1
$\delta M_H^{\mathrm{SM}}/M_H^{\mathrm{SM}}$ [%]	37		28		16

Table 1. *Current and anticipated future experimental uncertainties for* $\sin^2 \theta_{\mathrm{eff}}$, M_W *and* m_t. *Also shown is the relative precision of the indirect determination of* M_H^{SM} *[24]. Each column represents the combined results of all detectors and channels at a given collider, taking into account correlated systematic uncertainties. See Refs. [31–34] for details.*

This comparison will shed light on the basic theoretical components for generating the masses of the fundamental particles. On the other hand, an observed inconsistency would be a clear indication for the existence of a new physics scale.

3 The Higgs in Supersymmetry

3.1 Why SUSY?

Theories based on supersymmetry (SUSY) [3] are widely considered to be the theoretically most appealing extension of the SM. They are consistent with the approximate unification of the gauge coupling constants at the GUT scale and provide a way to cancel the quadratic divergences in the Higgs sector, hence stabilizing the huge hierarchy between the GUT and the Fermi scales. Furthermore, in SUSY theories the breaking of the electroweak symmetry is naturally induced at the Fermi scale, and the lightest supersymmetric particle can be neutral, weakly interacting and absolutely stable, providing therefore a natural solution for the dark matter problem.

The Minimal Supersymmetric Standard Model (MSSM) constitutes, hence its name, the minimal supersymmetric extension of the SM. The number of

SUSY generators is $N = 1$, the smallest possible value. In order to maintain anomaly cancellation, and contrary to the SM, a second Higgs doublet is needed [36]. All SM multiplets, including the two Higgs doublets, are extended to supersymmetric multiplets, resulting in scalar partners for quarks and leptons ("squarks" and "sleptons") and fermionic partners for the SM gauge boson and the Higgs bosons ("gauginos" and "gluinos"). So far, the direct search for SUSY particles has not been successful. One can only set lower bounds of $\mathcal{O}(100 \text{ GeV})$ on their masses [37].

3.2 The MSSM Higgs Sector

An excellent review on this subject is given in Ref. [38].

3.2.1 The Higgs Boson Sector at Tree Level

In the MSSM the Higgs potential [39] with two Higgs doublets is

$$
\begin{aligned}
V = & \ m_1^2 |\mathcal{H}_1|^2 + m_2^2 |\mathcal{H}_2|^2 - m_{12}^2 (\epsilon_{ab} \mathcal{H}_1^a \mathcal{H}_2^b + \text{h.c.}) \\
& + \frac{1}{8} (g^2 + g'^2) \left[|\mathcal{H}_1|^2 - |\mathcal{H}_2|^2 \right]^2 + \frac{1}{2} g^2 |\mathcal{H}_1^\dagger \mathcal{H}_2|^2,
\end{aligned}
\tag{60}
$$

and contains m_1, m_2, m_{12} as soft SUSY breaking parameters; g, g' are the $SU(2)$ and $U(1)$ gauge couplings, and $\epsilon_{12} = -1$.

The doublet fields H_1 and H_2 are decomposed in the following way:

$$
\mathcal{H}_1 = \begin{pmatrix} \mathcal{H}_1^0 \\ \mathcal{H}_1^- \end{pmatrix} = \begin{pmatrix} v_1 + \frac{1}{\sqrt{2}}(\phi_1^0 - i\chi_1^0) \\ -\phi_1^- \end{pmatrix},
$$

$$
\mathcal{H}_2 = \begin{pmatrix} \mathcal{H}_2^+ \\ \mathcal{H}_2^0 \end{pmatrix} = \begin{pmatrix} \phi_2^+ \\ v_2 + \frac{1}{\sqrt{2}}(\phi_2^0 + i\chi_2^0) \end{pmatrix}.
\tag{61}
$$

$\mathcal{H}_1$ gives mass to the down-type fermions, while $\mathcal{H}_2$ gives mass to the up-type fermions. The potential (60) can be described with the help of two independent parameters (besides g and g'): $\tan \beta = v_2/v_1$ and $M_A^2 = -m_{12}^2(\tan \beta + \cot \beta)$, where M_A is the mass of the $\mathcal{CP}$-odd Higgs boson A.

Which values can be expected for $\tan \beta$? One natural choice would be $\tan \beta \approx 1$, i.e., both vevs are about the same. On the other hand, one can argue that v_2 is responsible for the top quark mass, while v_1 gives rise to the bottom quark mass. Assuming that their mass differences come largely from the vevs, while their Yukawa couplings could be about the same, the natural value for $\tan \beta$ would then be $\tan \beta \approx m_t/m_b$. Consequently, one can expect

$$
1 \lesssim \tan \beta \lesssim 50.
\tag{62}
$$

The diagonalisation of the bilinear part of the Higgs potential, i.e., of the Higgs mass matrices, is performed via the orthogonal transformations

$$
\begin{pmatrix} H^0 \\ h^0 \end{pmatrix} = \begin{pmatrix} \cos\alpha & \sin\alpha \\ -\sin\alpha & \cos\alpha \end{pmatrix} \begin{pmatrix} \phi_1^0 \\ \phi_2^0 \end{pmatrix}, \tag{63}
$$

$$
\begin{pmatrix} G^0 \\ A^0 \end{pmatrix} = \begin{pmatrix} \cos\beta & \sin\beta \\ -\sin\beta & \cos\beta \end{pmatrix} \begin{pmatrix} \chi_1^0 \\ \chi_2^0 \end{pmatrix}, \tag{64}
$$

$$
\begin{pmatrix} G^\pm \\ H^\pm \end{pmatrix} = \begin{pmatrix} \cos\beta & \sin\beta \\ -\sin\beta & \cos\beta \end{pmatrix} \begin{pmatrix} \phi_1^\pm \\ \phi_2^\pm \end{pmatrix}. \tag{65}
$$

The mixing angle α is determined through

$$
\alpha = \arctan\left[\frac{-(M_A^2 + M_Z^2)\sin\beta\cos\beta}{M_Z^2\cos^2\beta + M_A^2\sin^2\beta - m_{h,\text{tree}}^2}\right], \quad -\frac{\pi}{2} < \alpha < 0 \tag{66}
$$

with $m_{h,\text{tree}}$ defined below in Eq. (70).

One gets the following Higgs spectrum:

$$
\begin{array}{rcl}
\text{2 neutral bosons, } \mathcal{CP} = +1 & : & h, H \\
\text{1 neutral boson, } \mathcal{CP} = -1 & : & A \\
\text{2 charged bosons} & : & H^+, H^- \\
\text{3 unphysical Goldstone bosons} & : & G, G^+, G^-.
\end{array} \tag{67}
$$

At tree level the mass matrix of the neutral $\mathcal{CP}$-even Higgs bosons is given in the ϕ_1-ϕ_2-basis in terms of M_Z, M_A, and $\tan\beta$ by

$$
\begin{aligned}
M_{\text{Higgs}}^{2,\text{tree}} &= \begin{pmatrix} m_{\phi_1}^2 & m_{\phi_1\phi_2}^2 \\ m_{\phi_1\phi_2}^2 & m_{\phi_2}^2 \end{pmatrix} \\
&= \begin{pmatrix} M_A^2\sin^2\beta + M_Z^2\cos^2\beta & -(M_A^2 + M_Z^2)\sin\beta\cos\beta \\ -(M_A^2 + M_Z^2)\sin\beta\cos\beta & M_A^2\cos^2\beta + M_Z^2\sin^2\beta \end{pmatrix},
\end{aligned} \tag{68}
$$

which by diagonalisation according to Eq. (63) yields the tree level Higgs boson masses

$$
M_{\text{Higgs}}^{2,\text{tree}} \xrightarrow{\alpha} \begin{pmatrix} m_{H,\text{tree}}^2 & 0 \\ 0 & m_{h,\text{tree}}^2 \end{pmatrix} \tag{69}
$$

with

$$
m_{H,h,\text{tree}}^2 = \frac{1}{2}\left[M_A^2 + M_Z^2 \pm \sqrt{(M_A^2 + M_Z^2)^2 - 4M_Z^2 M_A^2\cos^2 2\beta}\right]. \tag{70}
$$

From this formula the famous tree-level bound

$$m_{h,\text{tree}} \leq \min\{M_A, M_Z\} \cdot |\cos 2\beta| \leq M_Z \tag{71}$$

can be obtained. The charged Higgs boson mass is given by

$$m_{H^\pm}^2 = M_A^2 + M_W^2. \tag{72}$$

The masses of the gauge bosons are given in analogy to the SM:

$$M_W^2 = \frac{1}{2}g^2(v_1^2 + v_2^2), \qquad M_Z^2 = \frac{1}{2}(g^2 + g'^2)(v_1^2 + v_2^2), \qquad M_\gamma = 0. \tag{73}$$

The couplings of the Higgs bosons are modified from the corresponding SM couplings already at the tree-level. Some examples are

$$g_{hVV} = \sin(\beta - \alpha)\, g_{HVV}^{\text{SM}}, \quad V = W^\pm, Z, \tag{74}$$

$$g_{HVV} = \cos(\beta - \alpha)\, g_{HVV}^{\text{SM}}, \tag{75}$$

$$g_{hb\bar{b}}, g_{h\tau^+\tau^-} = -\frac{\sin\alpha}{\cos\beta}\, g_{Hb\bar{b}, H\tau^+\tau^-}^{\text{SM}}, \tag{76}$$

$$g_{ht\bar{t}} = \frac{\cos\alpha}{\sin\beta}\, g_{Ht\bar{t}}^{\text{SM}}, \tag{77}$$

$$g_{Ab\bar{b}}, g_{A\tau^+\tau^-} = \gamma_5 \tan\beta\, g_{Hb\bar{b}, H\tau^+\tau^-}^{\text{SM}}. \tag{78}$$

The following can be observed: The couplings of the $\mathcal{CP}$-even Higgs boson to SM gauge bosons are always suppressed with respect to the SM coupling. However, if g_{hVV}^2 is close to zero, g_{HVV}^2 is close to $(g_{HVV}^{\text{SM}})^2$ and vice versa, i.e., it is not possible to decouple both of them from the SM gauge bosons. The coupling of the h to down-type fermions can be *suppressed* or *enhanced* with respect to the SM value, depending on the size of $\sin\alpha/\cos\beta$. Especially for not-too-large values of M_A and large $\tan\beta$ one finds $|\sin\alpha/\cos\beta| \gg 1$, leading to a strong enhancement of this coupling. The same holds, in principle, for the coupling of the h to up-type fermions. However, for large parts of the MSSM parameter space the additional factor is found to be $|\cos\alpha/\sin\beta| < 1$. For the $\mathcal{CP}$-odd Higgs boson the additional factor is $\tan\beta$. According to Eq. (62) this can lead to a strongly enhanced coupling of the A boson to bottom quarks or τ leptons, resulting in new search strategies at the Tevatron and the LHC for the $\mathcal{CP}$-odd Higgs boson see Section 3.3.

For $M_A \gtrsim 150$ GeV the "decoupling limit" is reached. The couplings of the light Higgs boson become SM-like, i.e., the additional factors approach 1. The couplings of the heavy neutral Higgs bosons become similar, $g_{Axx} \approx g_{Hxx}$, and the masses of the heavy neutral and charged Higgs bosons fulfill $M_A \approx M_H \approx M_{H^\pm}$. As a consequence, search strategies for the A boson can also be applied to the H boson, and both are hard to disentangle at hadron colliders.

3.2.2 The Scalar Quark Sector

Since the most relevant squarks for the MSSM Higgs boson sector are the $\tilde{t}$ and $\tilde{b}$ particles, here we explicitly list their mass matrices in the basis of the gauge eigenstates $\tilde{t}_L, \tilde{t}_R$ and $\tilde{b}_L, \tilde{b}_R$:

$$\mathcal{M}_{\tilde{t}}^2$$

$$= \begin{pmatrix} M_{\tilde{t}_L}^2 + m_t^2 + \cos 2\beta(\tfrac{1}{2} - \tfrac{2}{3}s_{\mathrm{w}}^2)M_Z^2 & m_t X_t \\ m_t X_t & M_{\tilde{t}_R}^2 + m_t^2 + \tfrac{2}{3}\cos 2\beta s_{\mathrm{w}}^2 M_Z^2 \end{pmatrix},$$
$$(79)$$

$$\mathcal{M}_{\tilde{b}}^2$$

$$= \begin{pmatrix} M_{\tilde{b}_L}^2 + m_b^2 + \cos 2\beta(-\tfrac{1}{2} + \tfrac{1}{3}s_{\mathrm{w}}^2)M_Z^2 & m_b X_b \\ m_b X_b & M_{\tilde{b}_R}^2 + m_b^2 - \tfrac{1}{3}\cos 2\beta s_{\mathrm{w}}^2 M_Z^2 \end{pmatrix}.$$
$$(80)$$

$M_{\tilde{t}_L}$, $M_{\tilde{t}_R}$, $M_{\tilde{b}_L}$ and $M_{\tilde{b}_R}$ are the (diagonal) soft SUSY-breaking parameters. We furthermore have

$$m_t X_t = m_t(A_t - \mu \cot \beta), \quad m_b X_b = m_b (A_b - \mu \tan \beta). \qquad (81)$$

The soft SUSY-breaking parameters A_t and A_b denote the trilinear Higgs–stop and Higgs–sbottom coupling, and μ is the Higgs mixing parameter. $SU(2)$ gauge invariance requires the relation

$$M_{\tilde{t}_L} = M_{\tilde{b}_L}. \qquad (82)$$

Diagonalising $\mathcal{M}_{\tilde{t}}^2$ and $\mathcal{M}_{\tilde{b}}^2$ with the mixing angles $\theta_{\tilde{t}}$ and $\theta_{\tilde{b}}$, respectively, yields the physical $\tilde{t}$ and $\tilde{b}$ masses: $m_{\tilde{t}_1}$, $m_{\tilde{t}_2}$, $m_{\tilde{b}_1}$ and $m_{\tilde{b}_2}$.

3.2.3 Higher-Order Corrections to Higgs Boson Masses

A review about this subject can be found in Ref. [40]. In the Feynman diagrammatic (FD) approach the higher-order corrected $\mathcal{CP}$-even Higgs boson masses in the MSSM are derived by finding the poles of the (h, H)-propagator matrix. The inverse of this matrix is given by

$$(\Delta_{\mathrm{Higgs}})^{-1} = -i \begin{pmatrix} p^2 - m_{H,\mathrm{tree}}^2 + \hat{\Sigma}_{HH}(p^2) & \hat{\Sigma}_{hH}(p^2) \\ \hat{\Sigma}_{hH}(p^2) & p^2 - m_{h,\mathrm{tree}}^2 + \hat{\Sigma}_{hh}(p^2) \end{pmatrix}.$$
$$(83)$$

Determining the poles of the matrix Δ_{Higgs} in Eq. (83) is equivalent to solving the equation

$$\left[p^2 - m_{h,\mathrm{tree}}^2 + \hat{\Sigma}_{hh}(p^2)\right] \left[p^2 - m_{H,\mathrm{tree}}^2 + \hat{\Sigma}_{HH}(p^2)\right] - \left[\hat{\Sigma}_{hH}(p^2)\right]^2 = 0. \quad (84)$$

The very leading one-loop correction to M_h^2 is given by

$$\Delta M_h^2 = G_F m_t^4 \log \left(\frac{m_{\tilde{t}_1} m_{\tilde{t}_2}}{m_t^2} \right), \tag{85}$$

where G_F denotes the Fermi constant. Equation (85) shows two important aspects: First, the leading loop corrections go with m_t^4, which is a "very large number." Consequently, the loop corrections can strongly affect M_h and push the mass beyond the reach of LEP [21, 41]. Second, the scalar fermion masses (in this case the scalar top masses) appear in the log entering the loop corrections (acting as a "cut-off" where the new physics enter). In this way the light Higgs boson mass depends on all other sectors via loop corrections. This dependence is particularly pronounced for the scalar top sector due to the large mass of the top quark.

The status of the available results for the self-energy contributions to Eq. (83) can be summarized as follows. For the one-loop part, the complete result within the MSSM is known [42–45]. The by far dominant one-loop contribution is the $\mathcal{O}(\alpha_t)$ term due to top and stop loops; see also Eq. (85) ($\alpha_t \equiv h_t^2/(4\pi)$, h_t being the superpotential top coupling). Concerning the two-loop effects, their computation is quite advanced and has now reached a stage such that all the presumably dominant contributions are known. They include the strong corrections, usually indicated as $\mathcal{O}(\alpha_t \alpha_s)$, and Yukawa corrections, $\mathcal{O}(\alpha_t^2)$, to the dominant one-loop $\mathcal{O}(\alpha_t)$ term, as well as the strong corrections to the bottom/sbottom one-loop $\mathcal{O}(\alpha_b)$ term ($\alpha_b \equiv h_b^2/(4\pi)$), i.e., the $\mathcal{O}(\alpha_b \alpha_s)$ contribution. The latter can be relevant for large values of $\tan \beta$. Presently, the $\mathcal{O}(\alpha_t \alpha_s)$ [46–50], $\mathcal{O}(\alpha_t^2)$ [46, 51, 52] and the $\mathcal{O}(\alpha_b \alpha_s)$ [53, 54] contributions to the self-energies are known for vanishing external momenta. In the bottom/sbottom corrections the all-order resummation of the $\tan \beta$-enhanced terms, $\mathcal{O}(\alpha_b(\alpha_s \tan \beta)^n)$, is also performed [55, 56]. The $\mathcal{O}(\alpha_t \alpha_b)$ and $\mathcal{O}(\alpha_b^2)$ corrections were presented in Ref. [57]. A "nearly full" two-loop effective potential calculation (including even the momentum dependence for the leading pieces and the leading three-loop corrections) has been published [58]. Most recently another leading three-loop calculation, valid for certain SUSY mass combinations, became available [59]. Taking the available loop corrections into account, the upper limit of M_h is shifted to [60]

$$M_h \leq 135 \text{ GeV} \tag{86}$$

(as obtained with the code FeynHiggs [48, 60–62]). This limit takes into account the experimental uncertainty for the top quark mass, see Eq. (55), as well as the intrinsic uncertainties from unknown higher-order corrections [60, 69].

The charged Higgs boson mass is obtained by solving the equation

$$p^2 - m_{H^\pm}^2 - \hat{\Sigma}_{H^- H^+}(p^2) = 0. \tag{87}$$

The charged Higgs boson self-energy is known at the one-loop level [63, 64].

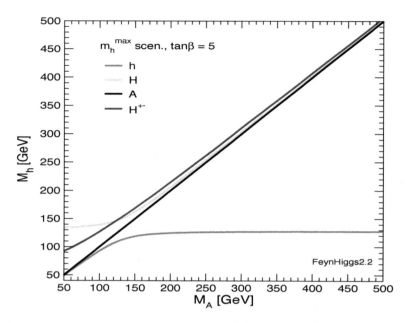

Figure 10. *The MSSM Higgs boson masses including higher-order correc-*
tions are shown as a function of M_A for $\tan\beta = 5$ in the m_h^{max} benchmark
scenario [65] (obtained with FeynHiggs *[48, 60–62]).*

3.3   MSSM Higgs Boson Searches at the LHC

The "decoupling limit" has been discussed for the tree-level couplings and
masses of the MSSM Higgs bosons in Section 3.2.1. This limit also persists
taking into account radiative corrections. The corresponding Higgs boson
masses are shown in Fig. 10 for $\tan\beta = 5$ in the m_h^{max} benchmark scenario [65]
obtained with FeynHiggs. For $M_A \gtrsim 150$ GeV the lightest Higgs boson mass
approaches its upper limit (depending on the SUSY parameters), and the
heavy Higgs boson masses are nearly degenerate. Furthermore, also the light
Higgs boson couplings including loop corrections approach their SM-value.
Consequently, for $M_A \gtrsim 150$ GeV the experimental searches for the lightest
MSSM Higgs boson, see Sect. 2.3, can be performed very similarly to the SM
Higgs boson searches (with $M_H^{\mathrm{SM}} = M_h$).

The various production cross sections at the LHC are shown in Fig. 11
(for $\sqrt{s} = 14$ TeV). For low masses the light Higgs cross sections are visible,
and for $M_H \gtrsim 130$ GeV the heavy $\mathcal{CP}$-even Higgs cross section is displayed,
while the cross sections for the $\mathcal{CP}$-odd A boson are given for the whole mass
range. As discussed in Section 3.2.1 the g_{Abb} coupling is enhanced by $\tan\beta$
with respect to the corresponding SM value. Consequently, the $b\bar{b}A$ cross
section is the largest or second-largest cross section for all M_A, despite the
relatively small value of $\tan\beta = 5$. For larger $\tan\beta$, see Eq. (62), this cross

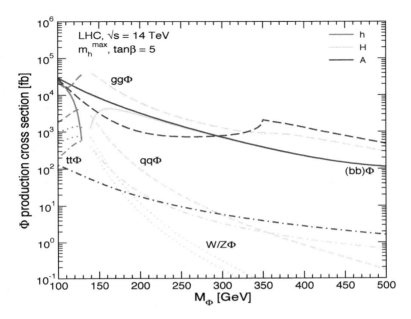

Figure 11. *Overview about the various neutral Higgs boson production cross sections at the LHC shown as a function of M_A for $\tan\beta = 5$ in the $m_h^{\max}$ scenario.*

section can become even more dominant. Furthermore, the coupling of the heavy $\mathcal{CP}$-even Higgs boson becomes very similar to the one of the A boson, and the two production cross sections, $b\bar{b}A$ and $b\bar{b}H$, are indistinguishable in the plot for $M_A > 200$ GeV.

Following the above discussion, the main search channel for heavy Higgs bosons at the LHC for $M_A \gtrsim 200$ GeV is the production in association with bottom quarks and the subsequent decay to tau leptons, $b\bar{b} \to b\bar{b} H/A \to b\bar{b}\,\tau^+\tau^-$. For heavy supersymmetric particles, with masses far above the Higgs boson mass scale, one has for the production and decay of the A boson [66]

$$\sigma(b\bar{b}A) \times \mathrm{BR}(A \to b\bar{b}) \simeq \sigma(b\bar{b}H)_{\mathrm{SM}} \frac{\tan^2\beta}{\left(1+\Delta_b\right)^2} \times \frac{9}{\left(1+\Delta_b\right)^2 + 9}, \quad (88)$$

$$\sigma(gg, b\bar{b} \to A) \times \mathrm{BR}(A \to \tau^+\tau^-) \simeq \sigma(gg, b\bar{b} \to H)_{\mathrm{SM}} \frac{\tan^2\beta}{\left(1+\Delta_b\right)^2 + 9}, \quad (89)$$

where $\sigma(b\bar{b}H)_{\mathrm{SM}}$ and $\sigma(gg, b\bar{b} \to H)_{\mathrm{SM}}$ denote the values of the corresponding

SM Higgs boson production cross sections for $M_H^{\rm SM} = M_A$. Δ_b is given by [55]

$$\Delta_b = \frac{2\alpha_s}{3\pi} \, m_{\tilde{g}} \, \mu \, \tan\beta \, \times \, I(m_{\tilde{b}_1}, m_{\tilde{b}_2}, m_{\tilde{g}}) + \frac{\alpha_t}{4\pi} \, A_t \, \mu \, \tan\beta \, \times \, I(m_{\tilde{t}_1}, m_{\tilde{t}_2}, |\mu|),$$
(90)

where the function I arises from the one-loop vertex diagrams and scales as $I(a, b, c) \sim 1/\max(a^2, b^2, c^2)$. Here $m_{\tilde{g}}$ is the gluino mass, and μ is the Higgs mixing parameter. As a consequence, the $b\bar{b}$ production rate depends sensitively on $\Delta_b \propto \mu \tan\beta$ because of the factor $1/(1 + \Delta_b)^2$, while this leading dependence on Δ_b cancels out in the $\tau^+\tau^-$ production rate. The formulas above apply, within a good approximation, also to the heavy $\mathcal{CP}$-even Higgs boson in the large $\tan\beta$ regime. Therefore, the production and decay rates of H are governed by formulas similar to those given above, leading to an approximate enhancement by a factor of 2 of the production rates with respect to the ones that would be obtained in the case of the single production of the $\mathcal{CP}$-odd Higgs boson as given in Eqs. (88) and (89).

Of particular interest is the "LHC wedge" region, i.e., the region in which only the light $\mathcal{CP}$-even MSSM Higgs boson, but none of the heavy MSSM Higgs bosons can be detected at the LHC at the $5\,\sigma$ level. It appears for $M_A \gtrsim 200$ GeV at intermediate $\tan\beta$ and widens to larger $\tan\beta$ values for larger M_A. Consequently, in the "LHC wedge" only an SM-like light Higgs boson can be discovered at the LHC. This region is bounded from above by the $5\,\sigma$ discovery contours for the heavy neutral MSSM Higgs bosons as described above. These discovery contours depend sensitively on the Higgs mass parameter μ. The dependence on μ enters in two different ways, on the one hand via higher-order corrections through $\Delta_b \propto \mu \tan\beta$, and on the other hand via the kinematics of Higgs decays into charginos and neutralinos, where μ enters in their respective mass matrices [3].

In Fig. 12 we show the $5\,\sigma$ discovery regions for the heavy neutral MSSM Higgs bosons in the channel $b\bar{b} \to b\bar{b}\, H/A, H/A \to \tau^+\tau^- \to$ jets [67]. As explained above, these discovery contours correspond to the upper bound of the "LHC wedge." A strong variation with the sign and the size of μ can be observed and should be taken into account in experimental and phenomenological analyses. The same higher-order corrections are relevant once a possible heavy Higgs boson signal at the LHC will be interpreted in terms of the underlying parameter space. From Eq. (90) it follows that an observed production cross section can be correctly connected to μ and $\tan\beta$ only if the scalar top and bottom masses, the gluino mass and the trilinear Higgs–stop coupling are measured and taken properly into account.

3.4   Electroweak Precision Observables

Also within SUSY one can attempt to fit the unknown parameters to the existing experimental data, in a similar fashion as was discussed in Section 2.4. However, fits within the MSSM differs from the SM fit in various ways. First, the number of free parameters is substantially larger in the MSSM, even when

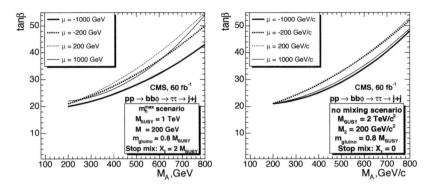

Figure 12. *The* 5σ *discovery regions (i.e., the upper bound of the "LHC wedge" region) for the heavy neutral Higgs bosons in the channel* $b\bar{b} \rightarrow b\bar{b} H/A, H/A \rightarrow \tau^+\tau^- \rightarrow$ *jets (taken from Ref. [67]).*

restricting to GUT-based models as discussed below. On the other hand, more observables can be taken into account, providing extra constraints on the fit. Within the MSSM the additional observables included are the anomalous magnetic moment of the muon $(g-2)_\mu$, B-physics observables such as BR($b \rightarrow s\gamma$) or BR($B_s \rightarrow \mu\mu$), and the relic density of cold dark matter (CDM), which can be provided by the lightest SUSY particle, the neutralino. These additional constraints would either have a minor impact on the best-fit regions or cannot be accommodated in the SM. Finally, as discussed in the previous subsections, whereas the light Higgs boson mass is a free parameter in the SM, it is a function of the other parameters in the MSSM. In this way, for example, the masses of the scalar tops and bottoms enter not only directly into the prediction of the various observables, but also indirectly via their impact on M_h.

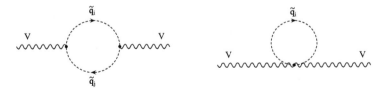

Figure 13. *Feynman diagrams for the contribution of scalar quark loops to the gauge boson self-energies at one-loop order,* $V = W, Z$, $\tilde{q} = \tilde{t}, \tilde{b}$.

Within the MSSM the dominant SUSY correction to electroweak precision observables arises from the scalar top and bottom contribution to the ρ parameter; see Eq. (53). The leading diagrams are shown in Fig. 13.

Generically one finds $\Delta\rho^{\text{SUSY}} > 0$, leading, for instance, to an upward shift in the prediction of M_W with respect to the SM prediction. The experimental result and the theory prediction of the SM and the MSSM for M_W are compared in Fig. 14 (updated from Ref. [68]). The predictions within the two models give rise to two bands in the m_t–M_W plane with only a relatively small overlap sliver (indicated by a dark-shaded area in Fig. 14). The allowed parameter region in the SM (the medium-shaded and dark-shaded bands, corresponding to the SM prediction in Fig. 6) arises from varying the only free parameter of the model, the mass of the SM Higgs boson, from $M_H^{\mathrm{SM}} = 114$ GeV, the LEP exclusion bound [21] (upper edge of the dark-shaded area), to 400 GeV (lower edge of the medium-shaded area). The light-shaded and the dark-shaded areas indicate allowed regions for the unconstrained MSSM, obtained from scattering the relevant parameters independently [68]. The decoupling limit with SUSY masses of $\mathcal{O}(2\text{ TeV})$ yields the lower edge of the dark-shaded area. Thus, the overlap region between the predictions of the two models corresponds in the SM to the region where the Higgs boson is light, i.e., in the MSSM allowed region ($M_h \lesssim 135$ GeV, see Eq. (86)). In the MSSM it corresponds to the case where all superpartners are heavy, i.e., the decoupling region of the MSSM. The current 68 and 95% C.L. experimental results for m_t, Eq. (55), and M_W, Eq. (54), are also indicated in the plot. As can be seen from Fig. 14, the current experimental 68% C.L. region for m_t and M_W exhibits a slight preference of the MSSM over the SM. This example indicates that the experimental measurement of M_W in combination with m_t prefers, within the MSSM, not too heavy SUSY mass scales.

As mentioned above, in order to restrict the number of free parameters in the MSSM one can resort to GUT-based models. Most fits have been performed in the Constrained MSSM (CMSSM), in which the input scalar masses m_0, gaugino masses $m_{1/2}$ and soft trilinear parameters A_0 are each universal at the GUT scale, $M_{\mathrm{GUT}} \approx 2 \times 10^{16}$ GeV, and in the non-universal Higgs mass model (NUHM1), in which a common SUSY-breaking contribution to the Higgs masses is allowed to be non-universal.

Here we follow the results obtained in Refs. [70–72], where an overview about different fitting techniques and an extensive list of references can be found in Ref. [72]. The computer code used for the fits shown below is `MasterCode` [70–73], which includes the following theoretical codes. For the RGE running of the soft SUSY-breaking parameters, it uses `SoftSUSY` [74], which is combined consistently with the codes used for the various low-energy observables: `FeynHiggs` [48, 60–62] is used for the evaluation of the Higgs masses and a_μ^{SUSY} (see also [75–78]); for the other electroweak precision data we have included a code based on [68, 79]; `SuFla` [80, 81] and `SuperIso` [82, 83] are used for flavour-related observables; and for dark-matter-related observables `MicrOMEGAs` [84] and `DarkSUSY` [85] are used. In the combination of the various codes, `MasterCode` makes extensive use of the SUSY Les Houches Accord [86, 87].

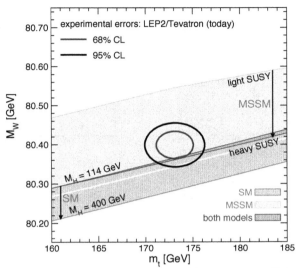

Figure 14. *Prediction for M_W in the MSSM and the SM (see text) as a function of m_t in comparison with the present experimental results for M_W and m_t (updated from Ref. [68]; see Ref. [69] for details).*

The global χ^2 likelihood function, which combines all theoretical predictions with experimental constraints, is now given as

$$\chi^2 = \sum_i^N \frac{(C_i - P_i)^2}{\sigma(C_i)^2 + \sigma(P_i)^2} + \sum_i^M \frac{(f_{\text{SM}_i}^{\text{obs}} - f_{\text{SM}_i}^{\text{fit}})^2}{\sigma(f_{\text{SM}_i})^2}$$
$$+ \chi^2(\text{BR}(B_s \to \mu\mu)) + \chi^2(\text{SUSY search limits}). \qquad (91)$$

Here N is the number of observables studied, C_i represents an experimentally measured value (constraint) and each P_i defines a prediction for the corresponding constraint that depends on the supersymmetric parameters. The experimental uncertainty, $\sigma(C_i)$, of each measurement is taken to be both statistically and systematically independent of the corresponding theoretical uncertainty, $\sigma(P_i)$, in its prediction (all the details can be found in Ref. [72]). $\chi^2(\text{BR}(B_s \to \mu\mu))$ denotes the χ^2 contributions from the one measurement for which only a one-sided bound is available so far. The lower limits from the direct searches for SUSY particles at LEP [88] are included as one-sided limits, denoted by "$\chi^2(\text{SUSY search limits})$" in Eq. (91). Furthermore, the three SM parameters $f_{\text{SM}} = \{\Delta\alpha_{\text{had}}, m_t, M_Z\}$ are included as fit parameters and allowed to vary with their current experimental resolutions $\sigma(f_{\text{SM}})$.

The results for the fits of M_h in the CMSSM and the NUHM1 are shown in Fig. 15 in the left and right plots, respectively. Also shown in Fig. 15 are the LEP exclusion on a SM Higgs (light shading) and the ranges that

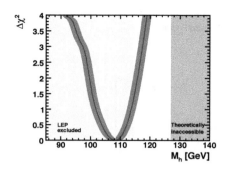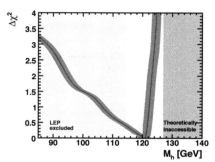

Figure 15. *The χ^2 functions for M_h in the CMSSM (left) and the NUHM1 (right) [72], including the theoretical uncertainties (dark bands). Also shown is the mass range excluded for an SM-like Higgs boson (light shading), and the ranges theoretically inaccessible in the supersymmetric models studied.*

are theoretically inaccessible in the supersymmetric models studied (medium shading). The LEP exclusion is directly applicable to the CMSSM, but cannot strictly be applied in the NUHM1; see Ref. [72] for details.

In the case of the CMSSM, we see in the left panel of Fig. 15 that the minimum of the χ^2 function occurs below the LEP exclusion limit. The fit result is still compatible at the 95% C.L. with the search limit, similar to the SM case. In the case of the NUHM1, shown in the right panel of Fig. 15, we see that the minimum of the χ^2 function occurs *above* the LEP lower limit on the mass of an SM Higgs. Thus, within the NUHM1 the combination of all other experimental constraints *naturally* evades the LEP Higgs constraints, and no tension between M_h and the experimental bounds exists.

Acknowledgements

I thank the organizers for their hospitality and for creating a very stimulating environment, as well as, together with the participants, for an exceptionally nice school dinner/farewell party.

References

[1] P. W. Higgs, Phys. Lett. **12** (1964) 132; Phys. Rev. Lett. **13** (1964) 508; Phys. Rev. **145** (1966) 1156;
F. Englert and R. Brout, Phys. Rev. Lett. **13** (1964) 321;
G. S. Guralnik, C. R. Hagen and T. W. B. Kibble, Phys. Rev. Lett. **13** (1964) 585.

[2] S.L. Glashow, Nucl. Phys. **B 22** (1961) 579;
S. Weinberg, Phys. Rev. Lett. **19** (1967) 19;

A. Salam, in: *Proceedings of the 8th Nobel Symposium*, Editor N. Svartholm, Stockholm, 1968.

[3] H. Nilles, Phys. Rept. **110** (1984) 1;
H. Haber and G. Kane, Phys. Rept. **117** (1985) 75;
R. Barbieri, Riv. Nuovo Cim. **11** (1988) 1.

[4] S. Weinberg, Phys. Rev. Lett. **37** (1976) 657;
J. Gunion, H. Haber, G. Kane and S. Dawson, *The Higgs Hunter's Guide* (Perseus Publishing, Cambridge, MA, 1990), and references therein.

[5] P. Fayet, Nucl. Phys. **B 90** (1975) 104; Phys. Lett. **B 64** (1976) 159; Phys. Lett. **B 69** (1977) 489; Phys. Lett. **B 84** (1979) 416;
H. Nilles, M. Srednicki and D. Wyler, Phys. Lett. **B 120** (1983) 346;
J. Frere, D. Jones and S. Raby, Nucl. Phys. **B 222** (1983) 11;
J. Derendinger and C. Savoy, Nucl. Phys. **B 237** (1984) 307;
J. Ellis, J. Gunion, H. Haber, L. Roszkowski and F. Zwirner, Phys. Rev. **D 39** (1989) 844;
M. Drees, Int. J. Mod. Phys. **A 4** (1989) 3635.

[6] N. Arkani-Hamed, A. Cohen and H. Georgi, Phys. Lett. **B 513** (2001) 232 [arXiv:hep-ph/0105239];
N. Arkani-Hamed, A. Cohen, T. Gregoire and J. Wacker, JHEP **0208** (2002) 020 [arXiv:/0202089].

[7] N. Arkani-Hamed, S. Dimopoulos and G. Dvali, Phys. Lett. **B 429** (1998) 263 [arXiv:hep-ph/9803315]; Phys. Lett. **B 436** (1998) 257 [arXiv:hep-ph/9804398];
I. Antoniadis, Phys. Lett. **B 246** (1990) 377;
J. Lykken, Phys. Rev. **D 54** (1996) 3693 [arXiv:hep-th/9603133];
L. Randall and R. Sundrum, Phys. Rev. Lett. **83** (1999) 3370 [arXiv:hep-ph/9905221].

[8] G. Weiglein et al. [LHC/ILC Study Group], Phys. Rept. **426** (2006) 47 [arXiv:hep-ph/0410364].

[9] A. De Roeck et al., Eur. Phys. J. **C66** (2010) 525 [arXiv:0909.3240 [hep-ph]].

[10] N. Cabibbo, L. Maiani, G. Parisi and R. Petronzio, Nucl. Phys. **B 158** (1979) 295;
R. Flores and M. Sher, Phys. Rev. **D 27** (1983) 1679;
M. Lindner, Z. Phys. **C 31** (1986) 295;
M. Sher, Phys. Rept. **179** (1989) 273;
J. Casas, J. Espinosa and M. Quiros, Phys. Lett. **342** (1995) 171. [arXiv:hep-ph/9409458].

[11] G. Altarelli and G. Isidori, Phys. Lett. **B 337** (1994) 141; J. Espinosa and M. Quiros, Phys. Lett. **353** (1995) 257 [arXiv:hep-ph/9504241].

[12] T. Hambye and K. Riesselmann, Phys. Rev. **D 55** (1997) 7255 [arXiv:hep-ph/9610272].

[13] T. Hahn, S. Heinemeyer, F. Maltoni, G. Weiglein and S. Willenbrock, arXiv:hep-ph/0607308.

[14] G. Aad et al. [The ATLAS Collaboration], arXiv:0901.0512.

[15] G. Bayatian et al. [CMS Collaboration], J. Phys. **G 34** (2007) 995.

[16] LEP Electroweak Working Group,
see: `lepewwg.web.cern.ch/LEPEWWG/Welcome.html`.

[17] Tevatron Electroweak Working Group, see: `tevewwg.fnal.gov`.

[18] The ALEPH, DELPHI, L3, OPAL, SLD Collaborations, the LEP Electroweak Working Group, the SLD Electroweak and Heavy Flavour Groups, Phys. Rept. **427** (2006) 257 [arXiv:hep-ex/0509008];
[The ALEPH, DELPHI, L3 and OPAL Collaborations, the LEP Electroweak Working Group], arXiv:hep-ex/0612034.

[19] A. Sirlin, Phys. Rev. **D 22** (1980) 971; W. Marciano and A. Sirlin, Phys. Rev. **D 22** (1980) 2695.

[20] M. Veltman, Nucl. Phys. **B 123** (1977) 89.

[21] LEP Higgs Working Group, Phys. Lett. **B 565** (2003) 61 [arXiv:hep-ex/0306033].

[22] ALEPH Collaboration, CDF Collaboration, D0 Collaboration, DELPHI Collaboration, L3 Collaboration, OPAL Collaboration, SLD Collaboration, LEP Electroweak Working Group, Tevatron Electroweak Working Group, SLD electroweak heavy flavour groups, arXiv:0911.2604 [hep-ex].

[23] Tevatron Electroweak Working Group and CDF Collaboration and D0 Collaboration, arXiv:0903.2503 [hep-ex].

[24] M. Grünewald, J. Phys. Conf. Ser. **110** (2008) 042008 [arXiv:0709.3744 [hep-ph]]; arXiv:0710.2838 [hep-ex].

[25] M. Grünewald, *priv. communication*.

[26] G. Montagna, O. Nicrosini, F. Piccinini and G. Passarino, Comput. Phys. Commun. **117** (1999) 278 [arXiv:hep-ph/9804211].

[27] D. Bardin et al., Comput. Phys. Commun. **133** (2001) 229 [arXiv:hep-ph/9908433]; A. Arbuzov et al., Comput. Phys. Commun. **174** (2006) 728 [arXiv:hep-ph/0507146].

[28] [CDF Collaboration and D0 Collaboration], arXiv:0903.4001 [hep-ex].

[29] H. Flacher, M. Goebel, J. Haller, A. Hocker, K. Moenig and J. Stelzer, Eur. Phys. J. **C 60** (2009) 543 [arXiv:0811.0009 [hep-ph]];
see: `cern.ch/gfitter`.

[30] [CDF Collaboration and D0 Collaboration], arXiv:0911.3930 [hep-ex].

[31] U. Baur, R. Clare, J. Erler, S. Heinemeyer, D. Wackeroth, G. Weiglein and D. Wood, arXiv:hep-ph/0111314.

[32] J. Erler, S. Heinemeyer, W. Hollik, G. Weiglein and P. Zerwas, Phys. Lett. **B 486** (2000) 125 [arXiv:hep-ph/0005024];
J. Erler and S. Heinemeyer, arXiv:hep-ph/0102083.

[33] R. Hawkings and K. Mönig, EPJdirect **C8** (1999) 1 [arXiv:hep-ex/9910022].

[34] G. Wilson, LC-PHSM-2001-009, see: `www.desy.de/~lcnotes/notes.html`.

[35] S. Heinemeyer et al., arXiv:hep-ph/0511332.

[36] S. Glashow and S. Weinberg, Phys. Rev. **D 15** (1977) 1958.

[37] C. Amsler et al. [Particle Data Group], Phys. Lett. **B 667** (2008) 1.

[38] A. Djouadi, Phys. Rept. **459** (2008) 1 [arXiv:hep-ph/0503173].

[39] J. Gunion, H. Haber, G. Kane and S. Dawson, *The Higgs Hunter's Guide*, Addison-Wesley, 1990.

[40] S. Heinemeyer, Int. J. Mod. Phys. **A 21** (2006) 2659 [arXiv:hep-ph/0407244].

[41] LEP Higgs working group, Eur. Phys. J. **C 47** (2006) 547 [arXiv:hep-ex/0602042].

[42] J. Ellis, G. Ridolfi and F. Zwirner, Phys. Lett. **B 257** (1991) 83;
Y. Okada, M. Yamaguchi and T. Yanagida, Prog. Theor. Phys. **85** (1991) 1;
H. Haber and R. Hempfling, Phys. Rev. Lett. **66** (1991) 1815.

[43] A. Brignole, Phys. Lett. **B 281** (1992) 284.

[44] P. Chankowski, S. Pokorski and J. Rosiek, Phys. Lett. **B 286** (1992) 307; Nucl. Phys. **B 423** (1994) 437 [arXiv:hep-ph/9303309].

[45] A. Dabelstein, Nucl. Phys. **B 456** (1995) 25 [arXiv:hep-ph/9503443]; Z. Phys. **C 67** (1995) 495 [arXiv:hep-ph/9409375].

[46] R. Hempfling and A. Hoang, Phys. Lett. **B 331** (1994) 99 [arXiv:hep-ph/9401219].

[47] S. Heinemeyer, W. Hollik and G. Weiglein, Phys. Rev. **D 58** (1998) 091701 [arXiv:hep-ph/9803277]; Phys. Lett. **B 440** (1998) 296 [arXiv:hep-ph/9807423].

[48] S. Heinemeyer, W. Hollik and G. Weiglein, Eur. Phys. Jour. **C 9** (1999) 343 [arXiv:hep-ph/9812472].

[49] R. Zhang, Phys. Lett. **B 447** (1999) 89 [arXiv:hep-ph/9808299];
J. Espinosa and R. Zhang, JHEP **0003** (2000) 026 [arXiv:hep-ph/9912236].

[50] G. Degrassi, P. Slavich and F. Zwirner, Nucl. Phys. **B 611** (2001) 403 [arXiv:hep-ph/0105096].

[51] J. Espinosa and R. Zhang, Nucl. Phys. **B 586** (2000) 3 [arXiv:hep-ph/0003246].

[52] A. Brignole, G. Degrassi, P. Slavich and F. Zwirner, Nucl. Phys. **B 631** (2002) 195 [arXiv:hep-ph/0112177].

[53] A. Brignole, G. Degrassi, P. Slavich and F. Zwirner, Nucl. Phys. **B 643** (2002) 79 [arXiv:hep-ph/0206101].

[54] S. Heinemeyer, W. Hollik, H. Rzehak and G. Weiglein, Eur. Phys. J. **C 39** (2005) 465 [arXiv:hep-ph/0411114].

[55] T. Banks, Nucl. Phys. **B 303** (1988) 172;
L. Hall, R. Rattazzi and U. Sarid, Phys. Rev. **D 50** (1994) 7048 [arXiv:hep-ph/9306309];
R. Hempfling, Phys. Rev. **D 49** (1994) 6168;
M. Carena, M. Olechowski, S. Pokorski and C. Wagner, Nucl. Phys. **B 426** (1994) 269 [arXiv:hep-ph/9402253].

[56] M. Carena, D. Garcia, U. Nierste and C. Wagner, Nucl. Phys. **B 577** (2000) 577 [arXiv:hep-ph/9912516].

[57] G. Degrassi, A. Dedes and P. Slavich, Nucl. Phys. **B 672** (2003) 144 [arXiv:hep-ph/0305127].

[58] S. Martin, Phys. Rev. **D 65** (2002) 116003 [arXiv:hep-ph/0111209]; Phys. Rev. **D 66** (2002) 096001 [arXiv:hep-ph/0206136]; Phys. Rev. **D 67** (2003) 095012 [arXiv:hep-ph/0211366]; Phys. Rev. **D 68** (2003) 075002 [arXiv:hep-ph/0307101]; Phys. Rev. **D 70** (2004) 016005 [arXiv:hep-ph/0312092]; Phys. Rev. **D 71** (2005) 016012 [arXiv:hep-ph/0405022]; Phys. Rev. **D 71** (2005) 116004 [arXiv:hep-ph/0502168]; Phys. Rev. **D 75** (2007) 055005 [arXiv:hep-ph/0701051];
S. Martin and D. Robertson, Comput. Phys. Commun. **174** (2006) 133 [arXiv:hep-ph/0501132].

[59] R. Harlander, P. Kant, L. Mihaila and M. Steinhauser, Phys. Rev. Lett. **100** (2008) 191602; Phys. Rev. Lett. **101** (2008) 039901] [arXiv:0803.0672 [hep-ph]].

[60] G. Degrassi, S. Heinemeyer, W. Hollik, P. Slavich and G. Weiglein, Eur. Phys.
     J. **C 28** (2003) 133 [arXiv:hep-ph/0212020].

[61] S. Heinemeyer, W. Hollik and G. Weiglein, Comput. Phys. Commun. **124**
     (2000) 76 [arXiv:hep-ph/9812320]; see: www.feynhiggs.de.

[62] M. Frank, T. Hahn, S. Heinemeyer, W. Hollik, H. Rzehak and G. Weiglein,
     JHEP **0702** (2007) 047 [arXiv:hep-ph/0611326].

[63] M. Diaz and H. Haber, Phys. Rev. **D 45** (1992) 4246.

[64] M. Frank, PhD thesis, University of Karlsruhe, 2002.

[65] M. Carena, S. Heinemeyer, C. Wagner and G. Weiglein, Eur. Phys. J. **C 26**
     (2003) 601 [arXiv:hep-ph/0202167].

[66] M. Carena, S. Heinemeyer, C. Wagner and G. Weiglein, Eur. Phys. J. **C 45**
     (2006) 797 [arXiv:hep-ph/0511023].

[67] S. Gennai, S. Heinemeyer, A. Kalinowski, R. Kinnunen, S. Lehti, A. Nikitenko
     and G. Weiglein, Eur. Phys. J. **C 52** (2007) 383 [arXiv:0704.0619 [hep-ph]];
     M. Hashemi, S. Heinemeyer, R. Kinnunen, A. Nikitenko and G. Weiglein,
     arXiv:0804.1228 [hep-ph].

[68] S. Heinemeyer, W. Hollik, D. Stockinger, A. M. Weber and G. Weiglein, JHEP
     0608 (2006) 052 [arXiv:hep-ph/0604147].

[69] S. Heinemeyer, W. Hollik and G. Weiglein, Phys. Rept. **425** (2006) 265
     [arXiv:hep-ph/0412214].

[70] O. Buchmueller et al., Phys. Lett. **B 657** (2007) 87 [arXiv:0707.3447 [hep-ph]].

[71] O. Buchmueller et al., JHEP **0809** (2008) 117 [arXiv:0808.4128 [hep-ph]].

[72] O. Buchmueller et al., Eur. Phys. J. **C 64** (2009) 391 [arXiv:0907.5568 [hep-ph]].

[73] See: cern.ch/mastercode.

[74] B. Allanach, Comput. Phys. Commun. **143** (2002) 305 [arXiv:hep-ph/0104145].

[75] T. Moroi, Phys. Rev. **D 53** (1996) 6565 [Erratum-ibid. **D 56** (1997) 4424]
     [arXiv:hep-ph/9512396].

[76] G. Degrassi and G. F. Giudice, Phys. Rev. **D 58** (1998) 053007 [arXiv:hep-
     ph/9803384].

[77] S. Heinemeyer, D. Stockinger and G. Weiglein, Nucl. Phys. **B 690** (2004) 62
     [arXiv:hep-ph/0312264].

[78] S. Heinemeyer, D. Stockinger and G. Weiglein, Nucl. Phys. **B 699** (2004) 103
     [arXiv:hep-ph/0405255].

[79] S. Heinemeyer, W. Hollik, A. M. Weber and G. Weiglein, JHEP **0804** (2008)
     039 [arXiv:0710.2972 [hep-ph]].

[80] G. Isidori and P. Paradisi, Phys. Lett. **B 639** (2006) 499 [arXiv:hep-
     ph/0605012].

[81] G. Isidori, F. Mescia, P. Paradisi and D. Temes, Phys. Rev. **D 75** (2007) 115019
     [arXiv:hep-ph/0703035], and references therein.

[82] F. Mahmoudi, Comput. Phys. Commun. **178** (2008) 745 [arXiv:0710.2067 [hep-
     ph]]; Comput. Phys. Commun. **180** (2009) 1579 [arXiv:0808.3144 [hep-ph]].

[83] D. Eriksson, F. Mahmoudi and O. Stal, JHEP **0811** (2008) 035
     [arXiv:0808.3551 [hep-ph]].

[84] G. Belanger, F. Boudjema, A. Pukhov and A. Semenov, Comput. Phys. Com-
     mun. **176** (2007) 367 [arXiv:hep-ph/0607059]; Comput. Phys. Commun. **149**

(2002) 103 [arXiv:hep-ph/0112278]; Comput. Phys. Commun. **174** (2006) 577 [arXiv:hep-ph/0405253].

[85] P. Gondolo, J. Edsjo, P. Ullio, L. Bergstrom, M. Schelke and E. Baltz, New Astron. Rev. **49** (2005) 149; JCAP **0407** (2004) 008 [arXiv:astro-ph/0406204].

[86] P. Skands et al., JHEP **0407** (2004) 036 [arXiv:hep-ph/0311123].

[87] B. Allanach et al., Comput. Phys. Commun. **180** (2009) 8 [arXiv:0801.0045 [hep-ph]].

[88] LEP Supersymmetry Working Group, see: lepsusy.web.cern.ch/lepsusy/.

B Physics in the LHC Era

Gino Isidori

INFN, National Laboratory of Frascati, Frascati, Italy
Institute for Advanced Study, Technical University of Munich, Munich, Germany

1   Introduction

In the last few years there has been great experimental progress in quark flavour physics. The validity of the Standard Model (SM) has been strongly reinforced by a series of challenging experimental tests in B, D and K decays. All the relevant SM parameters controlling quark-flavour dynamics (the quark masses and the angles of the Cabibbo–Kobayashi–Maskawa matrix [1,2]) have been determined with good accuracy. More importantly, several suppressed observables (such as Δm_{B_d}, Δm_{B_s}, $\mathcal{A}_{K\Psi}^{\mathrm{CP}}$, $B \to X_s\gamma$, ϵ_K, ...) that are potentially sensitive to physics beyond the SM have been measured with good accuracy and show no deviations from the SM. The situation is somewhat similar to the flavour-conserving electroweak precision observables after LEP: the SM works very well and genuine one-loop electroweak effects have been tested with relative accuracy in the 10%–30% range. As in the case of electroweak observables, non-standard effects in the quark flavour sector can only appear as small corrections to the leading SM contribution.

Despite the success of the SM in electroweak and quark-flavour physics, we also have clear indications that this theory is not complete: the phenomenon of neutrino oscillations and the evidence for dark matter cannot be explained within the SM. The SM is also affected by a serious theoretical problem because of the instability of the Higgs sector under quantum corrections. We do not yet have enough information to unambiguously determine how the SM Lagrangian should be modified; however, most realistic proposals point toward the existence of new degrees of freedom in the TeV range, possibly accessible at the high-p_T experiments at the LHC.

Assuming that the SM is not a complete theory, the precise tests of flavour dynamics performed so far imply a series of challenging constraints about the

new theory: if there are new degrees of freedom at the TeV scale, present
data tell us that they must possess a highly non-generic flavour structure.
This structure is far from being established and its deeper investigation is the
main goal of continuing high-precision flavour-physics in the LHC era. In these
lectures we focus on the interest and potential impact of future measurements
in the B-meson system in this perspective.

The lectures are organized as follows. In the first lecture we briefly recall the main features of flavour physics within the SM. We also address in
general terms the so-called *flavour problem*, namely the challenge to any SM
extension posed by the success of the SM in flavour physics. In the second
lecture we analyse in some detail the SM predictions for some of the most
interesting B physics observables to be measured in the LHC era. In the last
lecture we analyse flavour physics in various realistic beyond-the-SM scenarios, discussing how they can be tested by future experiments.

The presentation of these lectures is somehow original, but all the material
can be found, often with more details, in various sets of lectures [3] and review
articles [4–6] in the recent literature.

2 Flavour Physics within the SM and the Flavour Problem

2.1 The Flavour Sector of the SM

The Standard Model (SM) Lagrangian can be divided into two main parts,
the gauge and the Higgs (or symmetry breaking) sector. The gauge sector is
extremely simple and highly symmetric: it is completely specified by the local
symmetry $\mathcal{G}_{\text{local}}^{\text{SM}} = SU(3)_C \times SU(2)_L \times U(1)_Y$ and by the fermion content,

$$
\begin{aligned}
\mathcal{L}_{\text{gauge}}^{\text{SM}} &= \sum_{i=1\ldots3} \sum_{\psi=Q_L^i\ldots E_R^i} \bar{\psi} i \not{D} \psi \\
&\quad - \frac{1}{4} \sum_{a=1\ldots8} G_{\mu\nu}^a G_{\mu\nu}^a - \frac{1}{4} \sum_{a=1\ldots3} W_{\mu\nu}^a W_{\mu\nu}^a - \frac{1}{4} B_{\mu\nu} B_{\mu\nu}.
\end{aligned}
\tag{1}
$$

The fermion content consists of five fields with different quantum numbers
under the gauge groups,[1]

$$
Q_L^i(3,2)_{+1/6}, \quad U_R^i(3,1)_{+2/3}, \quad D_R^i(3,1)_{-1/3}, \quad L_L^i(1,2)_{-1/2}, \quad E_R^i(1,1)_{-1}, \tag{2}
$$

each of them appearing in three different replica or flavours ($i = 1, 2, 3$).

This structure gives rise to a large *global* flavour symmetry of $\mathcal{L}_{\text{gauge}}^{\text{SM}}$. Both
the local and the global symmetries of $\mathcal{L}_{\text{gauge}}^{\text{SM}}$ are broken by the introduction

---

[1] The notation used to indicate each field is $\psi(A, B)_Y$, where A and B denote the
representation under the $SU(3)_C$ and $SU(2)_L$ groups, respectively, and Y is the $U(1)_Y$
charge.

of an $SU(2)_L$ scalar doublet ϕ, or the Higgs field. The local symmetry is *spontaneously broken* by the vacuum expectation value of the Higgs field, $\langle\phi\rangle = v = (2\sqrt{2}G_F)^{-1/2} \approx 174$ GeV, while the global flavour symmetry is *explicitly broken* by the Yukawa interaction of ϕ with the fermion fields:

$$-\mathcal{L}^{\text{SM}}_{\text{Yukawa}} = Y_d^{ij}\bar{Q}_L^i\phi D_R^j + Y_u^{ij}\bar{Q}_L^i\tilde{\phi}U_R^j + Y_e^{ij}\bar{L}_L^i\phi E_R^j + \text{h.c.} \qquad (\tilde{\phi} = i\tau_2\phi^\dagger). \quad (3)$$

The large global flavour symmetry of $\mathcal{L}^{\text{SM}}_{\text{gauge}}$, corresponding to the independent unitary rotations in flavour space of the five fermion fields in Eq. (2), is a $U(3)^5$ group. This can be decomposed as follows:

$$\mathcal{G}_{\text{flavour}} = U(3)^5 \times \mathcal{G}_q \times \mathcal{G}_\ell, \quad (4)$$

where

$$\mathcal{G}_q = SU(3)_{Q_L} \times SU(3)_{U_R} \times SU(3)_{D_R}, \qquad \mathcal{G}_\ell = SU(3)_{L_L} \otimes SU(3)_{E_R}. \quad (5)$$

Three of the five $U(1)$ subgroups can be identified with the total baryon and lepton number, which are not broken by $\mathcal{L}_{\text{Yukawa}}$, and the weak hypercharge, which is gauged and broken only spontaneously by $\langle\phi\rangle \neq 0$. The subgroups controlling flavour-changing dynamics and flavour non-universality are the non-Abelian groups $\mathcal{G}_q$ and $\mathcal{G}_\ell$, which are explicitly broken by $Y_{d,u,e}$ not being proportional to the identity matrix.

The diagonalization of each Yukawa coupling requires, in general, two independent unitary matrices, $V_L Y V_R^\dagger = \text{diag}(y_1, y_2, y_3)$. In the lepton sector the invariance of $\mathcal{L}^{\text{SM}}_{\text{gauge}}$ under $\mathcal{G}_\ell$ allows us to freely choose the two matrices necessary to diagonalize Y_e without breaking gauge invariance, or without observable consequences. This is not the case in the quark sector, where we can freely choose only three of the four unitary matrices necessary to diagonalize both Y_d and Y_u. Choosing the basis where Y_d is diagonal (and eliminating the right-handed diagonalization matrix of Y_u) we can write

$$Y_d = \lambda_d, \qquad Y_u = V^\dagger\lambda_u, \quad (6)$$

where

$$\lambda_d = \text{diag}(y_d, y_s, y_b), \quad \lambda_u = \text{diag}(y_u, y_c, y_t), \qquad y_q = \frac{m_q}{v}. \quad (7)$$

Alternatively, we could choose a gauge-invariant basis where $Y_d = V\lambda_d$ and $Y_u = \lambda_u$. Since the flavour symmetry does not allow the diagonalization from the left of both Y_d and Y_u, in both cases we are left with a non-trivial unitary mixing matrix, V, which is nothing but the Cabibbo–Kobayashi–Maskawa (CKM) mixing matrix [1,2].

A generic unitary 3×3 $[N \times N]$ complex unitary matrix depends on three $[N(N-1)/2]$ real rotational angles and six $[N(N+1)/2]$ complex phases. Having chosen a quark basis where the Y_d and Y_u have the form in (6) leaves us with a residual invariance under the flavour group which allows us to eliminate five of the six complex phases in V (the relative phases of the various quark fields). As a result, the physical parameters in V are four: three real angles and one complex CP-violating phase. The full set of parameters controlling the breaking of the quark flavour symmetry in the SM is composed of the six quark masses in $\lambda_{u,d}$ and the four parameters of V.

For practical purposes it is often convenient to work in the mass eigenstate basis of both up- and down-type quarks. This can be achieved by rotating independently the up- and down-components of the quark doublet Q_L, or by moving the CKM matrix from the Yukawa sector to the charged weak current in $\mathcal{L}_{\text{gauge}}^{\text{SM}}$:

$$J_W^\mu\big|_{\text{quarks}} = \bar{u}_L^i \gamma^\mu d_L^i \quad \overset{u,d \text{ mass-basis}}{\longrightarrow} \quad \bar{u}_L^i V_{ij} \gamma^\mu d_L^j. \tag{8}$$

However, it must be stressed that V originates from the Yukawa sector (in particular by the misalignment of Y_u and Y_d in the $SU(3)_{Q_L}$ subgroup of $\mathcal{G}_q$): in the absence of Yukawa couplings we can always set $V_{ij} = \delta_{ij}$.

To summarize, quark-flavour physics within the SM is characterized by a large flavour symmetry, $\mathcal{G}_q$, defined by the gauge sector, whose only breaking sources are the two Yukawa couplings Y_d and Y_u. The CKM matrix arises through the misalignment of Y_u and Y_d in flavour space.

2.2   Some Properties of the CKM Matrix

The standard parameterisation of the CKM matrix [7] in terms of three rotational angles (θ_{ij}) and one complex phase (δ) is

$$V = \begin{pmatrix} V_{ud} & V_{us} & V_{ub} \\ V_{cd} & V_{cs} & V_{cb} \\ V_{td} & V_{ts} & V_{tb} \end{pmatrix}$$

$$= \begin{pmatrix} c_{12}c_{13} & s_{12}c_{13} & s_{13}e^{-i\delta} \\ -s_{12}c_{23} - c_{12}s_{23}s_{13}e^{i\delta} & c_{12}c_{23} - s_{12}s_{23}s_{13}e^{i\delta} & s_{23}c_{13} \\ s_{12}s_{23} - c_{12}c_{23}s_{13}e^{i\delta} & -s_{23}c_{12} - s_{12}c_{23}s_{13}e^{i\delta} & c_{23}c_{13} \end{pmatrix}, \tag{9}$$

where $c_{ij} = \cos\theta_{ij}$ and $s_{ij} = \sin\theta_{ij}$ ($i, j = 1, 2, 3$).

The off-diagonal elements of the CKM matrix show a strongly hierarchical pattern: $|V_{us}|$ and $|V_{cd}|$ are close to 0.22, the elements $|V_{cb}|$ and $|V_{ts}|$ are of order 4×10^{-2} whereas $|V_{ub}|$ and $|V_{td}|$ are of order 5×10^{-3}. The Wolfenstein parameterisation, namely the expansion of the CKM matrix elements in powers of the small parameter $\lambda \doteq |V_{us}| \approx 0.22$, is a convenient way to exhibit this hierarchy in a more explicit way [8]:

$$V = \begin{pmatrix} 1 - \frac{\lambda^2}{2} & \lambda & A\lambda^3(\varrho - i\eta) \\ -\lambda & 1 - \frac{\lambda^2}{2} & A\lambda^2 \\ A\lambda^3(1 - \varrho - i\eta) & -A\lambda^2 & 1 \end{pmatrix} + \mathcal{O}(\lambda^4), \tag{10}$$

where A, ϱ, and η are free parameters of order 1. Because of the smallness of λ and the fact that for each element the expansion parameter is actually λ^2, this is a rapidly converging expansion.

The Wolfenstein parameterisation is certainly more transparent than the standard parameterisation. However, if one requires sufficient level of accuracy, the terms of $\mathcal{O}(\lambda^4)$ and $\mathcal{O}(\lambda^5)$ have to be included in phenomenological applications. This can be achieved in many different ways, according to the convention adopted. The simplest (and nowadays commonly adopted) choice is obtained by *defining* the parameters $\{\lambda, A, \varrho, \eta\}$ in terms of the angles of the exact parameterisation in Eq. (9) as follows:

$$\lambda \doteq s_{12}, \qquad A\lambda^2 \doteq s_{23}, \qquad A\lambda^3(\varrho - i\eta) \doteq s_{13}e^{-i\delta}. \qquad (11)$$

The change of variables $\{s_{ij}, \delta\} \to \{\lambda, A, \varrho, \eta\}$ in Eq. (9) leads to an exact parameterisation of the CKM matrix in terms of the Wolfenstein parameters. Expanding this expression up to $\mathcal{O}(\lambda^5)$ leads to

$$\begin{pmatrix} 1 - \frac{1}{2}\lambda^2 - \frac{1}{8}\lambda^4 & \lambda + \mathcal{O}(\lambda^7) & A\lambda^3(\varrho - i\eta) \\ -\lambda + \frac{1}{2}A^2\lambda^5[1 - 2(\varrho + i\eta)] & 1 - \frac{1}{2}\lambda^2 - \frac{1}{8}\lambda^4(1 + 4A^2) & A\lambda^2 + \mathcal{O}(\lambda^8) \\ A\lambda^3(1 - \bar{\varrho} - i\bar{\eta}) & -A\lambda^2 + \frac{1}{2}A\lambda^4[1 - 2(\varrho + i\eta)] & 1 - \frac{1}{2}A^2\lambda^4 \end{pmatrix},$$

$$(12)$$

where

$$\bar{\varrho} = \varrho\left(1 - \frac{\lambda^2}{2}\right) + \mathcal{O}(\lambda^4), \qquad \bar{\eta} = \eta\left(1 - \frac{\lambda^2}{2}\right) + \mathcal{O}(\lambda^4). \qquad (13)$$

The advantage of this generalization of the Wolfenstein parameterisation is the absence of relevant corrections to V_{us}, V_{cd}, V_{ub} and V_{cb}, and a simple change in V_{td}, which facilitates the implementation of experimental constraints.

The unitarity of the CKM matrix implies the following relations between its elements:

$$\textbf{I)} \quad \sum_{k=1...3} V_{ik}^* V_{ki} = 1, \qquad \textbf{II)} \quad \sum_{k=1...3} V_{ik}^* V_{kj \neq i} = 0. \qquad (14)$$

These relations are a distinctive feature of the SM, where the CKM matrix is the only source of quark flavour mixing. Their experimental verification is therefore a useful tool to set bounds, or possibly reveal, new sources of flavour symmetry breaking. Among the relations of type **II**, the one obtained for $i = 1$ and $j = 3$, namely

$$V_{ud}V_{ub}^* + V_{cd}V_{cb}^* + V_{td}V_{tb}^* = 0 \qquad (15)$$

or $\quad \dfrac{V_{ud}V_{ub}^*}{V_{cd}V_{cb}^*} + \dfrac{V_{td}V_{tb}^*}{V_{cd}V_{cb}^*} + 1 = 0 \qquad \leftrightarrow \qquad [\bar{\varrho} + i\bar{\eta}] + [(1 - \bar{\varrho}) - i\bar{\eta}] + 1 = 0,$ (16)

is particularly interesting since it involves the sum of three terms all of the same order in λ and is usually represented as a unitarity triangle in the complex plane, as shown in Fig. 1. It is worth stressing that Eq. (15) is invariant

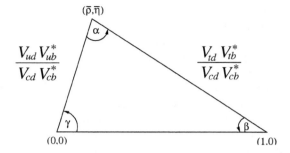

Figure 1. *The CKM unitarity triangle.*

under any phase transformation of the quark fields. Under such transformations the triangle in Fig. 1 is rotated in the complex plane, but its angles and the sides remain unchanged. Both angles and sides of the unitary triangle are indeed observable quantities which can be measured in suitable experiments.

2.3 Present Status of CKM Fits

The values of $|V_{us}|$ and $|V_{cb}|$, or λ and A in the parameterisation (12), are determined with good accuracy from $K \to \pi \ell \nu$ and $B \to X_c \ell \nu$ decays, respectively. According to the recent analysis in Ref. [4], their numerical values are

$$\lambda = 0.2259 \pm 0.0009, \qquad A = 0.802 \pm 0.015. \tag{17}$$

Using these results, all the other constraints on the elements of the CKM matrix can be expressed as constraints on $\bar{\varrho}$ and $\bar{\eta}$ (or constraints on the CKM unitarity triangle in Fig. 1). The list of the most sensitive observables used to determine $\bar{\varrho}$ and $\bar{\eta}$ in the SM includes:

- The rates of inclusive and exclusive charmless semileptonic B decays, which depend on $|V_{ub}|$ and provide a constraint on $\bar{\varrho}^2 + \bar{\eta}^2$.

- The time-dependent CP asymmetry in $B \to \psi K_S$ decays ($\mathcal{A}_{K\Psi}^{\mathrm{CP}}$), which depends on the phase of the B_d–$\bar{B}_d$ mixing amplitude relative to the decay amplitude (see Section 3.2). Within the SM this translates into a constraint on $\sin 2\beta$.

- The rates of various $B \to DK$ decays, which provide constraints on the angle γ (see Section 3.3.2).

- The rates of various $B \to \pi\pi, \rho\pi, \rho\rho$ decays, which constrain $\alpha = \pi - \beta - \gamma$.

- The ratio between the mass splittings in the neutral B and B_s systems, which depends on $|V_{td}/V_{ts}|^2 \propto [(1 - \bar{\varrho})^2 + \bar{\eta}^2]$.

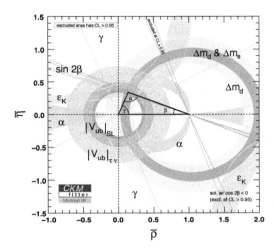

Figure 2. *Allowed region in the $\bar{\varrho}, \bar{\eta}$ plane, from [10]. Superimposed are the individual constraints from charmless semileptonic B decays ($|V_{ub}|$), mass differences in the B_d (Δm_d) and B_s (Δm_s) systems, CP violation in the neutral kaon (ε_K) and in the B_d systems (sin 2β), the combined constraints on α and γ from various B decays.*

- The indirect CP violating parameter of the kaon system (ϵ_K), which determines a hyperbola in the $\bar{\varrho}$ and $\bar{\eta}$ plane (see Ref. [4] for more details).

The resulting constraints are shown in Fig. 2. As can be seen, they are all consistent with a unique value of $\bar{\varrho}$ and $\bar{\eta}$ [4]:

$$\rho = +0.158 \pm 0.021, \qquad \eta = +0.343 \pm 0.013. \tag{18}$$

The compatibility of different constraints on the CKM unitarity triangle is a powerful consistency test of the SM in describing flavour-changing phenomena. From the plot in Fig. 2 it is quite clear, at least in a qualitative way, that there is little room for non-SM contributions in flavour changing transitions. A more quantitative evaluation of this statement is presented in the next section.

2.4 The SM as an Effective Theory

As anticipated in the introduction, despite the impressive phenomenological success of the SM in flavour and electroweak physics, there are various convincing arguments which motivate us to consider this model only as the low-energy limit of a more complete theory.

Assuming that the new degrees of freedom which complete the theory are heavier than the SM particles, we can integrate them out and describe physics beyond the SM in full generality by means of an *effective theory* approach. The SM Lagrangian becomes the renormalizable part of a more general local Lagrangian which includes an infinite tower of operators with dimension $d > 4$, constructed in terms of SM fields and suppressed by inverse powers of an effective scale Λ. These operators are the residual effect of having integrated out the new heavy degrees of freedom, whose mass scale is parameterised by the effective scale $\Lambda > m_W$.

As we will discuss in more detail in Section 3.1, integrating out heavy degrees of freedom is a procedure often adopted also within the SM: it allows us to simplify the evaluation of amplitudes which involve different energy scales. This approach is indeed a generalization of the Fermi theory of weak interactions, where the dimension-six four-fermion operators describing weak decays are the results of having integrated out the W field. The only difference when applying this procedure to physics beyond the SM is that in this case, as in the original work by Fermi, we don't know the nature of the degrees of freedom we are integrating out. This implies we are not able to determine a priori the values of the effective couplings of the higher-dimensional operators. The advantage of this approach is that it allows us to analyse all realistic extensions of the SM in terms of a limited number of parameters (the coefficients of the higher-dimensional operators). The drawback is the impossibility of establishing correlations of new physics (NP) effects at low and high energies.

Assuming for simplicity that there is a single elementary Higgs field, responsible for the $SU(2)_L \times U(1)_Y \to U(1)_Q$ spontaneous breaking, the Lagrangian of the SM considered as an effective theory can be written as

$$\mathcal{L}_{\text{eff}} = \mathcal{L}_{\text{gauge}}^{\text{SM}} + \mathcal{L}_{\text{Higgs}}^{\text{SM}} + \mathcal{L}_{\text{Yukawa}}^{\text{SM}} + \Delta\mathcal{L}_{d>4}, \tag{19}$$

where $\Delta\mathcal{L}_{d>4}$ denotes the series of higher-dimensional operators invariant under the SM gauge group:

$$\Delta\mathcal{L}_{d>4} = \sum_{d>4} \sum_{n=1}^{N_d} \frac{c_n^{(d)}}{\Lambda^{d-4}} \mathcal{O}_n^{(d)}(\text{SM fields}). \tag{20}$$

If NP appears at the TeV scale, as we expect from the stabilization of the mechanism of electroweak symmetry breaking, the scale Λ cannot exceed a few TeV. Moreover, if the underlying theory is natural (no fine-tuning in the coupling constants), we expect $c_i^{(d)} = O(1)$ for all the operators which are not forbidden (or suppressed) by symmetry arguments. The observation that this expectation is *not* fulfilled by several dimension-six operators contributing to flavour-changing processes is often denoted as the *flavour problem*.

If the SM Lagrangian were invariant under some flavour symmetry, this problem could easily be circumvented. For instance, in the case of baryon- or lepton-number violating processes, which are exact symmetries of the SM

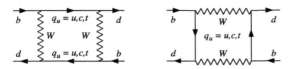

Figure 3. *Box diagrams contributing to B_d-$\bar{B}_d$ mixing in the unitary gauge.*

Lagrangian, we can avoid the tight experimental bounds by promoting B and L to be exact symmetries of the new dynamics at the TeV scale. The peculiar aspects of flavour physics is that there is no exact flavour symmetry in the low-energy theory. In this case it is not sufficient to invoke a flavour symmetry for the underlying dynamics. We also need to specify how this symmetry is broken in order to describe the observed low-energy spectrum and, at the same time, be in agreement with the precise experimental tests of flavour-changing processes.

2.4.1 Bounds on the Scale of New Physics from $\Delta F = 2$ Processes

The best way to quantify the flavour problem is by looking at the consistency of the tree- and loop-mediated constraints on the CKM matrix discussed in Section 2.3.

In the first approximation we can assume that NP effects are negligible in processes which are dominated by tree-level amplitudes. Following this assumption, the values of $|V_{us}|$, $|V_{cb}|$ and $|V_{ub}|$, as well as the constraints on α and γ are essentially NP free. As can be seen in Fig. 2, this implies we can determine completely the CKM matrix assuming generic NP effects in loop-mediated amplitudes. We can then use the measurements of observables which are loop-mediated within the SM to bound the couplings of effective NP operators in Eq. (20) which contribute to these observables at the tree level.

The loop-mediated constraints shown in Fig. 2 are those from the mixing of B_d, B_s and K^0 with the corresponding anti-particles (generically denoted as $\Delta F = 2$ amplitudes). Within the SM, these amplitudes are generated by box amplitudes of the type in Fig. 3 (and similarly for B_s, and K^0) and are affected by small hadronic uncertainties (with the exception of Δm_K). We will come back to the evaluation of these amplitudes in more detail in Section 3.2. For the moment it is sufficient to note that the leading contribution is obtained with the top-quark running inside the loop, giving rise to the highly suppressed result

$$\mathcal{M}_{\Delta F=2}^{\text{SM}} \approx \frac{G_F^2 m_t^2}{16\pi^2} \, V_{3i}^* V_{3j} \, \langle \bar{M} | (\bar{d}_L^i \gamma^\mu d_L^j)^2 | M \rangle \times F\left(\frac{m_t^2}{m_W^2}\right), \qquad (21)$$

for $M = K^0, B_d, B_s$ and where F is a loop function of order one (i, j denote the flavour indexes of the meson valence quarks).

The magnitude and phase of all these mixing amplitudes have been determined with good accuracy from experiments with the exception of the CP-violating phase in B_s–$\bar{B}_s$ mixing. As shown in Fig. 2, in all cases where the experimental information is precise, the magnitude of the new physics amplitude cannot exceed, in size, the SM contribution.

To translate this information into bounds on the scale of new physics, let's consider the following set of $\Delta F = 2$ dimension-six operators

$$\mathcal{O}^{ij}_{\Delta F=2} = (\bar{Q}^i_L \gamma^\mu Q^j_L)^2, \qquad Q^i_L = \begin{pmatrix} u^i_L \\ d^i_L \end{pmatrix}, \tag{22}$$

where i, j are flavour indexes in the basis defined by Eq. (6). These operators contribute at the tree level to the meson-antimeson mixing amplitudes. Denoting the couplings of the non-standard operators in (22) by c_{ij}, the condition $|\mathcal{M}^{NP}_{\Delta F=2}| < |\mathcal{M}^{SM}_{\Delta F=2}|$ implies[2]

$$\Lambda < \frac{3.4 \text{ TeV}}{|V^*_{3i}V_{3j}|/|c_{ij}|^{1/2}} < \begin{cases} 9 \times 10^3 \text{ TeV} \times |c_{21}|^{1/2} & \text{from} \quad K^0 - \bar{K}^0 \\ 4 \times 10^2 \text{ TeV} \times |c_{31}|^{1/2} & \text{from} \quad B_d - \bar{B}_d. \\ 7 \times 10^1 \text{ TeV} \times |c_{32}|^{1/2} & \text{from} \quad B_s - \bar{B}_s \end{cases} \tag{23}$$

The main messages of these bounds are the following:

- New physics models with a generic flavour structure (c_{ij} of order 1) at the TeV scale are ruled out. If we want to keep Λ in the TeV range, physics beyond the SM must have a highly non-generic flavour structure.

- In the specific case of the $\Delta F = 2$ operators in (22), in order to keep Λ in the TeV range, we must find a symmetry argument such that $|c_{ij}| \lesssim |V^*_{3i}V_{3j}|^2$.

The strong constraining power of $\Delta F = 2$ observables is a consequence of their strong suppression within the SM. They are suppressed not only by the typical $1/(4\pi)^2$ factor of loop amplitudes, but also by the GIM mechanism [12] and by the hierarchy of the CKM matrix ($|V_{3i}| \ll 1$, for $i \neq 3$). Reproducing a similar structure beyond the SM is a highly non-trivial task. As we will discuss in the last lecture, only in a few cases this can be implemented in a natural way.

To conclude, we stress that the good agreement of SM and experiments for B_d and K^0 mixing does not imply that further studies of flavour physics are not interesting. On the one hand, even for $|c_{ij}| \approx |V^*_{3i}V_{3j}|$, which can be considered the most pessimistic case, as we will discuss in Section 4.1, we are presently constraining physics at the TeV scale. Therefore, improving these

---

[2] A more refined analysis, with complete statistical treatment and separate bounds forming the real and the imaginary parts of the various amplitudes, leading to slightly more stringent bounds, can be found in [11].

bounds, if possible, would be extremely valuable. One the other hand, as we will discuss in the next lecture, there are various interesting observables which have not been deeply investigated yet, whose study could reveal additional key features about the flavour structure of physics beyond the SM.

3   *B* Physics Phenomenology: Mixing, CP Violation, and Rare Decays

As we have seen in the previous lecture, the exploration of the mechanism of quark-flavour mixing has entered a new era. The precise measurements of mixing-induced CP violation and tree level-allowed semileptonic transition have provided an important consistency check of the SM, and a precise determination of the CKM matrix. The next goal is to understand if there is still room for new physics or, more precisely, if there is still room for new sources of flavour symmetry breaking close to the electroweak scale. From this perspective, a few rare *B* decays mediated by flavour-changing neutral-current (FCNC) amplitudes, and also the CP violating phase of B_s mixing, represent a fundamental tool.

Beside the experimental sensitivity, the conditions which allow us to perform significant NP searches in rare decays can be summarized as follows: (1) decay amplitudes dominated by electroweak dynamics, and thus with enhanced sensitivity to non-standard contributions; (2) small theoretical error within the SM, or good control of both perturbative and non-perturbative corrections. In the next section we analyse how and in which cases these two conditions can be achieved.

3.1   Theoretical Tools: Low-Energy Effective Lagrangians

The decays of *B* mesons are processes which involve at least two different energy scales: the electroweak scale, characterized by the *W* boson mass, which determines the fl-changing transition at the quark level, and the scale of strong interactions Λ_{QCD}, related to the hadron formation. The presence of these two widely separated scales makes the calculation of the decay amplitudes starting from the full SM Lagrangian quite complicated: large logarithms of the type $\log(m_W/\Lambda_{\text{QCD}})$ may appear, leading to a breakdown of ordinary perturbation theory.

This problem can be substantially simplified by integrating out the heavy SM fields (*W* and *Z* bosons, as well as the top quark) at the electroweak scale, and constructing an appropriate low-energy effective theory where only the light SM fields appear. The weak effective Lagrangians thus obtained contain local operators of dimension six (and higher), written in terms of light SM fermions, photon and gluon fields, suppressed by inverse powers of the *W* mass.

To be concrete, let's consider the example of charged-current semileptonic weak interactions. The basic building block in the full SM Lagrangian is

$$\mathcal{L}_W^{\text{full SM}} = \frac{g}{\sqrt{2}} J_W^\mu(x) W_\mu^+(x) + \text{h.c.}, \tag{24}$$

where

$$J_W^\mu(x) = V_{ij}\, \bar{u}_L^i(x)\gamma^\mu d_L^j(x) + \bar{e}_L^j(x)\gamma^\mu \nu_L^j(x) \tag{25}$$

is the weak charged current already introduced in Eq. (8). Integrating out the W field at the tree level we contract two vertexes of this type generating the non-local transition amplitude

$$i\mathcal{T} = -i\frac{g^2}{2} \int d^4x D_{\mu\nu}(x, m_W)\, T\left[J_W^\mu(x), J_W^{\nu\dagger}(0)\right], \tag{26}$$

which involves only light fields. Here $D_{\mu\nu}(x, m_W)$ is the W propagator in coordinate space: expanding it in inverse powers of m_W,

$$D_{\mu\nu}(x, m_W) = \int \frac{d^4q}{(2\pi)^4} e^{-iq\cdot x} \frac{-ig_{\mu\nu} + \mathcal{O}(q_\mu, q_\nu)}{q^2 - m_W^2 + i\varepsilon} = \delta(x)\frac{ig_{\mu\nu}}{m_W^2} + \ldots, \tag{27}$$

the leading contribution to $\mathcal{T}$ can be interpreted as the tree-level contribution of the following effective local Lagrangian

$$\mathcal{L}_{\text{eff}}^{(0)} = -\frac{4G_F}{\sqrt{2}} g_{\mu\nu} J_W^\mu(x) J_W^{\nu\dagger}(x), \tag{28}$$

where $G_F/\sqrt{2} = g^2/(8m_W^2)$ is the Fermi coupling. If we select the product of one quark and one leptonic current,

$$\mathcal{L}_{\text{eff}}^{\text{semi-lept}} = -\frac{4G_F}{\sqrt{2}} V_{ij}\, \bar{u}_L^i(x)\gamma^\mu d_L^j(x)\, \bar{\nu}_L(x)\gamma_\mu e_L(x) + \text{h.c.}, \tag{29}$$

we obtain an effective Lagrangian which provides an excellent description of semileptonic weak decays. The neglected terms in the expansion of Eq. (27) correspond to corrections of $\mathcal{O}(m_B^2/m_W^2)$ to the decay amplitudes. In principle, these corrections could be taken into account by adding appropriate dimension-eight operators in the effective Lagrangian. However, in most cases they are safely negligible.

The case of charged semileptonic decays is particularly simple since we can ignore QCD effects: the operator (Eq. (29)) is not renormalized by strong interactions. The situation is slightly more complicated in the case of non-leptonic or flavour-changing neutral-current processes, where QCD corrections and higher-order weak interactions cannot be neglected, but the basic strategy is the same. First of all we need to identify a complete basis of local operators that also includes those generated beyond the tree level. In general, given a fixed order in the $1/m_W^2$ expansion of the amplitudes, we need to consider all operators of corresponding dimension (e.g., dimension six at the first order in the $1/m_W^2$ expansion) compatible with the symmetries of the system. Then we must introduce an artificial scale in the problem, the renormalization scale μ, which is needed to regularize QCD (or QED) corrections in the effective theory.

The effective Lagrangian for generic $\Delta F = 1$ processes assumes the form

$$\mathcal{L}_{\Delta F=1} = -4\frac{G_F}{\sqrt{2}} \sum_i C_i(\mu)Q_i, \tag{30}$$

where the sum runs over the complete basis of operators. As explicitly indicated, the effective couplings $C_i(\mu)$ (known as Wilson coefficients) depend, in general, on the renormalization scale. The dependence from this scale cancels when evaluating the matrix elements of the effective Lagrangian for physical processes, that we can generically indicate as

$$\mathcal{M}(i \to f) = -4\frac{G_F}{\sqrt{2}} \sum_i C_i(\mu)\langle f|Q_i(\mu)|i\rangle. \tag{31}$$

The independence of $\mathcal{M}$ from μ holds for any initial and final state, including partonic states at high energies. This implies that the $C_i(\mu)$ obey a series of renormalization group equations (RGE), whose structure is completely determined by the anomalous dimensions of the effective operators. These equations can be solved using standard RG techniques, allowing the resummation of all large logs of the type $\alpha_s(\mu)^{n+m} \log(m_W/\mu)^n$ to all orders in n (working at order $m + 1$ in perturbation theory). The scale μ acts as a separator of short- and long-distance virtual corrections: short-distance effects are included in the $C_i(\mu)$, whereas long-distance effects are left as explicit degrees of freedom in the effective theory.[3]

In practice, the problem reduces to the following three well-defined and independent steps:

1. the evaluation of the *initial conditions* of the $C_i(\mu)$ at the electroweak scale ($\mu \approx m_W$);

2. the evaluation of the anomalous dimension of the effective operators, and the corresponding *RGE evolution* of the $C_i(\mu)$ from the electroweak scale down to the energy scale of the physical process ($\mu \approx m_B$);

3. the evaluation of the *matrix elements* of the effective Lagrangian for the physical hadronic processes (which involve energy scales from m_B down to Λ_{QCD}).

---

[3] This statement would be correct if the theory were regularized using a dimensional cut-off. It is not fully correct if μ is the scale appearing in the (often adopted) dimensional-regularization + minimal-subtraction (MS) renormalization scheme.

The first step is the one where physics beyond the SM may appear: if we assume NP is heavy, it may modify the initial conditions of the Wilson coefficients at the high scale, while it cannot affect the two subsequent steps. While the RGE evolution and the hadronic matrix elements are not directly related to NP, they may influence the sensitivity to NP of physical observables. In particular, the evaluation of hadronic matrix elements is potentially affected by non-perturbative QCD effects: these are often a large source of theoretical uncertainty which can obscure NP effects. RGE effects do not induce sizable uncertainties since they can be fully handled within perturbative QCD; however, the sizable logs generated by the RGE running may *dilute* the interesting short-distance information encoded in the value of the Wilson coefficients at the high scale. As we will discuss in the following, only in specific observables are these two effects small and under good theoretical control.

A deeper discussion about the construction of low-energy effective Lagrangians, with a detailed discussions of the first two steps mentioned above, can be found in Ref. [13].

3.1.1 Effective Operators for Rare Decays and $\Delta F = 2$ Amplitudes

Let's take a closer look at processes where the underlying parton process is $b \to s + \bar{q}q$. In this case the relevant effective Lagrangian can be written as

$$\mathcal{L}_{b \to s}^{\text{non-lept}} = -4 \frac{G_F}{\sqrt{2}} \left(\sum_{q=u,c} \lambda_q^s \sum_{i=1,2} C_i(\mu) Q_i^q(\mu) - \lambda_t^s \sum_{i=3}^{10} C_i(\mu) Q_i(\mu) \right), \quad (32)$$

where $\lambda_q^s = V_{qb}^* V_{qs}$, and the operator basis is

$$
\begin{aligned}
Q_1^q &= \bar{b}_L^\alpha \gamma^\mu q_L^\alpha \, \bar{q}_L^\beta \gamma_\mu s_L^\beta, & Q_2^q &= \bar{b}_L^\alpha \gamma^\mu q_L^\beta \, \bar{q}_L^\beta \gamma_\mu s_L^\alpha, \\
Q_3 &= \bar{b}_L^\alpha \gamma^\mu s_L^\alpha \sum_q \bar{q}_L^\beta \gamma_\mu q_L^\beta, & Q_4 &= \bar{b}_L^\alpha \gamma^\mu s_L^\beta \sum_q \bar{q}_L^\beta \gamma_\mu q_L^\alpha, \\
Q_5 &= \bar{b}_L^\alpha \gamma^\mu s_L^\alpha \sum_q \bar{q}_R^\beta \gamma_\mu q_R^\beta, & Q_6 &= \bar{b}_L^\alpha \gamma^\mu s_L^\beta \sum_q \bar{q}_R^\beta \gamma_\mu q_R^\alpha, \quad (33) \\
Q_7 &= \tfrac{3}{2} \bar{b}_L^\alpha \gamma^\mu s_L^\alpha \sum_q e_q \bar{q}_R^\beta \gamma_\mu q_R^\beta, & Q_8 &= \tfrac{3}{2} \bar{b}_L^\alpha \gamma^\mu s_L^\beta \sum_q e_q \bar{q}_R^\beta \gamma_\mu q_R^\alpha, \\
Q_9 &= \tfrac{3}{2} \bar{b}_L^\alpha \gamma^\mu s_L^\alpha \sum_q e_q \bar{q}_L^\beta \gamma_\mu q_L^\beta, & Q_{10} &= \tfrac{3}{2} \bar{b}_L^\alpha \gamma^\mu s_L^\beta \sum_q e_q \bar{q}_L^\beta \gamma_\mu q_L^\alpha,
\end{aligned}
$$

with $\{\alpha, \beta\}$ and e_q denoting the color indexes and the electric charge of the quark q, respectively.

Out of these operators, only Q_1^c and Q_1^u are generated at the tree-level by the W exchange. Indeed, comparing with the tree level structure in (28), we find

$$C_1^{u,c}(m_W) = 1 + \mathcal{O}(\alpha_s, \alpha), \qquad C_{2-10}^{u,c}(m_W) = 0 + \mathcal{O}(\alpha_s, \alpha). \quad (34)$$

However, after including RGE effects and running down to $\mu \approx m_b$, both $C_1^{u,c}$ and $C_2^{u,c}$ become $\mathcal{O}(1)$, while C_{3-6} become $\mathcal{O}(\alpha_s(m_b))$. In all these cases there is little hope to identify NP effects: the leading initial condition is the tree-level W exchange, which is hardly modified by NP. In principle, the coefficients of the electroweak penguin operators, Q_7–Q_{10}, are more interesting:

their initial conditions are related to electroweak penguin and box diagrams. However, it is hard to distinguish their contribution from those of the other four-quark operators in non-leptonic processes. Moreover, for C_{7-10} the relative contribution from long-distance physics (running down from m_W to m_b) is sizable and dilutes the interesting short-distance information.

For $b \to s$ transitions with a photon or a lepton pair in the final state, additional dimension-six operators must be included in the basis,

$$\mathcal{L}^{\text{rare}}_{b \to s} = \mathcal{L}^{\text{non-lept}}_{b \to s} + 4\frac{G_F}{\sqrt{2}}\lambda^s_t \left(C_{7\gamma}Q_{7\gamma} + C_{8g}Q_{8g} + C_{9V}Q_{9V} + C_{10A}Q_{10A}\right), \tag{35}$$

where

$$Q_{7\gamma} = \frac{e}{16\pi^2}m_b\bar{b}^\alpha_R\sigma^{\mu\nu}F_{\mu\nu}s^\alpha_L, \qquad Q_{8g} = \frac{g_s}{16\pi^2}m_b\bar{b}^\alpha_R\sigma^{\mu\nu}G^A_{\mu\nu}T^As^\alpha_L,$$

$$Q_{9V} = \frac{1}{2}\bar{b}^\alpha_L\gamma^\mu s^\alpha_L\,\bar{l}\gamma_\mu l, \qquad Q_{10A} = \frac{1}{2}\bar{b}^\alpha_L\gamma^\mu s^\alpha_L\,\bar{l}\gamma_\mu\gamma_5 l, \tag{36}$$

and $G^A_{\mu\nu}$ ($F_{\mu\nu}$) is the gluon (photon) field strength tensor. The initial conditions of these operators are particularly sensitive to NP: within the SM they are generated by one-loop penguin and box diagrams dominated by the top-quark exchange. The most theoretically clean is C_{10A}, which does not mix with any of the four-quark operators listed above and which has a vanishing anomalous dimension:

$$C^{\text{SM}}_{10A}(m_W) = \frac{g^2}{8\pi^2}\frac{x_t}{8}\left[\frac{4 - x_t}{1 - x_t} + \frac{3x_t}{(1 - x_t)^2}\ln x_t\right], \qquad x_t = \frac{m^2_t}{m^2_W}. \tag{37}$$

NP effects at the TeV scale could easily modify this result, and this deviation would directly show up in low-energy observables sensitive to C_{10A}, such as $\mathcal{A}_{\text{FB}}(B \to K^*\ell^+\ell^-)$ and $\mathcal{B}(B \to \ell^+\ell^-)$ (see Sections 3.3.3 and 3.3.4). We finally note that while the operators in Eqs. (33) and (36) form a complete basis within the SM, this is not necessarily the case beyond the SM. In particular, within specific scenarios right-handed current operators (e.g., those obtained from (36) for $q_{L(R)} \to q_{R(L)}$) may also appear.

The $\Delta F = 2$ effective weak Lagrangians are simpler than the $\Delta F = 1$ ones: the SM operator basis includes one operator only. The Lagrangian relevant for B^0_d–$\bar{B}^0_d$ and B^0_s–$\bar{B}^0_s$ mixing is conventionally written as ($q = \{d, s\}$):

$$\mathcal{L}^{\text{SM}}_{\Delta B=2} = \frac{G^2_F}{4\pi^2}m^2_W(V^*_{tb}V_{tq})^2\,\eta_B(\mu)\,S_0(x_t)\,(\bar{b}_L\gamma_\mu q_L\,\bar{b}_L\gamma^\mu q_L), \tag{38}$$

where the initial condition of the Wilson coefficient is the loop function $S_0(x_t)$, corresponding to the box diagrams in Fig. 3. The effect of the QCD corrections is a multiplicative correction factor, $\eta_B(\mu)$, which can be computed with high accuracy and turns out to be of order one. The explicit expression of the loop function, dominated by the top-quark exchange, is

$$S_0(x_t) = \frac{4x_t - 11x^2_t + x^3_t}{4(1 - x_t)^2} - \frac{3x^3_t \ln x_t}{2(1 - x_t)^3}. \tag{39}$$

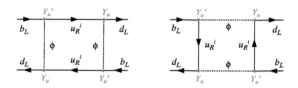

Figure 4. *One-loop contributions $\Delta F = 2$ amplitudes in the gaugeless limit.*

3.1.2   The Gaugeless Limit of FCNC and $\Delta F = 2$ Amplitudes

An interesting aspect which is common to the electroweak loop functions in Eqs. (37) and (39) is the fact they diverge in the limit $m_t/m_W \to \infty$. This behavior is apparently strange: it contradicts the expectation that contributions of heavy particles at low energy decouple in the limit where their masses increase. The origin of this effect can be understood by noting that the leading contributions to both amplitudes are generated only by the Yukawa interaction. These contributions can be better isolated in the *gaugeless* limit of the SM, i.e., if we send to zero the gauge couplings. In this limit $m_W \to 0$ and the derivation of the effective Lagrangian discussed in Section 3.1 does not make sense. However, the leading contributions to the effective Lagrangians for $\Delta F = 2$ and rare decays are unaffected. Indeed, the leading contributions to these processes are generated by Yukawa interactions of the type in Fig. 4, where the scalar fields are the Goldstone-bosons components of the Higgs field (which are not eaten up by the W in the limit $g \to 0$). Since the top is still heavy, we can integrate it out, obtaining the following result for $\mathcal{L}_{\Delta B=2}$:

$$\mathcal{L}_{\Delta B=2}^{\text{SM}}\Big|_{g_i \to 0} = \frac{G_F^2 m_t^2}{16\pi^2}(V_{tb}^* V_{tq})^2 (\bar{b}_L \gamma_\mu q_L)^2 = \frac{[(Y_u Y_u^*)_{bq}]^2}{128\pi^2 m_t^2}(\bar{b}_L \gamma_\mu q_L)^2. \quad (40)$$

Taking into account that $S_0(x) \to x/4$ for $x \to \infty$, it is easy to verify that this result is equivalent to the one in Eq. (39) in the large m_t limit. A similar structure holds for the $\Delta F = 1$ amplitude contributing to the axial operator Q_{10A}.

The last expression in Eq. (40), which holds in the limit where we neglect the charm Yukawa coupling, shows that the decoupling of the amplitude with the mass of the top is compensated by four powers of the top Yukawa coupling at the numerator. The divergence for $m_t \to \infty$ can thus be understood as the divergence of one of the fundamental couplings of the theory. Note also that in the gaugeless limit there is no GIM mechanism: the contributions of the various up-type quarks inside the loops do not cancel each other: they are directly weighted by the corresponding Yukawa couplings, and this is why the top-quark contribution is the dominant one.

This exercise illustrates the key role of the Yukawa coupling in determining the main properties flavour physics within the SM, as advertised in the first lecture. It also illustrates the interplay of flavour and electroweak symmetry breaking in determining the structure of short-distance-dominated flavour-changing processes in the SM.

3.1.3 Hadronic Matrix Elements

As anticipated, all non-perturbative effects are confined to the hadronic matrix elements of the operators of the effective Lagrangians. As far as the evaluation of the matrix elements is concerned, we can divide B-physics observables into three main categories: (1) inclusive decays, (2) one-hadron final states, (3) multi-hadron processes.

The heavy-quark expansion [14] forms a solid theoretical framework to evaluate the hadronic matrix elements for inclusive processes by relating the inclusive hadronic rates to those of free b quarks, calculable in perturbation theory, by means of a systematic expansion in inverse powers of Λ_{QCD}/m_b. Thanks to quark–hadron duality, the lowest-order terms in this expansion are the pure partonic rates, and for sufficiently inclusive observables higher-order corrections are usually very small. This technique has been very successful in the past in the case of charged-current semileptonic decays, as well as $B \to X_s\gamma$. However, it has a limited domain of applicability, due to the difficulty of selecting and reconstructing hadronic inclusive states. It cannot be used at hadronic machines, and even at B factories it cannot be applied to very rare decays.

For processes with a single hadron in the final state, the hadronic effects are often (although not always) confined to the matrix elements of a single quark current. These can be expressed in terms of the meson decay constants,

$$\langle 0|b\gamma_\mu\gamma_5 q|B_q(p)\rangle = ip_\mu F_{B_q}, \tag{41}$$

or appropriate $B \to H$ hadronic form factors. Lattice QCD is the best tool to evaluate these non-perturbative quantities from first principles, at least in the kinematical region where the form factors are real (no re-scattering phase allowed). At present not all the form-factors relevant for B-physics phenomenology are computed on the lattice with good accuracy, but the field is evolving rapidly (see, e.g., Refs. [15, 16]). To this category also belong the so-called bag parameters for $\Delta B = 2$ mixing, $B_{d,s}$, defined by

$$\eta_B(\mu)\langle \bar{B}_q|(\bar{b}_L\gamma_\mu q_L)^2|B_q\rangle = \frac{2}{3}f_{B_q}^2 m_{B_q}^2 \eta_B(\mu)B_q(\mu) = \frac{2}{3}f_{B_q}^2 m_{B_q}^2 \hat{\eta}_B\hat{B}_q, \tag{42}$$

where both $\hat{B}_q$ and $\hat{\eta}_B$ are scale-independent quantities ($\hat{\eta}_B = 0.55 \pm 0.01$). For later convenience, we report here the lattice averages for meson decay constants and bag parameters obtained in Ref. [16]:

$$F_{B_s} = 245 \pm 25 \text{ GeV}, \qquad \hat{B}_s = 1.22 \pm 0.12, \tag{43}$$

$$\frac{F_{B_s}}{F_{B_d}} = 1.21 \pm 0.04, \qquad \frac{\hat{B}_{B_s}}{\hat{B}_{B_d}} = 1.00 \pm 0.03. \tag{44}$$

As can be seen, the absolute determinations are affected by $\mathcal{O}(10\%)$ errors, while the ratios, which are sensitive to $SU(3)$ breaking effects only, are known with a better precision. This is why the ratio $\Delta m_{B_d}/\Delta m_{B_s}$ gives more significant constraint in Fig. 1 rather than Δm_{B_d} only.

The last class of hadronic matrix elements is the one of multi-hadron final states, such as the two-body non-leptonic decays $B \to \pi\pi$ and $B \to K\pi$, as well as many other processes with more than one hadron in the final state. These are the most difficult ones to be estimated from first principles with high accuracy. A lot of progress in the recent pass has been achieved thanks to QCD factorization [17] and the SCET [18] approaches, which provide factorization formulae to relate these hadronic matrix elements to two-body hadronic form factors in the large m_b limit. However, it is fair to say that the errors associated with the $\Lambda_{\rm QCD}/m_b$ corrections are still quite large. This subject is quite interesting in itself, but it is beyond the scope of these lectures, where we focus on clean B-physics observables for NP studies. To this purpose, the only interesting non-leptonic channels are those where, with suitable ratios, or using $SU(2)$ relations among hadronic matrix elements, we can eliminate completely all hadronic unknowns. Examples of this type are the $B \to DK$ channels discussed in Section 3.3.2.

3.2   Time Evolution of $B_{d,s}$ States

The non-vanishing amplitude mixing a B^0 meson (B_d^0 or B_s^0) with the corresponding anti-meson, described within the SM by the effective Lagrangian in Eq. (38), induces a time-dependent oscillation between B^0 and $\bar{B}^0$ states: an initially produced B^0 or $\bar{B}^0$ evolves in time into a superposition of B^0 and $\bar{B}^0$. Denoting by $|B^0(t)\rangle$ (or $|\bar{B}^0(t)\rangle$) the state vector of a B meson which is tagged as a B^0 (or $\bar{B}^0$) at time $t = 0$, the time evolution of these states is governed by the following equation:

$$i\frac{d}{dt}\left(\begin{array}{c} |B(t)\rangle \\ |\bar{B}(t)\rangle \end{array} \right) = \left(M - i\frac{\Gamma}{2} \right)\left(\begin{array}{c} |B(t)\rangle \\ |\bar{B}(t)\rangle \end{array} \right), \qquad (45)$$

where the mass-matrix M and the decay-matrix Γ are t-independent, Hermitian 2×2 matrices. CPT invariance implies that $M_{11} = M_{22}$ and $\Gamma_{11} = \Gamma_{22}$, while the off-diagonal element $M_{12} = M_{21}^*$ is the one we can compute using the effective Lagrangian $\mathcal{L}_{\Delta B=2}$.

The mass eigenstates are the eigenvectors of $M - i\Gamma/2$. We express them in terms of the fl eigenstates as

$$|B_L\rangle = p\,|B^0\rangle + q\,|\bar{B}^0\rangle, \qquad |B_H\rangle = p\,|B^0\rangle - q\,|\bar{B}^0\rangle, \qquad (46)$$

with $|p|^2 + |q|^2 = 1$. Note that, in general, $|B_L\rangle$ and $|B_H\rangle$ are not orthogonal to each other. The time evolution of the mass eigenstates is governed by the two eigenvalues $M_H - i\Gamma_H/2$ and $M_L - i\Gamma_L/2$:

$$|B_{H,L}(t)\rangle = e^{-(iM_{H,L}+\Gamma_{H,L}/2)t}\,|B_{H,L}(t = 0)\rangle. \qquad (47)$$

For later convenience it is also useful to define

$$m = \frac{M_H + M_L}{2}, \quad \Gamma = \frac{\Gamma_L + \Gamma_H}{2}, \quad \Delta m = M_H - M_L, \quad \Delta \Gamma = \Gamma_L - \Gamma_H.$$

$$(48)$$

With these conventions the time evolution of initially tagged B^0 or $\bar{B}^0$ states is

$$|B^0(t)\rangle = e^{-imt} e^{-\Gamma t/2} \left[f_+(t) \, |B^0\rangle + \frac{q}{p} f_-(t) \, |\bar{B}^0\rangle \right],$$

$$|\bar{B}^0(t)\rangle = e^{-imt} e^{-\Gamma t/2} \left[\frac{p}{q} f_-(t) \, |B^0\rangle + f_+(t) \, |\bar{B}^0\rangle \right], \qquad (49)$$

where

$$f_+(t) = \cosh \frac{\Delta \Gamma t}{4} \cos \frac{\Delta m t}{2} - i \sinh \frac{\Delta \Gamma t}{4} \sin \frac{\Delta m t}{2}, \qquad (50)$$

$$f_-(t) = -\sinh \frac{\Delta \Gamma t}{4} \cos \frac{\Delta m t}{2} + i \cosh \frac{\Delta \Gamma t}{4} \sin \frac{\Delta m t}{2}, \qquad (51)$$

In both B_s and B_d systems the following hierarchies hold: $|\Gamma_{12}| \ll |M_{12}|$ and $\Delta \Gamma \ll \Delta m$. They are experimentally verified and can be traced back to the fact that $|\Gamma_{12}|$ is a genuine long-distance $\mathcal{O}(G_F^2)$ effect (it is indeed related to the absorptive part of the box diagrams in Fig. 3) which does not share the large m_t enhancement of $|M_{12}|$ (which is a short-distance-dominated quantity). Taking into account this hierarchy leads to the following approximate expressions for the quantities appearing in the time-evolution formulae in terms of M_{12} and Γ_{12}:

$$\Delta m = 2 |M_{12}| \left[1 + \mathcal{O} \left(\left| \frac{\Gamma_{12}}{M_{12}} \right|^2 \right) \right], \qquad (52)$$

$$\Delta \Gamma = 2 |\Gamma_{12}| \cos \phi \left[1 + \mathcal{O} \left(\left| \frac{\Gamma_{12}}{M_{12}} \right|^2 \right) \right], \qquad (53)$$

$$\frac{q}{p} = -e^{-i\phi_B} \left[1 - \frac{1}{2} \left| \frac{\Gamma_{12}}{M_{12}} \right| \sin \phi + \mathcal{O} \left(\left| \frac{\Gamma_{12}}{M_{12}} \right|^2 \right) \right], \qquad (54)$$

where $\phi = \arg(-M_{12}/\Gamma_{12})$ and ϕ_B is the phase of M_{12}. Note that ϕ_B thus defined is not measurable and depends on the phase convention adopted, while ϕ is a phase-convention-independent quantity which can be measured in experiments.

Taking into account the above results, the time-dependent decay rates of an initially tagged B^0 or $\bar{B}^0$ state into some final state f can be written as

$$\Gamma[B^0(t=0) \to f(t)] = \mathcal{N}_0 |A_f|^2 e^{-\Gamma t} \left\{ \frac{1 + |\lambda_f|^2}{2} \cosh \frac{\Delta \Gamma t}{2} \right.$$

$$+\frac{1-|\lambda_f|^2}{2}\cos(\Delta mt) - \mathrm{Re}\,\lambda_f \sinh\frac{\Delta\Gamma t}{2} - \mathrm{Im}\,\lambda_f\sin(\Delta mt)\Big\},$$

$$\Gamma[\bar{B}^0(t=0)\to f(t)] = \mathcal{N}_0|A_f|^2\left(1+\left|\frac{\Gamma_{12}}{M_{12}}\right|\sin\phi\right)e^{-\Gamma t}\left\{\frac{1+|\lambda_f|^2}{2}\times\right.$$

$$\times\cosh\frac{\Delta\Gamma t}{2} - \frac{1-|\lambda_f|^2}{2}\cos(\Delta mt) - \mathrm{Re}\,\lambda_f\sinh\frac{\Delta\Gamma t}{2} + \mathrm{Im}\,\lambda_f\sin(\Delta mt)\Big\},$$

where $\mathcal{N}_0$ is the flux normalization and, following the standard notation, we have defined

$$\lambda_f = \frac{q}{p}\frac{\bar{A}_f}{A_f} \approx -e^{-i\phi_B}\frac{\bar{A}_f}{A_f}\left[1-\frac{1}{2}\left|\frac{\Gamma_{12}}{M_{12}}\right|\sin\phi\right] \tag{55}$$

in terms of the decay amplitudes

$$A_f = \langle f|\mathcal{L}_{\Delta F=1}|B^0\rangle, \qquad \bar{A}_f = \langle f|\mathcal{L}_{\Delta F=1}|\bar{B}^0\rangle. \tag{56}$$

From the above expressions it is clear that the key quantity we can access experimentally in the time-dependent study of B decays is the combination λ_f. Both real and imaginary parts of λ_f can be measured, and indeed this is a phase-convention independent quantity: the phase convention in ϕ_B is compensated by the phase convention in the decay amplitudes. In other words, what we can measure is the weak-phase difference between M_{12} and the decay amplitudes.

For generic final states, λ_f is a quantity that is difficult to evaluate. However, it becomes particularly simple in the case where f is a CP eigenstate, $\mathrm{CP}\,|f\rangle = \eta_f|f\rangle$, and the weak phase of the decaying amplitude is known. In this case $\bar{A}_f/A_f$ is a pure phase factor ($|\bar{A}_f/A_f| = 1$), determined by the weak phase of the decaying amplitude:

$$\lambda_f|_{\mathrm{CP-eigen.}} = \eta_f\frac{q}{p}e^{-2i\phi_A}, \qquad A_f = |A_f|e^{i\phi_A}, \qquad \eta_f = \pm 1. \tag{57}$$

The cleanest example of this type is the $|\psi K_S\rangle$ final state for B_d decays. In this case the final state is a CP eigenstate and the decay amplitude is real (to a very good approximation) in the standard CKM phase convention. Indeed the underlying partonic transition is dominated by the Cabibbo-allowed tree-level process $b\to c\bar{c}s$, which has a vanishing phase in the standard CKM phase convention. In addition the leading one-loop corrections (top-quark penguins) have the same vanishing weak phase. Since in the B_d system we can safely neglect Γ_{12}/M_{12}, this implies

$$\lambda_{\psi K_s}^{B_b} = -e^{-i\phi_{B_d}}, \qquad \mathrm{Im}\left(\lambda_{\psi K_s}^{B_b}\right)_{\mathrm{SM}} = \sin(2\beta), \tag{58}$$

where the SM expression of ϕ_{B_d} is nothing but the phase of the CKM combination $(V_{tb}^*V_{td})^2$ appearing in Eq. (38). Given the smallness of $\Delta\Gamma_d$, this quantity is easily extracted by the ratio

$$\frac{\Gamma[\bar{B}_d(t=0) \to \psi K_s(t)] - \Gamma[B^0(t=0) \to f\psi K_s(t)]}{\Gamma[\bar{B}_d(t=0) \to \psi K_s(t)] + \Gamma[B^0(t=0) \to f\psi K_s(t)]}$$
$$\approx \mathrm{Im}\left(\lambda_{\psi K_s}^{B_b}\right)\sin(\Delta m_{B_d}t),$$

which can be considered the golden measurement of B factories.

Another class of interesting final states are CP-conjugate channels $|f\rangle$ and $|\bar{f}\rangle$ which are accessible only to B^0 or $\bar{B}^0$ states, such that $|A_f| = |\bar{A}_{\bar{f}}|$ and $\bar{A}_f = A_{\bar{f}}=0$. Typical examples of this type are the charged semileptonic channels. In this case the asymmetry

$$\frac{\Gamma[\bar{B}^0(t=0) \to f(t)] - \Gamma[B^0(t=0) \to \bar{f}(t)]}{\Gamma[\bar{B}^0(t=0) \to f(t)] + \Gamma[B^0(t=0) \to \bar{f}(t)]} = \left|\frac{\Gamma_{12}}{M_{12}}\right|\sin\phi\left[1 + \mathcal{O}\left(\frac{\Gamma_{12}}{M_{12}}\right)\right]$$

turns out to be time independent and a clean way to determine the indirect CP-violating phase ϕ.

3.3 A Selection of Particularly Interesting Observables in the LHC Era

Given the general arguments about the sensitivity to physics beyond the SM presented at the end of the first lecture, and the theoretical tools discussed above, in the following we analyse some interesting measurements which could be performed in the LHC era, and particularly at the Tevatron and at the LHCb experiment. The list presented here is far from exhaustive (for a more complete analysis we refer to the reviews in Refs. [4–6,19]), but it should serve as an illustration of the interesting potential of B physics at hadron colliders.

3.3.1 CP Violation in B_s Mixing

The CP violating phase of B_s–$\bar{B}_s$ mixing is the last missing ingredient of down-type $\Delta F = 2$ observables. The fact we have not found any deviations from the SM in B_d–$\bar{B}_d$ and Δm_{B_s} should not discourage the measurement of this missing piece of the $\Delta F = 2$ puzzle.

The golden channel for the measurement of the CP-violating phase of B_s–$\bar{B}_s$ mixing is the time-dependent analysis of the $B_s(\bar{B}_s) \to \psi\phi$ decay. At the quark level $B_s \to \psi\phi$ shares the same virtues of $B_d \to \psi K$ (partonic amplitude of the type $b \to c\bar{c}s$), which is used to extract the phase of B_d–$\bar{B}_d$ mixing. However, there are a few points which makes this measurement much more challenging:

- The B_s oscillations are much faster $(\Delta m_{B_s}/\Delta m_{B_d} \approx |V_{t_s}/V_{td}|^2 \approx 30)$, making the time-dependent analysis quite difficult (and essentially inaccessible at B factories).

- Contrary to $|\psi K\rangle$, which has a single angular momentum and is a pure CP eigenstate, the vector–vector state $|\psi \phi\rangle$ produced by the B_s decay has different angular momenta, corresponding to different CP eigenstates. These must be disentangled with a proper angular analysis of the four-body final state $|(\ell^+\ell^-)_\psi (K^+K^-)_\phi\rangle$. To avoid contamination from the nearby $|\psi f_0\rangle$ state, the fit should also include a $|(\ell^+\ell^-)_\psi (K^+K^-)_{S-\text{wave}}\rangle$ component, for a total of ten independent (and unknown) weak amplitudes.[4]

- Contrary to the B_d system, the width difference cannot be neglected in the B_s case, leading to an additional key parameter to be included in the fit.

Modulo the experimental difficulties listed above, the process is theoretically clean and a complete fit of the decay distributions should allow the extraction of

$$\lambda_{\psi\phi}^{B_s} \approx -e^{-i\phi_{B_s}}, \qquad (59)$$

where the SM prediction is

$$\phi_{B_s}^{\text{SM}} = -\arg\frac{(V_{tb}^*V_{ts})^2}{|V_{tb}^*V_{ts}|^2} \approx -0.03. \qquad (60)$$

The tiny value of $\phi_{B_s}^{\text{SM}}$ implies that, within the SM, no CP asymmetry should be observed in the near future. The present status of the combined fit of $\Delta\Gamma_s$ and ϕ_{B_s} from the CDF [21] and D0 [22] experiments at Tevatron is shown in Fig. 5. As can be noted, the agreement with the SM is not good, but the errors are still large.

As we will see in Section 4.1 a clear evidence for $\phi_{B_s} \neq 0$ at Tevatron, or even within the first year of LHCb (which realistically would imply $|\phi_{B_s}| > 0.1$), would not only signal the presence of physics beyond the SM, but would also rule out the whole class of MFV models.

3.3.2   CP Violation in Charged B Decays

Among non-leptonic channels $B^\pm \to DK$ decays are particularly interesting since, via appropriate asymmetries, they allow us to extract the CKM angle γ in a very clean way. The extraction of γ involves only tree-level B decay amplitudes, and is virtually free from hadronic uncertainties (which are eliminated directly by data). It is therefore an essential element for a precise determination of the SM Yukawa couplings in the presence of NP.

The main strategy is based on the following two observations:

---

[4] The formalism is essentially the same adopted in the four-body angular analysis of the $B_d \to J/\psi K\pi$ decay at B factories [20], with the only difference that $\Delta\Gamma$ cannot be neglected in the B_s case.

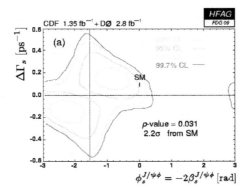

Figure 5. *Combined fit of CDF and D0 results on $\Delta\Gamma$ and ϕ_{B_s} from the time-dependent analysis of $B_s \to \psi\phi$ decays, from Ref. [23].*

- The partonic amplitudes for $B^- \to \bar{D}K^-$ ($b \to c\bar{u}s$) and $B^- \to \bar{D}K^-$ ($b \to u\bar{c}s$) are pure tree-level amplitudes (no penguins allowed given the four different quark flavours). As a result, their weak phase difference is completely determined and is $\gamma = \arg\left(-V_{ud}V_{ub}^*/V_{cd}V_{cb}^*\right)$.

- Thanks to D–$\bar{D}$ mixing, there are several final states f accessible to both D and $\bar{D}$, where the two tree-level amplitudes can interfere. By combining the four final states $B^\pm \to fK^\pm$ and $B^\pm \to \bar{f}K^\pm$, we can extract γ and all the relevant hadronic unknowns of the system.

The first strategy, proposed by Gronau, London and Wyler [24], was based on the selection of $D(\bar{D})$ decays to two-body CP eigenstates. But it has later been realized that any final state accessible to both D and $\bar{D}$ (such as the $K^\pm\pi^\mp$ channels [25] or multibody final states [26]) may work as well.

Let's start from the case of $D(\bar{D})$ decays to CP eigenstates, where the formalism is particularly transparent. The key quantity is the ratio

$$r_B e^{i\delta_B} = \frac{A(B^+ \to D^0 K^+)}{A(B^+ \to \bar{D}^0 K^+)},\tag{61}$$

where δ is a strong phase. Denoting CP-even and CP-odd final states f_+ and f_-, we then have

$$
\begin{aligned}
A\left(B^- \to f_+ K^-\right) &= A_0 \times \left[1 + r_B e^{i(\delta_B - \gamma)}\right], \\
A\left(B^- \to f_- K^-\right) &= A_0 \times \left[1 - r_B e^{i(\delta_B - \gamma)}\right], \\
A\left(B^+ \to f_+ K^+\right) &= A_0 \times \left[1 + r_B e^{i(\delta_B + \gamma)}\right], \\
A\left(B^+ \to f_- K^+\right) &= A_0 \times \left[1 - r_B e^{i(\delta_B + \gamma)}\right].
\end{aligned}\tag{62}
$$

It is clear that combining the four rates we can extract the three hadronic unknowns (A_0, r_B and δ) as well as γ. It is also clear that the sensitivity to γ vanishes in the limit $r_B \to 0$, and indeed the main limitation of this method is that r_B turns out to be very small.

The formalism is essentially unchanged if we consider final states that are not CP eigenstates, such as the $K^\pm \pi^\mp$ states. These have the advantage that the suppression of r_B is partially compensated by the CKM suppression of the corresponding $D(\bar{D}) \to K^\pm \pi^\mp$ decays. Indeed the effective relevant ratio becomes

$$r_{\text{eff}} e^{i\delta_{\text{eff}}} = \frac{A(B^+ \to D^0 K^+)}{A(B^+ \to \bar{D}^0 K^+)} \times \frac{A(D^0 \to K^- \pi^+)}{A(\bar{D}^0 \to K^- \pi^+)}, \tag{63}$$

which is substantially larger than r_B.

Once r_B and δ_B (or r_{eff} and δ_{eff}) are determined from data, the extraction of γ essentially has no theoretical uncertainty. In principle a theoretical error could be induced by the neglected CP-violating effects in charm mixing. In practice, the experimental bounds on charm mixing make this effect totally negligible. The key issue is only collecting high statistics on these highly suppressed decay modes: a clear target for B physics at hadron machines.

3.3.3   The Forward-Backward Asymmetry in $B \to K^* \ell^+ \ell^-$

Theoretical predictions for exclusive FCNC decays are not easy. Even if the final state involves only one hadron, in most of the kinematical region re-scattering effects of the type $B \to K^* H \bar{H} \to K^* \ell^+ \ell^-$ are possible, making it difficult to estimate precisely the decay rate. However, the largest source of uncertainty is typically the normalization of the hadronic form factors. The theoretical error can be substantially reduced in appropriate ratios or differential distributions. A clean example of this type is the normalized forward-backward asymmetry in $B \to K^* \ell^+ \ell^-$.

The observable is defined as

$$\mathcal{A}_{FB}(s) = \frac{1}{d\Gamma(B \to K^* \mu^+ \mu^-)/ds} \int_{-1}^{1} d\cos\theta \, \frac{d^2\Gamma(B \to K^* \mu^+ \mu^-)}{ds \, d\cos\theta} \text{sgn}(\cos\theta), \tag{64}$$

where θ is the angle between the momenta of μ^+ and $\bar{B}$ in the dilepton centre-of-mass frame. Assuming that the leptonic current has only a vector (V) or axial-vector (A) structure (as in the SM), the FB asymmetry provides a direct measure of the A–V interference. Indeed, at the lowest order one can write

$$\mathcal{A}_{\text{FB}}(q^2) \propto \text{Re}\left\{ C_{10A}^* \left[\frac{q^2}{m_b^2} C_{9V}^{\text{eff}} + r(q^2) \frac{m_b C_{7\gamma}}{m_B} \right] \right\},$$

where $r(q^2)$ is an appropriate ratio of $B \to K^*$ vector and tensor form factors [27]. There are three main features of this observable that provide clear and independent short-distance information:

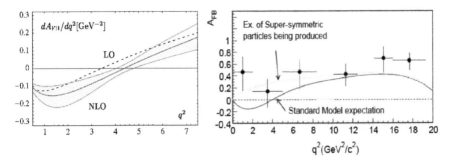

Figure 6. *Left: SM prediction for the zero of the forward-backward asymmetry in $B^- \to K^{*-}\ell^+\ell^-$ at LO and NLO (the band indicates the theoretical uncertainty) [28]. Right: Recent measurement of the forward-backward asymmetry in $B^- \to K^{*-}\ell^+\ell^-$ by BELLE [30].*

1. The position of the zero (q_0) of $\mathcal{A}_{\mathrm{FB}}(q^2)$ in the low-q^2 region (see Fig. 6) [27]: as shown by means of a full NLO calculation [28], the experimental measurement of q_0^2 could allow a determination of C_7/C_9 at the 10% level.

2. The sign of $\mathcal{A}_{\mathrm{FB}}(q^2)$ around the zero. This is fixed unambiguously in terms of the relative sign of C_{10} and C_9: within the SM one expects $\mathcal{A}_{\mathrm{FB}}(q^2 > q_0^2) > 0$ for $|\bar{B}\rangle \equiv |b\bar{d}\rangle$ mesons.

3. The relation $\mathcal{A}[\bar{B}]_{\mathrm{FB}}(q^2) = -\mathcal{A}[B]_{\mathrm{FB}}(q^2)$. This follows from the CP-odd structure of $\mathcal{A}_{\mathrm{FB}}$ and holds at the 10^{-3} level within the SM [29], where C_{10} has a negligible CP-violating phase.

The present status of the measurement at BELLE [30] is shown in Fig. 6. In this case, similarly to B_s–$\bar{B}_s$ mixing, the agreement with the SM is not good, leaving open the room for speculation about sizable non-standard effects. However, the errors are clearly too large to draw definite conclusions.

3.3.4 The Leptonic Decay $B \to \ell^+\ell^-$

The purely leptonic decays constitute a special case among exclusive transitions. Within the SM only the axial-current operator, Q_{10A}, induces a non-vanishing contribution to these decays. As a result, the short-distance contribution is not *diluted* by the mixing with four-quark operators. Moreover, the hadronic matrix element involved is the simplest we can consider, namely the B-meson decay constant in Eq. (41). As we have seen, present estimates of F_{B_d} and F_{B_s} from lattice QCD are already at the 10% level, and in the future the error is expected to go below 5%.

The price to pay for this theoretically clean amplitude is a strong helicity suppression for $\ell = \mu$ (and $\ell = e$) which are the channels with the best

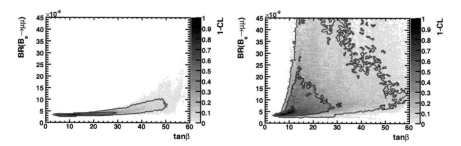

Figure 7. *The correlation between branching ratio for $B_s \to \mu^+\mu^-$ and $\tan\beta$ in the CMSSM (left panel) and in the CMSSM with non-universal Higgs masses (right panel, see Section 4.2.2 for more details), from Ref. [32].*

experimental signature. Employing the full NLO expression of C_{10A}^{SM}, we can write

$$\mathcal{B}(B_s \to \mu^+\mu^-)^{SM} = 3.1 \times 10^{-9} \left(\frac{|V_{ts}|}{0.04}\right)^2 \times \left(\frac{F_{B_s}}{0.21 \text{ GeV}}\right)^2 \quad (65)$$

$$\frac{\mathcal{B}(B_s \to \tau^+\tau^-)^{SM}}{\mathcal{B}(B_s \to \mu^+\mu^-)^{SM}} = 215, \qquad \frac{\mathcal{B}(B_s \to e^+e^-)^{SM}}{\mathcal{B}(B_s \to \mu^+\mu^-)^{SM}} = 2.4 \times 10^{-5}. (66)$$

The corresponding B_d modes are both suppressed by an additional factor $|V_{td}/V_{ts}|^2 \approx 1/30$. The present experimental bound closest to SM expectations is the one obtained by CDF for $B_s \to \mu^+\mu^-$ [31]

$$\mathcal{B}(B_s \to \mu^+\mu^-) < 5.8 \times 10^{-8} \quad (95\% \text{ CL}), \quad (67)$$

which is about one order of magnitude above the SM level.

The strong helicity suppression and the theoretical cleanness make these modes excellent probes of several new-physics models and, particularly, of scalar FCNC amplitudes. Scalar FCNC operators, such as $\bar{b}_R s_L \bar{\mu}_R \mu_L$, are present within the SM but are negligible because of the smallness of down-type Yukawa couplings. On the other hand, these amplitudes could be non-negligible in models with an extended Higgs sector (see Section 4.1.2). In particular, within the MSSM, where two Higgs doublets are coupled separately to up- and down-type quarks, a strong enhancement of scalar FCNCs can occur at large $\tan\beta = v_u/v_d$. This effect is very small in non-helicity-suppressed B decays (because of the small Yukawa couplings), but could easily enhance $B \to \ell^+\ell^-$ rates by one order of magnitude.

An illustration of the possible enhancement in a constrained version of the MSSM (that will be discussed in Section 4.2.2) is shown in Fig. 7. This figure shows that the present search for $B_s \to \mu^+\mu^-$ at CDF is already quite interesting, even if the sensitivity is well above the SM level. In a long-term perspective, the discovery and precise measurement of all the accessible

$B \to \ell^+\ell^-$ channels is definitely one of the most interesting items of the
B-physics program at hadron colliders.

4 Flavour Physics beyond the SM: Models and Predictions

4.1 Minimal Flavour Violation

The main idea of minimal flavour violation (MFV) is that flavour-violating
interactions are linked to the known structure of Yukawa couplings also beyond the SM. In a more quantitative way, the MFV construction consists of
identifying the flavour symmetry and symmetry-breaking structure of the SM
and enforcing it beyond the SM.

The MFV hypothesis consists of two ingredients [33]: (1) a *flavour symmetry* and (2) a set of *symmetry-breaking terms*. The symmetry is nothing
but the large global symmetry $\mathcal{G}_{\text{flavour}}$ of the SM Lagrangian in the absence
of Yukawa couplings shown in Eq. (4). Since this global symmetry, and particularly the $SU(3)$ subgroups controlling quark flavour-changing transitions,
is already broken within the SM, we cannot promote it to be an exact symmetry of the NP model. Some breaking would appear at the quantum level
because of the SM Yukawa interactions. The most restrictive assumption we
can make to *protect* in a consistent way quark-flavour mixing beyond the SM
is to assume that Y_d and Y_u are the only sources of flavour symmetry breaking
in the NP model. To implement and interpret this hypothesis in a consistent
way, we can assume that $\mathcal{G}_q$ is a good symmetry and promote $Y_{u,d}$ to be non-
dynamical fields (*spurions*) with non-trivial transformation properties under
$\mathcal{G}_q$:

$$Y_u \sim (3, \bar{3}, 1), \qquad Y_d \sim (3, 1, \bar{3}). \tag{68}$$

If the breaking of the symmetry occurs at very high-energy scales, at low
energies we would only be sensitive to the background values of the Y, i.e., to
the ordinary SM Yukawa couplings. The role of the Yukawa in breaking the
flavour symmetry becomes similar to the role of the Higgs in the breaking of
the gauge symmetry. However, in the case of the Yukawa we don't know (and
we do not attempt to construct) a dynamical model which gives rise to this
symmetry breaking.

Within the effective theory approach to physics beyond the SM introduced
in Section 2.4, we can say that an effective theory satisfies the criterion of Minimal Flavour Violation in the quark sector if all higher-dimensional operators,
constructed from SM and Y fields, are invariant under CP and (formally)
under the flavour group $\mathcal{G}_q$ [33].

According to this criterion one should in principle consider operators with
arbitrary powers of the (dimensionless) Yukawa fields. However, a strong simplification arises with the observation that all the eigenvalues of the Yukawa

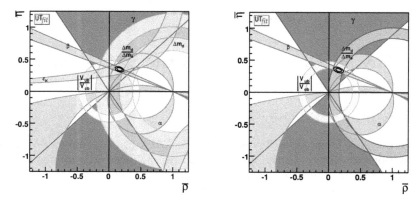

Figure 8. *Fit of the CKM unitarity triangle (in 2008) within the SM (left) and in generic extensions of the SM satisfying the MFV hypothesis (right) [11].*

matrices are small, but for the top one, and that the off-diagonal elements of the CKM matrix are very suppressed. Working in the basis of Eq. (6) we have

$$\left[Y_u(Y_u)^\dagger\right]^n_{i\neq j} \;\approx\; y_t^n V_{it}^* V_{tj}. \tag{69}$$

As a consequence, in the limit where we neglect light quark masses, the leading $\Delta F = 2$ and $\Delta F = 1$ FCNC amplitudes get exactly the same CKM suppression as in the SM:

$$\mathcal{A}(d^i \to d^j)_{\mathrm{MFV}} \;=\; (V_{ti}^* V_{tj})\; \mathcal{A}_{\mathrm{SM}}^{(\Delta F=1)}\left[1 + a_1 \frac{16\pi^2 M_W^2}{\Lambda^2}\right], \tag{70}$$

$$\mathcal{A}(M_{ij} - \bar{M}_{ij})_{\mathrm{MFV}} \;=\; (V_{ti}^* V_{tj})^2 \mathcal{A}_{\mathrm{SM}}^{(\Delta F=2)}\left[1 + a_2 \frac{16\pi^2 M_W^2}{\Lambda^2}\right]. \tag{71}$$

where the $\mathcal{A}_{\mathrm{SM}}^{(i)}$ are the SM loop amplitudes and the a_i are $\mathcal{O}(1)$ real parameters. The a_i depend on the specific operator considered but are flavour independent. This implies the same relative correction in $s \to d$, $b \to d$, and $b \to s$ transitions of the same type: a key prediction which can be tested in experiment.

As pointed out in Ref. [34], within the MFV framework several of the constraints used to determine the CKM matrix (and in particular the unitarity triangle) are not affected by NP. In this framework, NP effects are negligible not only in tree-level processes but also in a few clean observables sensitive to loop effects, such as the time-dependent CPV asymmetry in $B_d \to \psi K_{L,S}$. Indeed the structure of the basic flavour-changing coupling in Eq. (71) implies that the weak CPV phase of B_d–$\bar{B}_d$ mixing is $\arg[(V_{td}V_{tb}^*)^2]$, exactly as in the SM. This construction provides a natural (a posteriori) justification of why no NP effects have been observed in the quark sector: by construction, most of the clean observables measured at B factories are insensitive to NP effects in the MFV framework. A comparison of the CKM fits in the SM and in generic MFV models is shown in Fig. 8. Essentially only ϵ_K and Δm_{B_d} (but not the ratio $\Delta m_{B_d}/\Delta m_{B_s}$) are sensitive to non-standard effects within MFV models.

Given the built-in CKM suppression, the bounds on higher-dimensional operators in the MFV framework turn out to be in the TeV range. This can easily be understood by the discussion in Section 2.4.1: the MFV bounds on operators contributing to ϵ_K and Δm_{B_d} are obtained from Eq. (23) setting $|c_{ij}| = |y_t^2 V_{3i}^* V_{3j}|^2$. From this perspective we could say that the MFV hypothesis provides a solution to the flavour problem.

4.1.1 General Considerations

The idea that the CKM matrix rules the strength of FCNC transitions beyond the SM is a concept that has been implemented and discussed in several works, especially after the first results of the B factories (see, e.g., Refs. [35, 36]). However, it is worth stressing that the CKM matrix represents only one part of the problem: a key role in determining the structure of FCNCs is also played by quark masses, or better by the Yukawa eigenvalues. In this respect, the above MFV criterion provides the maximal protection of FCNCs (or the minimal violation of flavour symmetry), since the full structure of Yukawa matrices is preserved. Moreover, contrary to other approaches, the above MFV criterion is based on a renormalization-group-invariant symmetry argument, which can easily be extended to TeV-scale-effective theories where new degrees of freedoms, such as extra Higgs doublets or SUSY partners of the SM fields are included.

In particular, it is worth stressing that the MFV hypothesis can also be implemented in the so-called Higgsless models, i.e., assuming that the breaking of the electroweak symmetry is induced by some strong dynamics at the TeV scale (similar to the breaking of chiral symmetry in QCD). In this case Eq. (3) is replaced by an effective interaction between fermion fields and the Goldstone bosons associated to the spontaneous breaking of the gauge symmetry. From the point of view of the flavour symmetry, this interaction is identical to the one in (3), and this allows us to proceed as in the case with the explicit Higgs field (see, e.g., Ref. [37]). The only difference between weakly and strongly interacting theories at the TeV scale is that in the latter case the expansion in powers of the Yukawa spurions is not necessarily a rapidly convergent series. If this is the case, then a resummation of the terms involving the top-quark Yukawa coupling needs to be performed [38]

This model-independent structure does not hold in most of the alternative definitions of MFV models that can be found in the literature. For instance, the definition of Ref. [36] (denoted constrained MFV, or CMFV) contains the additional requirement that the effective FCNC operators playing a significant role within the SM are the only relevant ones beyond the SM. This condition is realized only in weakly coupled theories at the TeV scale with only one light Higgs doublet, such as the MSSM with small $\tan \beta$. It does not hold in several other frameworks, such as Higgsless models, or the MSSM with large $\tan \beta$.

Although the MFV seems to be a natural solution to the flavour problem, it should be stressed that we are still very far from having proved the validity of this hypothesis from data. A proof of the MFV hypothesis can be achieved only with positive evidence of physics beyond the SM exhibiting the flavour-universality pattern (same relative correction in $s \to d$, $b \to d$, and $b \to s$ transitions of the same type) predicted by the MFV assumption. While this goal is quite difficult to achieve, the MFV framework is quite predictive and thus could easily be falsified. Some of the most interesting predictions which could be tested in the near future are the following:

- No new CPV phases in B_s mixing, hence $|\phi_{B_s}| < 0.05$ from $\mathcal{A}_{\text{CP}}(B_s \to \psi\phi)$.

- Ratio of B_s and B_d decays into $\ell^+\ell^-$ pairs determined by the CKM matrix: $\mathcal{B}(B_d \to \ell^+\ell^-)/\mathcal{B}(B_s \to \ell^+\ell^-) \approx |V_{td}/V_{ts}|^2$.

- No new CPV phases in $b \to s\gamma$, hence vanishingly small CP asymmetries in $B \to K^*\gamma$ and $B \to K^*\ell^+\ell^-$.

Violations of these bounds would not only imply physics beyond the SM, but also a clear signal of new sources of flavour symmetry breaking beyond the Yukawa couplings.

4.1.2 MFV at Large $\tan \beta$

If the Yukawa Lagrangian contains more than one Higgs field, we can still assume that the Yukawa couplings are the only irreducible breaking sources of $\mathcal{G}_q$, but we can change their overall normalization. A particularly interesting scenario is the two-Higgs-doublet model where the two Higgses are coupled separately to up- and down-type quarks:

$$-\mathcal{L}_Y^{2HDM} = \bar{Q}_L \lambda_d D_R \phi_D + \bar{Q}_L Y_u U_R \phi_U + \bar{L}_L Y_e E_R \phi_D \; + \; \text{h.c.} \qquad (72)$$

This Lagrangian is invariant under an extra $U(1)$ symmetry with respect to the one-Higgs Lagrangian in Eq. (3): a symmetry under which the only charged fields are D_R and E_R (charge $+1$) and ϕ_D (charge -1). This symmetry, denoted U_{PQ}, prevents tree-level FCNCs and implies that $Y_{u,d}$ are the only sources of $\mathcal{G}_q$ breaking appearing in the Yukawa interaction (similar to the one-Higgs-doublet scenario). Coherently with the MFV hypothesis, we can then assume that $Y_{u,d}$ are the only relevant sources of $\mathcal{G}_q$ breaking appearing in all the low-energy effective operators. This is sufficient to ensure that flavour mixing is still governed by the CKM matrix, and naturally guarantees a good agreement with present data in the $\Delta F = 2$ sector. However, the extra symmetry of the Yukawa interaction allows us to change the overall normalization of $Y_{u,d}$ with interesting phenomenological consequences in specific rare modes.

The normalization of the Yukawa couplings is controlled by the ratio of the vacuum expectation values of the two Higgs fields, or by the parameter $\tan\beta = \langle\phi_U\rangle/\langle\phi_D\rangle = v_u/v_d$. Defining the eigenvalues $\lambda_{u,d}$ as in Eq. (6),

$$\lambda_u = \frac{1}{v_u}\text{diag}(m_u, m_c, m_t),$$

$$\lambda_d = \frac{1}{v_d}\text{diag}(m_d, m_s, m_b) = \frac{\tan\beta}{v_u}\text{diag}(m_d, m_s, m_b). \quad (73)$$

For $\tan\beta \gg 1$ the smallness of the b quark mass can be attributed to the smallness of v_d with respect to $v_u \approx v$, rather than to the smallness of the corresponding Yukawa coupling. As a result, for $\tan\beta \gg 1$ we can no longer neglect down-type Yukawa couplings. Since the b-quark Yukawa coupling becomes $\mathcal{O}(1)$, the large-$\tan\beta$ regime is particularly interesting for all the helicity-suppressed observables in B physics (i.e., the observables suppressed within the SM by the smallness of the b quark Yukawa coupling).

Another important aspect of this scenario is that the $U(1)_{\text{PQ}}$ symmetry cannot be exact: it has to be broken at least in the scalar potential in order to avoid the presence of a massless pseudoscalar Higgs boson. Even if the breaking of $U(1)_{\text{PQ}}$ and $\mathcal{G}_q$ are decoupled, the presence of $U(1)_{\text{PQ}}$ breaking sources can have important implications for the structure of the Yukawa interaction, especially if $\tan\beta$ is large [40,41]. We can indeed consider new dimension-four operators such as

$$\epsilon\,\bar{Q}_L\lambda_d D_R\tilde{\phi}_U \quad \text{or} \quad \epsilon\,\bar{Q}_L\lambda_u\lambda_u^\dagger\lambda_d D_R\tilde{\phi}_U, \quad (74)$$

where ϵ denotes a generic MFV-invariant $U(1)_{\text{PQ}}$-breaking source. Even if $\epsilon \ll 1$, the product $\epsilon \times \tan\beta$ can be $\mathcal{O}(1)$, inducing large corrections to the down-type Yukawa sector:

$$\epsilon\,\bar{Q}_L\lambda_d D_R\tilde{\phi}_U \xrightarrow{\text{vev}} \epsilon\,\bar{Q}_L\lambda_d D_R\langle\phi_U\rangle = (\epsilon \times \tan\beta)\,\bar{Q}_L\lambda_d D_R\langle\phi_D\rangle. \quad (75)$$

This is what happens in supersymmetry, where the operators in Eq. (74) are generated at the one-loop level $[\epsilon \sim 1/(16\pi^2)]$, and the large $\tan\beta$ solution is particularly welcome in the context of Grand Unified models [42].

One of the clearest phenomenological consequences is a suppression (typically in the 10–50% range) of the $B \to \ell\nu$ decay rate with respect to its SM expectation [43]. But the most striking signature could arise from the rare decays $B_{s,d} \to \ell^+\ell^-$ whose rates could be enhanced over the SM expectations by more than one order of magnitude [44] as already shown in Fig. 7 in the context of supersymmetric models. An enhancement of both $B_s \to \ell^+\ell^-$ and $B_d \to \ell^+\ell^-$ respecting the MFV relation $\Gamma(B_s \to \ell^+\ell^-)/\Gamma(B_d \to \ell^+\ell^-) \approx |V_{ts}/V_{td}|^2$ would be an unambiguous signature of MFV at large $\tan\beta$ [39].

4.2 Flavour Breaking in the MSSM

The Minimal Supersymmetric extension of the SM (MSSM) is one of the best-motivated and definitely the most-studied extension of the SM at the TeV scale. For a detailed discussion of this model we refer to the review in Ref. [45] and to the lectures by J. Ellis at this school. Here we restrict ourselves to the analysis of some properties of this model relevant to flavour physics.

The particle content of the MSSM consists of the SM gauge and fermion fields plus a scalar partner for each quark and lepton (squarks and sleptons) and a spin-1/2 partner for each gauge field (gauginos). The Higgs sector has two Higgs doublets with the corresponding spin-1/2 partners (higgsinos) and a Yukawa coupling of the type in Eq. (72). While gauge and Yukawa interactions of the model are completely specified in terms of the corresponding SM couplings, the so-called soft-breaking sector[5] of the theory contains several new free parameters, most of which are related to flavour-violating observables. For instance, after the up-type Higgs field gets a vev ($\phi_U \to \langle \phi_U \rangle$), the 6×6 mass matrix of the up-type squarks has the following structure,

$$
\tilde{M}_U^2 = \begin{pmatrix} \tilde{m}_{Q_L}^2 & A_U \langle \phi_U \rangle \\ A_U^\dagger \langle \phi_U \rangle & \tilde{m}_{U_R}^2 \end{pmatrix} + \mathcal{O}\left(m_Z, m_{\text{top}}\right), \tag{76}
$$

where $\tilde{m}_{Q_L}$, $\tilde{m}_{U_R}$ and A_U are 3×3 unknown matrices. Indeed the adjective *minimal* in the MSSM acronyms refers to the particle content of the model but does not specify its flavour structure.

Because of this large number of free parameters, we cannot discuss the implications of the MSSM in flavour physics without specifying in more detail the flavour structure of the model. The versions of the MSSM analysed in the literature range from the so-called Constrained MSSM (CMSSM), where the complete model is specified in terms of only four free parameters (in addition to the SM couplings), to the MSSM without R parity and generic flavour structure, which contains a few hundred new free parameters.

Throughout the large body of work in the past decades it has became clear that the MSSM with generic flavour structure and squarks in the TeV range is not compatible with precision tests in flavour physics. This is true even if we impose R parity, the discrete symmetry which forbids single s-particle production, usually advocated to prevent a too-fast proton decay. In this case we have no tree-level FCNC amplitudes, but the loop-induced contributions are still too large compared to the SM ones unless the squarks are highly degenerate or have very small intra-generation mixing angles. This

---

[5] Supersymmetry must be broken in order to be consistent with observations (we do not observe degenerate spin partners in nature). The soft breaking terms are the most general supersymmetry-breaking terms which preserve the nice ultraviolet properties of the model. They can be divided into two main classes: (1) mass terms which break the mass degeneracy of the spin partners (e.g., sfermion or gaugino mass terms); (2) trilinear couplings among the scalar fields of the theory (e.g., sfermion-sfermion-Higgs couplings).

is nothing but a manifestation in the MSSM context of the general flavour problem illustrated in the first lecture.

The flavour problem of the MSSM is an important clue about the underlying mechanism of supersymmetry breaking. On general grounds, mechanisms of SUSY breaking with flavour universality (such as gauge mediation) or with heavy squarks (especially in the case of the first two generations) tends to be favoured. However, several options are still open. These range from the very restrictive CMSSM case, which is a special case of MSSM with MFV, to more general scenarios with new small but non-negligible sources of flavour symmetry breaking.

4.2.1 Flavour Universality, MFV and RGE in the MSSM

Since the squark fields have well-defined transformation properties under the SM quark-flavour group $\mathcal{G}_q$, the MFV hypothesis can easily be implemented in the MSSM framework following the general rules outlined in Section 4.1.

We need to consider all possible interactions compatible with (1) softly-broken supersymmetry; (2) the breaking of $\mathcal{G}_q$ via the spurion fields $Y_{U,D}$. This allows the squark mass terms and the trilinear quark-squark-Higgs couplings to be expressed as follows [33, 46]:

$$
\begin{aligned}
\tilde{m}_{Q_L}^2 &= \tilde{m}^2 \left(a_1 \mathbb{1} + b_1 Y_U Y_U^\dagger + b_2 Y_D Y_D^\dagger + b_3 Y_D Y_D^\dagger Y_U Y_U^\dagger + \ldots \right), \\
\tilde{m}_{U_R}^2 &= \tilde{m}^2 \left(a_2 \mathbb{1} + b_5 Y_U^\dagger Y_U + \ldots \right), \\
A_U &= A \left(a_3 \mathbb{1} + b_6 Y_D Y_D^\dagger + \ldots \right) Y_U,
\end{aligned}
\tag{77}
$$

and similarly for the down-type terms. The dimensional parameters $\tilde{m}$ and A, expected to be in the range of a few 100 GeV – 1 TeV, set the overall scale of the soft-breaking terms. In Eq. (77) we have explicitly shown all independent flavour structures which cannot be absorbed into a redefinition of the leading terms (up to tiny contributions that are quadratic in the Yukawas of the first two families), when $\tan\beta$ is not too large and the bottom Yukawa coupling is small, the quadratic terms in Y_D can be dropped.

In a bottom-up approach, the dimensionless coefficients a_i and b_i should be considered to be free parameters of the model. Note that this structure is renormalization-group invariant: the values of a_i and b_i change according to the Renormalization Group (RG) flow, but the general structure of Eq. (77) is unchanged. This is not the case if the b_i are set to zero, corresponding to the so-called hypothesis of *flavour universality*. In several explicit mechanisms of supersymmetry breaking, the condition of flavour universality holds at some high-scale M, such as the scale of Grand Unification in the CMSSM (see below) or the mass scale of the messenger particles in gauge mediation (see Ref. [47]). In this case non-vanishing $b_i \sim (1/4\pi)^2 \ln M^2/\tilde{M}^2$ are generated by the RG evolution. As recently pointed out in Ref. [48] the RG flow in the MSSM-MFV framework exhibits quasi infra-red fixed points: even if we start with all the $b_i = \mathcal{O}(1)$ at some high scale, the only non-negligible terms at the TeV scale are those associated with the $Y_U Y_U^\dagger$ structures.

If we are interested only in low-energy processes we can integrate out the supersymmetric particles at one loop and project this theory into the general MFV effective theory approach discussed before. In this case the coefficients of the dimension-six effective operators written in terms of SM and Higgs fields are computable in terms of the supersymmetric soft-breaking parameters. The typical effective scale suppressing these operators (assuming an overall coefficient $1/\Lambda^2$) is $\Lambda \sim 4\pi\tilde{m}$. Since the bounds on Λ within MFV are in the few TeV range, we then conclude that if MFV holds, the present bounds on FCNCs do not exclude squarks in the few hundred GeV mass range, i.e., well within the LHC reach.

4.2.2   The CMSSM Framework

The CMSSM, also known as mSUGRA, is the supersymmetric extension of the SM with the minimal particle content and the maximal number of universality conditions on the soft-breaking terms. At the scale of Grand Unification ($M_{\rm GUT} \sim 10^{16}$ GeV) it is assumed that there are only three independent soft-breaking terms: the universal gaugino mass ($\tilde{m}_{1/2}$), the universal trilinear term (A) and the universal sfermion mass ($\tilde{m}_0$). The model has two additional free parameters in the Higgs sector (the so-called μ and B terms), which control the vacuum expectation values of the two Higgs fields (also determined by the RG running from the unification scale down to the electroweak scale). Imposing the correct W- and Z-boson masses allow us to eliminate one of these Higgs-sector parameters, and the remaining one is usually chosen to be $\tan\beta$. As a result, the model is fully specified in terms of the three high-energy parameters $\{\tilde{m}_{1/2}, \tilde{m}_0, A\}$ and the low-energy parameter $\tan\beta$.[6] This constrained version of the MSSM is an example of a SUSY model with MFV. Note, however, that the model is much more constrained than the general MSSM with MFV: in addition to being flavour universal, the soft-breaking terms at the unification scale obey various additional constraints (e.g., in Eq. (77) we have $a_1 = a_2$ and $b_i = 0$).

In the MSSM with R parity we can distinguish five main classes of one-loop diagrams contributing to FCNC and CP-violating processes with external down-type quarks. They are distinguished according to the virtual particles running inside the loops: W and up-quarks (i.e., the leading SM amplitudes), charged-Higgs and up-quarks, charginos and up-squarks, neutralinos and down-squarks, gluinos and down-squarks. Within the CMSSM, the charged-Higgs and chargino exchanges yield the dominant non-standard contributions.

Given the low number of free parameters, the CMSSM is very predictive and phenomenologically constrained by the precision measurements in flavour

---

[6]More precisely, for each choice of $\{\tilde{m}_{1/2}, \tilde{m}_0, A, \tan\beta\}$ there is a discrete ambiguity related to the sign of the μ term.

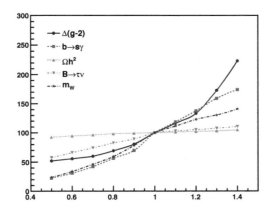

Figure 9. *Relative sizes of the 95% C.L. preferred area in the* $(m_0, \tan\beta)$ *plane of the CMSSM, as a function hypothetical variations of the present errors on* $(g-2)_\mu$, *the cold dark matter abundance* (Ωh^2), $\mathcal{B}(B \to X_s\gamma)$, $\mathcal{B}(B^+ \to \tau^+\nu)$, *and* m_W *[49]. The error scaling is relative to the current combined theory and experimental error.*

physics. The most powerful low-energy constraint comes from $B \to X_s\gamma$. For large values of $\tan\beta$, strong constraints are also obtained from $B_s \to \mu^+\mu^-$ and $B^+ \to \tau^+\nu$. If these observables are within the present experimental bounds, the constrained nature of the model implies essentially no observable deviations from the SM in other flavour-changing processes. Interestingly enough, the CMSSM satisfy at the same time the flavour constraints and those from electroweak precision observables for squark masses below 1 TeV [49]. An illustration of the constraining power of the flavour observables in determining the allowed parameter space of the model is shown in Fig. 9. As can be seen, a reduction of the error on $B \to X_s\gamma$ by 50% would imply a substantial reduction of the allowed region in the CMSSM parameter space.

It is worth stressing that as soon as we relax the strong universality assumptions of the CMSSM, the phenomenology of the model can vary substantially. An illustration of this statement is provided by Fig. 7, where we compare the predictions for $B_s \to \mu^+\mu^-$ in the CMSSM and in its minimal variation, the so-called Non-Universal Higgs Mass (NUHM) scenario. In the latter case only the condition of universality for the soft breaking terms in the Higgs sectors is relaxed, increasing by one unit the number of free parameters of the model. As can be noted, the difference is substantial (in both cases all existing constraints are satisfied). This also illustrates how precise data from flavour physics are essential to discriminate different versions of the MSSM. A recent detailed analysis of the discriminating power of flavour observables with respect to different versions of the MSSM, even beyond MFV, can be found in Ref. [50].

4.2.3   The Mass Insertion Approximation in the General MSSM

Flavour universality at the GUT scale is not a general property of the MSSM, even if the model is embedded in a Grand Unified Theory. If this assumption is relaxed, new interesting phenomena can occur in flavour physics. The most general one is the appearance of gluino-mediated one-loop contributions to FCNC amplitudes [51].

The main problem when going beyond simplifying assumptions, such as flavour universality or MFV, is the proliferation in the number of free parameters. A useful model-independent parameterisation to describe the new phenomena occurring in the general MSSM with R parity conservation is the so-called mass insertion (MI) approximation [52]. Selecting a flavour basis for fermion and sfermion states where all the couplings of these particles to neutral gauginos are flavour diagonal, the new flavour-violating effects are parameterised in terms of the non-diagonal entries of the sfermion mass matrices. More precisely, denoting by Δ the off-diagonal terms in the sfermion mass matrices (i.e., the mass terms relating sfermions of the same electric charge, but different flavour), the sfermion propagators can be expanded in terms of $\delta = \Delta/\tilde{m}^2$, where $\tilde{m}$ is the average sfermion mass. As long as Δ is significantly smaller than $\tilde{m}^2$ (as suggested by the absence of sizable deviations from the SM), one can truncate the series to the first term of this expansion and the experimental information concerning FCNC and CP-violating phenomena translates into upper bounds on δ [53].

The major advantage of the MI method is that it is not necessary to perform a full diagonalization of the sfermion mass matrices, obtaining a substantial simplification in the comparison of flavour-violating effects in different processes. There exist four types of mass insertions connecting flavours i and j along a sfermion propagator: $(\Delta_{ij})_{LL}$, $(\Delta_{ij})_{RR}$, $(\Delta_{ij})_{LR}$ and $(\Delta_{ij})_{RL}$. The indexes L and R refer to the helicity of the fermion partners.

In most cases the leading non-standard amplitude is the gluino-exchange one, which is enhanced by one or two powers of the ratio $(\alpha_{\text{strong}}/\alpha_{\text{weak}})$ with respect to neutralino- or chargino-mediated amplitudes. When analysing the bounds, it is customary to consider one non-vanishing MI at a time, barring accidental cancellations. This procedure is justified a posteriori by observing that the MI bounds have typically a strong hierarchy, making the destructive interference among different MIs rather unlikely. The bounds thus obtained from recent measurements in B and K physics are reported in Table 1. The bounds mainly depend on the gluino and on the average squark mass, scaling as the inverse mass (the inverse mass square) for bounds derived from $\Delta F = 2$ ($\Delta F = 1$) observables.

The only clear pattern emerging from these bounds is that there is no room for sizable new sources of flavour-symmetry breaking. However, it is too early

$|(\delta^d_{12})_{LL,RR}| < 1 \cdot 10^{-2}$	$|(\delta^d_{12})_{LL=RR}| < 2 \cdot 10^{-4}$	$|(\delta^d_{12})_{LR}| < 5 \cdot 10^{-4}$
$|(\delta^d_{13})_{LL,RR}| < 7 \cdot 10^{-2}$	$|(\delta^d_{13})_{LL=RR}| < 5 \cdot 10^{-3}$	$|(\delta^d_{13})_{LR}| < 1 \cdot 10^{-2}$
$|(\delta^d_{23})_{LL}| < 2 \cdot 10^{-1}$	$|(\delta^d_{23})_{LL=RR}| < 5 \cdot 10^{-2}$	$|(\delta^d_{23})_{LR,RL}| < 5 \cdot 10^{-3}$

Table 1. *Upper bounds at 95% C.L. on the dimensionless down-type mass-insertion parameters (see text) for squark and gluino masses of 350 GeV (from Ref. [5]).*

to draw definite conclusions since some of the bounds, especially those in the 2-3 sector, are still rather weak. As suggested by various authors (see, e.g., Ref. [54]), the possibility of sizable deviations from the SM in the 2-3 sector could fit well with the large 2-3 mixing of light neutrinos, in the context of a unification of quark and lepton sectors and could possibly also explain a large CP-violating phase in B_s–$\bar{B}_s$ mixing.

4.3 Flavour Protection from Hierarchical Fermion Profiles

So far we have assumed that the suppression of flavour-changing transitions beyond the SM can be attributed to a flavour symmetry, and a specific form of the symmetry-breaking terms. An interesting alternative is the possibility of a generic *dynamical suppression* of flavour-changing interactions, related to the weak mixing of the light SM fermions with the new dynamics at the TeV scale. A mechanism of this type is the so-called RS-GIM mechanism occurring in models with a warped extra dimension. In this framework the hierarchy of fermion masses, which is attributed to the different localization of the fermions in the bulk [55], implies that the lightest fermions are those localized far from the infra-red (SM) brane. As a result, the suppression of FCNCs involving light quarks is a consequence of the small overlap of the light fermions with the lightest Kaluza–Klein excitations [56].

As pointed out in [57], the general features of this class of models can also be described by means of an effective theory approach. The two main assumptions of this approach are the following:

- There exists a (non-canonical) basis for the SM fermions where their kinetic terms exhibit a rather hierarchical form:

$$\mathcal{L}^{\text{quarks}}_{\text{kin}} = \sum_{\Psi = Q_L, U_R, D_R} \overline{\Psi} Z_\psi^{-2} \slashed{D} \Psi,$$

$$Z_\psi = \text{diag}(z_\psi^{(1)}, z_\psi^{(2)}, z_\psi^{(3)}), \qquad z_\psi^{(1)} \ll z_\psi^{(2)} \ll z_\psi^{(3)} \lesssim 1. \quad (78)$$

- In such a basis there is no flavour symmetry and all the flavour-violating interactions, including the Yukawa couplings, are $\mathcal{O}(1)$.

Once the fields are transformed into the canonical basis, the hierarchical kinetic terms act as a distorting lens, through which all interactions are seen as approximately aligned on the magnification axes of the lens. As anticipated, this construction provides an effective four-dimensional description of a wide class of models with a warped extra dimension. However, it should be stressed that this mechanism is not a general feature of models with extra dimensions: as discussed in [58], in extra-dimensional models it is also possible to postulate the existence of additional symmetries and, for instance, recover an MFV structure.

The dynamical mechanism of hierarchical fermion profiles is quite effective in suppressing FCNCs beyond the SM. In particular, it can be shown that all the dimension-six FCNC left-left operators (such as the $\Delta F = 2$ terms in Eq. (22)), have the same suppression as in MFV [57]. However, a residual problem is present in the left-right operators contributing to CP-violating observables in the kaon system: ϵ_K [11,59] and ϵ'/ϵ_K [60], with potentially visible effects also in rare K decays [61]. As a result, and contrary to most of the models discussed before, in this framework one expects no significant NP effects in the B system, while possible improved measurements and predictions in the K system could reveal some deviation from the SM.

5   Conclusions

The absence of significant deviations from the SM in quark-flavour physics provides key information about any extension of the SM. Only models with a highly non-generic flavour structure can both stabilize the electroweak sector and, at the same time, be compatible with flavour observables. In such models we expect new particles within the LHC reach; however, the structure of the new theory cannot be determined using only the high-p_T data from LHC. As illustrated in these lectures, there are still various open questions about the flavour structure of the model that can be addressed only at low energies, and in particular through B decays.

If we are interested only in physics beyond the SM, the set of B-physics observables to be measured with higher precision, and the rare transitions to be searched for are limited. But it is far from being a small set. As shown in these lectures, we still know very little about CP violation in the B_s system, and about FCNC transitions of the type $B \to K^*\ell^+\ell^-$ and $B_{s,d} \to \ell^+\ell^-$. A systematic reduction in the determination of the SM Yukawa couplings, such as the determination of γ from $B \to DK$ decays, could possibly yet reveal non-standard effects also in observables which we have already measured well, such as ϵ_K or the B_d mixing phase.

Acknowledgements

I wish to thank the organizers of SUSSP65 for the invitation to this interesting and well-organized school, as well as all the participants who contributed to a very enjoyable and stimulating atmosphere.

References

[1] N. Cabibbo, Phys. Rev. Lett. **10** (1963) 531.

[2] M. Kobayashi and T. Maskawa, Prog. Theor. Phys. **49** (1973) 652.

[3] A. J. Buras, arXiv:hep-ph/0505175, M. Neubert, arXiv:hep-ph/0512222; Y. Nir, arXiv:0708.1872 [hep-ph].

[4] M. Antonelli *et al.*, Phys. Rept. **494** (2010) 197 [arXiv:0907.5386 [hep-ph]]

[5] M. Artuso *et al.*, Eur. Phys. J. C **57** (2008) 309 [arXiv:0801.1833 [hep-ph]].

[6] A. J. Buras, PoS **EPS-HEP2009** (2009) 024 [arXiv:0910.1032 [hep-ph]].

[7] L. L. Chau and W. Y. Keung, Phys. Rev. Lett. **53** (1984) 1802.

[8] L. Wolfenstein, Phys. Rev. Lett. **51** (1983) 1945.

[9] A. J. Buras, M. E. Lautenbacher, and G. Ostermaier, Phys. Rev. D **50** (1994) 3433 [arXiv:hep-ph/9403384].

[10] J. Charles *et al.* [CKMfitter Collaboration], Eur. Phys. J. **C41** (2005) 1 [hep-ph/0406184]; [updates at http://www.slac.stanford.edu/xorg/ckmfitter/]

[11] M. Bona *et al.* [UTfit Collaboration], JHEP **0803** (2008) 049 [arXiv:0707.0636 [hep-ph]]; [updates at http://www.utfit.org/].

[12] S. L. Glashow, J. Iliopoulos and L. Maiani, Phys. Rev. D **2** (1970) 1285.

[13] G. Buchalla, A. J. Buras and M. E. Lautenbacher, Rev. Mod. Phys. **68** (1996) 1125 [arXiv:hep-ph/9512380].

[14] H. Georgi, Phys. Lett. B **240** (1990) 447.

[15] E. Gamiz, PoS **LATTICE2008** (2008) 014. [arXiv:0811.4146 [hep-lat]]

[16] V. Lubicz and C. Tarantino, Nuovo Cim. **123B** (2008) 674 [arXiv:0807.4605 [hep-lat]].

[17] M. Beneke, G. Buchalla, M. Neubert and C. T. Sachrajda, Nucl. Phys. B **591** (2000) 313 [arXiv:hep-ph/0006124].

[18] C. W. Bauer, S. Fleming, D. Pirjol and I. W. Stewart, Phys. Rev. D **63** (2001) 114020 [arXiv:hep-ph/0011336].

[19] K. Anikeev *et al.*, arXiv:hep-ph/0201071.

[20] B. Aubert *et al.* [BABAR Collaboration], Phys. Rev. D **71** (2005) 032005 [arXiv:hep-ex/0411016].

[21] T. Aaltonen *et al.* [CDF Collaboration], Phys. Rev. Lett. **100** (2008) 161802 [arXiv:0712.2397 [hep-ex]].

[22] V. M. Abazov *et al.* [D0 Collaboration], Phys. Rev. Lett. **101** (2008) 241801 [arXiv:0802.2255 [hep-ex]].

[23] E. Barberio *et al.* [Heavy Flavour Averaging Group], arXiv:0808.1297 [hep-ex].

[24] M. Gronau and D. London, Phys. Lett. B **253** (1991) 483; M. Gronau and D. Wyler, Phys. Lett. B **265** (1991) 172.

[25] D. Atwood, I. Dunietz and A. Soni, Phys. Rev. Lett. **78** (1997) 3257 [arXiv:hep-ph/9612433].

[26] A. Giri, Y. Grossman, A. Soffer and J. Zupan, Phys. Rev. D **68** (2003) 054018 [arXiv:hep-ph/0303187].

[27] G. Burdman, Phys. Rev. D **57** (1998) 4254 [hep-ph/9710550].

[28] M. Beneke, T. Feldmann and D. Seidel, Nucl. Phys. B **612** (2001) 25 [hep-ph/0106067].

[29] G. Buchalla, G. Hiller and G. Isidori, Phys. Rev. D **63** (2001) 014015 [hep-ph/0006136].

[30] J. T. Wei *et al.* [BELLE Collaboration], Phys. Rev. Lett. **103** (2009) 171801 [arXiv:0904.0770 [hep-ex]].

[31] M. Aoki [CDF Collaboration and D0 Collaboration], arXiv:0906.3320 [hep-ex].

[32] O. Buchmueller *et al.*, Eur. Phys. J. C **64** (2009) 391 [arXiv:0907.5568 [hep-ph]].

[33] G. D'Ambrosio, G. F. Giudice, G. Isidori and A. Strumia, Nucl. Phys. B **645** (2002) 155 [hep-ph/0207036].

[34] A. J. Buras, P. Gambino, M. Gorbahn, S. Jager and L. Silvestrini, Phys. Lett. B **500** (2001) 161 [arXiv:hep-ph/0007085].

[35] A. Ali and D. London, Eur. Phys. J. C **9** (1999), 687 [hep-ph/9903535]; S. Laplace, Z. Ligeti, Y. Nir and G. Perez, Phys. Rev. D **65** (2002) 094040 [hep-ph/0202010].

[36] A. J. Buras, Acta Phys. Polon. B **34** (2003) 5615 [arXiv:hep-ph/0310208].

[37] R. Barbieri, G. Isidori and D. Pappadopulo, JHEP **0902** (2009) 029 [arXiv:0811.2888 [hep-ph]].

[38] T. Feldmann and T. Mannel, Phys. Rev. Lett. **100** (2008) 171601 [arXiv:0801.1802 [hep-ph]]; A. L. Kagan, G. Perez, T. Volansky and J. Zupan, Phys. Rev. D **80** (2009) 076002 [arXiv:0903.1794 [hep-ph]].

[39] T. Hurth, G. Isidori, J. F. Kamenik and F. Mescia, Nucl. Phys. B **808** (2009) 326 [arXiv:0807.5039 [hep-ph]].

[40] L. J. Hall, R. Rattazzi and U. Sarid, Phys. Rev. D **50** (1994) 7048 [hep-ph/9306309]; T. Blazek, S. Raby and S. Pokorski, Phys. Rev. D **52** (1995) 4151 [hep-ph/9504364].

[41] K. S. Babu and C. F. Kolda, Phys. Rev. Lett. **84** (2000) 228 [hep-ph/9909476]; G. Isidori and A. Retico, JHEP **0111** (2001) 001 [hep-ph/0110121].

[42] M. Olechowski and S. Pokorski, Phys. Lett. B **214** (1988) 393; P. Krawczyk and S. Pokorski, Phys. Rev. Lett. **60** (1988) 182.

[43] W. S. Hou, Phys. Rev. D **48** (1993) 2342; A. G. Akeroyd and S. Recksiegel, J. Phys. G **29** (2003) 2311 [hep-ph/0306037]; G. Isidori and P. Paradisi, Phys. Lett. B **639** (2006) 499 [hep-ph/0605012].

[44] C. Hamzaoui, M. Pospelov and M. Toharia, Phys. Rev. D **59** (1999) 095005 [hep-ph/9807350]; S. R. Choudhury and N. Gaur, Phys. Lett. B **451** (1999) 86 [arXiv:hep-ph/9810307].

[45] S. P. Martin, arXiv:hep-ph/9709356.

[46] L. J. Hall and L. Randall, Phys. Rev. Lett. **65** (1990) 2939.

[47] G. F. Giudice and R. Rattazzi, Phys. Rept. **322** (1999) 419 [arXiv:hep-ph/9801271].

[48] P. Paradisi, M. Ratz, R. Schieren and C. Simonetto, Phys. Lett. B **668** (2008) 202 [arXiv:0805.3989 [hep-ph]]; G. Colangelo, E. Nikolidakis and C. Smith, Eur. Phys. J. C **59** (2009) 75 [arXiv:0807.0801 [hep-ph]].

[49] O. Buchmueller *et al.*, Phys. Lett. B **657** (2007) 87 [arXiv:0707.3447 [hep-ph]].

[50] W. Altmannshofer, A. J. Buras, S. Gori, P. Paradisi and D. M. Straub, Nucl. Phys. B **830** (2010) 17 [arXiv:0909.1333 [hep-ph]].

[51] J. R. Ellis and D. V. Nanopoulos, Phys. Lett. B **110** (1982) 44; R. Barbieri and R. Gatto, Phys. Lett. B **110** (1982) 211.

[52] L. J. Hall, V. A. Kostelecky and S. Raby, Nucl. Phys. B **267** (1986) 415.

[53] F. Gabbiani, E. Gabrielli, A. Masiero and L. Silvestrini, Nucl. Phys. B **477** (1996) 321 [arXiv:hep-ph/9604387].

[54] M. Ciuchini, A. Masiero, P. Paradisi, L. Silvestrini, S. K. Vempati and O. Vives, Nucl. Phys. B **783** (2007) 112 [arXiv:hep-ph/0702144].

[55] N. Arkani-Hamed and M. Schmaltz, Phys. Rev. D **61** (2000) 033005 [arXiv:hep-ph/9903417]; T. Gherghetta and A. Pomarol, Nucl. Phys. B **586** (2000) 141 [arXiv:hep-ph/0003129]. S. J. Huber and Q. Shafi, Phys. Lett. B **498** (2001) 256 [arXiv:hep-ph/0010195];

[56] K. Agashe, G. Perez and A. Soni, Phys. Rev. D **71** (2005) 016002 [arXiv:hep-ph/0408134]; Phys. Rev. Lett. **93** (2004) 201804 [arXiv:hep-ph/0406101]; R. Contino, T. Kramer, M. Son and R. Sundrum, JHEP **0705** (2007) 074 [arXiv:hep-ph/0612180].

[57] S. Davidson, G. Isidori and S. Uhlig, Phys. Lett. B **663** (2008) 73 [arXiv:0711.3376 [hep-ph]].

[58] G. Cacciapaglia, C. Csaki, J. Galloway, G. Marandella, J. Terning and A. Weiler, JHEP **0804** (2008) 006 [arXiv:0709.1714 [hep-ph]]; C. Csaki, A. Falkowski and A. Weiler, Phys. Rev. **D80** (2009) 016001 [arXiv:0806.3757 [hep-ph]].

[59] C. Csaki, A. Falkowski and A. Weiler, JHEP **0809** (2008) 008 [arXiv:0804.1954 [hep-ph]]; M. Blanke *et al.*, JHEP **0903** (2009) 001 [arXiv:0809.1073 [hep-ph]]; M. Bauer *et al.*, Phys. Rev. **D79** (2009) 076001 [arXiv:0811.3678 [hep-ph]].

[60] O. Gedalia, G. Isidori and G. Perez, Phys. Lett. **B682** (2009) 200 [arXiv:0905.3264 [hep-ph]]

[61] M. Blanke, A. J. Buras, B. Duling, K. Gemmler and S. Gori, JHEP **0903** (2009) 108 [arXiv:0812.3803 [hep-ph]].

BSM Phenomenology

John Ellis

CERN, Geneva, Switzerland

1 The Standard Model and the Higgs Boson

The electroweak sector of the Standard Model (SM) (for more detailed accounts, see, e.g., [1–3]) grew out of experimental information on charged-current weak interactions, and of the realisation that the four-point Fermi description ceases to be valid above $\sqrt{s} = 600$ GeV [3]. Electroweak theory was able to predict the existence of neutral-current interactions, as discovered by the Gargamelle Collaboration in 1973 [4]. One of its greatest subsequent successes was the detection in 1983 of the $W^{\pm}$ and Z^0 bosons [5–8], whose existences it had predicted. Over time, thanks to the accumulating experimental evidence, the $SU(2)_L \otimes U(1)_Y$ electroweak theory and $SU(3)_C$ quantum chromodynamics, collectively known as the Standard Model, have come to be regarded as the correct description of electromagnetic, weak and strong interactions up to the energies that have been probed so far. However, although the SM has many successes, it also has some shortcomings, as discussed in subsequent lectures; see also [9–11].

Within the SM, the electromagnetic and weak interactions are described by a Lagrangian that is symmetric under local weak isospin and hypercharge gauge transformations, described using the $SU(2)_L \otimes U(1)_Y$ group (the L subindex refers to the fact that the weak $SU(2)$ group acts only the left-handed projections of fermion states; Y is the hypercharge). We can write the $SU(2)_L \otimes U(1)_Y$ part of the SM Lagrangian as

$$
\begin{aligned}
\mathcal{L} = \ & -\frac{1}{4}\mathbf{F}^a_{\mu\nu}\mathbf{F}^{a\mu\nu} \\
& + \ i\bar{\psi}\slashed{D}\psi + h.c. \\
& + \ \psi_i y_{ij}\psi_j\phi + h.c. \\
& + \ |D_\mu\phi|^2 - V(\phi).
\end{aligned}
\tag{1}
$$

This is short enough to write on a T-shirt!

However, whereas the first two lines of (1) have been confirmed in many different experiments, there is **no** experimental evidence for the last two lines, and one of the main objectives of the LHC is to discover whether it is right, needs modification or is simply wrong. These lines describe the Higgs boson and its interactions with fermions, its kinetic term and its effective potential $V(\phi^*\phi) = \mu^2(\phi^*\phi) + \lambda(\phi^*\phi)^2$. If $\mu^2 < 0$, minimizing this effective potential gives a non-zero v.e.v., and thereby masses to the fermions and the neutral and charged weak-interaction bosons. The latter are related through

$$m_Z = m_W \sqrt{1 + g'^2/g^2}. \tag{2}$$

Experimentally, the weak gauge boson masses are known to high accuracy to be [12]

$$m_W = 80.399 \pm 0.023 \text{ GeV}, \qquad m_Z = 91.1875 \pm 0.0021 \text{ GeV}. \tag{3}$$

For the Higgs mass, after minimization of the potential we find

$$m_H = -2\mu^2. \tag{4}$$

A priori, however, there is no theoretical prediction within the SM, since μ is not determined by any of the known parameters of the SM. Later we will see various ways in which experiments constrain the Higgs mass, and then proceed to discuss alternative ideas for electroweak symmetry breaking, as well as other ideas beyond the SM.

1.1 Parameters of the Standard Model

The transformation from being one of the possible explanations of electromagnetic, weak and strong phenomena into a description in outstanding agreement with experiments is reflected in the dozens of electroweak precision measurements available today [12–14]. These are sensitive to quantum corrections at and beyond the one-loop level, which are essential for obtaining agreement with the data. The calculations of these corrections rely upon the renormalisability (calculability) of the SM,[1] and depend on the masses of heavy virtual particles, such as the top quark and the Higgs boson and possibly other particles beyond the SM. The consistency with the data may be used to constrain the masses of these particles.

Many of these observables have quadratic sensitivity to the mass of the top quark, e.g.,

$$s_W^2 \equiv 1 - m_W^2/m_Z^2 \ni -\frac{2\alpha}{16\pi \sin^2(\theta_W)} \frac{m_t^2}{m_Z^2}. \tag{5}$$

This effect was used before the discovery of the top quark to predict successfully its mass [15], and the consistency of the prediction with experiment can

---

[1] A crucial aspect of this is cancellation of anomalous triangle diagrams between quarks and leptons, which may be a hint of an underlying Grand Unified Theory. See Section 4.

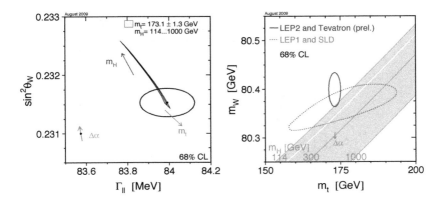

Figure 1. *Left: LEP and SLD measurements of* $\sin^2 \theta_W$ *and the leptonic decay width of the* Z^0, Γ_{ll}, *compared with the SM prediction for different values of* m_t *and* m_H. *Right: The predictions for* m_t *and* m_W *made in the SM using LEP1 and SLD data (dotted mango-shaped contour) for different values of* m_H, *compared with the LEP2 and Tevatron measurements (ellipse). The arrows show the additional effects of the uncertainty in the value of* α_{em} *at the* Z^0 *peak [13].*

be used to constrain possible new physics beyond the SM, particularly mass-squared differences between isospin partner particles, that would contribute analogously to Eq. (5). Many electroweak observables are also logarithmically sensitive to the mass of the Higgs boson, e.g.,

$$s_W^2 \ni \frac{5\alpha}{24\pi} \ln \left(\frac{m_H^2}{m_W^2} \right) \tag{6}$$

when $m_H \gg m_W$. If there were no Higgs boson, or nothing to do its job,[2] radiative corrections such as (6) would diverge, and the SM calculations would become meaningless. Two examples of precision electroweak observables, namely the coupling of the Z^0 boson to leptons and the mass of the W boson, are shown in Fig. 1.

The agreement of the precision electroweak observables with the SM can be used to predict m_H, just as it was used previously to predict m_t. Since the early 1990s [16], this method has been used to tighten the vise on the Higgs, providing ever-stronger indications that it is probably relatively light, as hinted in Fig. 2. The latest estimate of the Higgs mass is [13]

$$m_H = 89^{+35}_{-26} \text{ GeV.} \tag{7}$$

Although the central value is somewhat below the lower limit of 114.4 GeV set by direct searches at LEP [17], there is consistency at the 1σ level, and no

---

[2]See Section 2 for a discussion of possible alternatives.

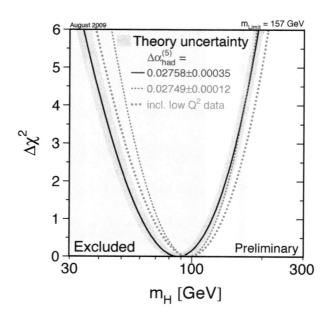

Figure 2. *The χ^2 likelihood function for m_H in a global electroweak fit. The blue band around the (almost) parabolic solid curve represents the theoretical uncertainty: the other curves indicate the effects of different calculations of the renormalisation of α_{em} and of including low-energy data. The shaded regions are those excluded by LEP and by the Tevatron [13].*

significant discrepancy. *A priori*, the relatively light mass range (7) suggests that the Higgs boson interacts relatively weakly, with a small quartic coupling λ, though there is no theoretical consensus on this: see the discussion in the next section.

1.2 Bounds on the Standard Model Higgs Boson Mass

1.2.1 Upper Bounds from Unitarity

As already emphasized, if there were no Higgs boson, and nothing analogous to replace it, the SM would not be a calculable, renormalisable theory. This incompleteness is reflected in the behaviours of physical quantities as the Higgs mass increases. The most basic example of this is W^+W^- scattering [18], whose high-energy s-wave amplitude grows with m_H:

$$T \sim -\frac{4G_F}{\sqrt{2}} m_H^2, \qquad (8)$$

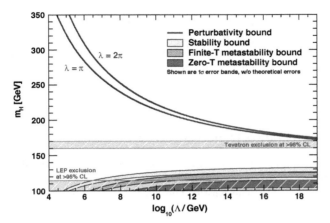

Figure 3. *The scale Λ at which the two-loop RGEs drive the quartic SM Higgs coupling non-perturbative (upper curves), and the scale Λ at which the RGEs create an instability in the electroweak vacuum (lower curves). The widths of the bands reflect the uncertainties in m_t and $\alpha_s(m_Z)$ (added quadratically). The perturbativity upper bound (sometimes referred to as the "triviality" bound) is given for $\lambda = \pi$ (lower bold line) and $\lambda = 2\pi$ (upper bold line). Their difference indicates the theoretical uncertainty in this bound. The absolute vacuum stability bound is displayed by the light-shaded band, while the less restrictive finite-temperature and zero-temperature metastability bounds are medium- and dark-shaded, respectively. The grey-hatched areas indicate the LEP [17] and Tevatron [19] exclusion domains. Figure taken from [20].*

Imposing the unitarity bound $|T| < 1$, one finds the upper limit $M_H^2 < 4\pi\sqrt{2}/G_F$, which is strengthened to

$$M_H^2 < \frac{8\pi\sqrt{2}}{3G_F} \sim 1 \text{ TeV}^2 \tag{9}$$

when one makes a coupled analysis including the $Z^0 Z^0$ channel.

A related effect is seen in the behaviour of the quartic self-coupling λ of the Higgs field. Like any of the SM parameters, λ is subject to renormalisation via loop corrections. Loops of fermions, most importantly the top quark, tend to *decrease* λ as the renormalisation scale Λ increases, whereas loops of bosons tend to *increase* λ. In particular, if the Higgs mass $\gtrsim m_t$, the positive renormalisation due to the Higgs self-coupling itself is dominant, and λ increases uncontrollably with Λ. The larger the value of m_H, the larger the low-energy value of λ, and the smaller the value of Λ at which λ blows up. In general, one should regard the limiting value of Λ, also for smaller m_H, as a scale where novel non-perturbative dynamics must set in. This behaviour is seen in the upper part of Fig. 3, where we see, for example, that if $m_H = 170$ GeV, then $\Lambda \sim 10^{19}$ GeV, whereas if $m_H = 300$ GeV, the coupling λ blows up at a scale $\Lambda \sim 10^6$ GeV. One may ask: under what circumstances does $m_H \sim \Lambda$ itself? The answer is when $m_H \sim 700$ GeV: if the Higgs boson were heavier than this mass, the Higgs self-coupling would blow up at a scale smaller than its mass. In this case, Higgs physics would necessarily be described by some new strongly interacting theory, cf. the technicolour theories described in Section 2.

1.2.2   Lower Bounds from Vacuum Stability

Looking at lower values of m_H in Fig. 3, we see an uneventful range of m_H extending down to $m_H \sim 130$ GeV, where (as far as we know) the SM could continue to be valid all the way to the Planck scale. At lower m_H, there is a band below which the present electroweak vacuum becomes unstable at some scale $\Lambda < 10^{19}$ GeV. For example, if the Higgs is slightly above the present experimental lower limit from LEP, $m_H \sim 115$ GeV, the present electroweak vacuum is unstable against decay into a vacuum with $\langle|\phi|\rangle \sim 10^7$ GeV. This instability is due to the negative renormalisation of λ by the top quark, which overcomes the positive renormalisation due to λ itself, and drives $\lambda < 0$ [3].

If m_H is only slightly below the top band, and above the middle band, it is true that the present electroweak vacuum is in principle unstable against decay into a state with $\langle|\phi|\rangle > \Lambda$, but it would not have decayed during the conventional thermal expansion of the Universe at finite temperatures. Below the middle band but above the lowest band, the vacuum would have decayed to a correspondingly large value of $\langle|\phi|\rangle$ at some finite temperature, but its present-day (low-temperature) lifetime is longer than the age of the Universe. Below the lowest band, the lifetime for decay to a vacuum with $\langle|\phi|\rangle > \Lambda$ would be less than the present age of the Universe at low temperatures, and we should really watch out!

In fact, as we will see shortly, such low values of m_H are almost excluded by LEP searches for the SM Higgs boson, as also seen in Fig. 3.

One could in principle avoid this vacuum instability by introducing some new physics at an energy scale $< \Lambda$: what type of physics [21]? One needs to overcome the negative effects of renormalisation of λ by loops with the top quark circulating. The sign of renormalisation could be reversed by loops with some boson circulating, potentially restoring the stability of the electroweak vacuum. However, then one should consider the renormalisation of the quartic coupling between the Higgs and the new boson. It turns out that the renormalisation of this coupling is in turn very unstable, and that the best way to stabilize this coupling would be to introduce a new fermion.

These new scalars and fermions look very much like the partners of the top quark and Higgs bosons, respectively, that are predicted by supersymmetry [21]. In Section 3 we will study in more detail the renormalisation of mass and vacuum parameters in a supersymmetric theory.

---

[3]The widths of the boundary bands indicate the uncertainties in these calculations.

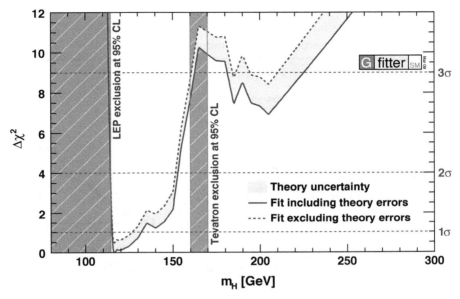

Figure 4. *Dependence on m_H of the $\Delta\chi^2$ function obtained from the global fit of the SM parameters to precision electroweak data [20], excluding (left) or including (right) the results from direct searches at LEP and the Tevatron.*

1.2.3 Results from Searches at LEP and the Tevatron

As seen in Fig. 1, searches for the reaction $e^+e^- \to Z^0 + H$ at LEP established a lower limit on the possible mass of an SM Higgs boson [17]:

$$m_H > 114.4 \text{ GeV} \tag{10}$$

at the 95% confidence level. The lower limit (10) is somewhat higher than the central value of the SM Higgs mass preferred by the global precision electroweak fit (7), but there is no significant tension between these two pieces of information. Fig. 4 shows the χ^2 likelihood function obtained by combining the LEP search and the global electroweak fit. At the 95% confidence level, one finds [17]

$$m_H < 157 \text{ GeV}, 186 \text{ GeV}, \tag{11}$$

depending on whether one uses precision electroweak data alone, or also includes the lower limit (10) from the direct search at LEP. The χ^2 function obtained by combining the LEP limit (10) with the precision electroweak fit is shown in Fig. 4. Notice the little blip at $m_H \sim 115$ GeV, reflecting the hint of a signal found in the last run at the highest LEP energies: this was only at the 1.7σ level, insufficient to claim any evidence.

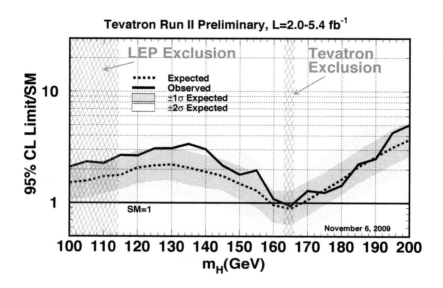

Figure 5. *Combined 95% confidence level upper limit from searches by CDF and D0 for the Higgs boson at the Tevatron collider [19], compared with the SM expectation.*

Searches at the Fermilab Tevatron collider have recently started to exclude a region of mass for the SM Higgs boson, as also seen in Figs. 1, 3 and 4. At the time of writing, these searches exclude [19]

$$163 \text{ GeV} < m_H < 166 \text{ GeV} \qquad (12)$$

at the 95% confidence level, as seen in Fig. 5. At smaller masses, the Tevatron 95% confidence level upper limit on Higgs production and decay is within a factor of a few of the SM calculation, and the integrated luminosity is expected to double over the next couple of years.

Fig. 4 also includes the effect on the χ^2 likelihood function of combining the Tevatron search with the global electroweak fit and the LEP search. We see from this that the 'blow-up' region $m_H > 180$ GeV is strongly disfavoured: above the 99% confidence level if the Tevatron data are included, compared with 96% if they are dropped [20]. The combination of all the data yields a 68% confidence level range [14]

$$m_H = 116^{+16}_{-1.3} \text{ GeV}. \qquad (13)$$

The Tevatron continued running until September 2011, accumulating $\mathcal{O}(10)/\text{fb}$

of integrated luminosity. That could be sufficient to exclude an SM Higgs boson over all the mass range between Eqs. (10) and (12), which would exclude all the preferred range given in Eq. (11) – a very intriguing possibility! Alternatively, perhaps the Tevatron will find some evidence for a Higgs boson with a mass within this range? We can, in any case, expect that the LHC will either discover an SM Higgs boson or prove that it does not exist.

1.3 Issues beyond the Standard Model

The success of the SM is very impressive. However, our rejoicing is muted by the fact that in order to specify the SM we need at least 19 input parameters in order to calculate physical processes, namely:

- three coupling parameters, which we can choose to be the strong coupling constant, α_s, the fine structure constant, α_{em}, and the weak mixing angle, $\sin^2(\theta_W)$;

- two parameters that specify the shape of the Higgs potential, μ^2 and λ (or, equivalently, m_H and m_W or m_Z);

- six quark masses (or the six Yukawa couplings for the quarks);

- four parameters (three mixing angles and one weak CP-violating angle) for the Cabibbo–Kobayashi–Maskawa matrix [see Eq. (19) below];

- three charged-lepton masses (or the corresponding Yukawa couplings);

- one parameter to allow for non-perturbative CP violation in QCD, θ_{QCD}.

Moreover, because we now know that neutrinos have mass and that they mix, the SM must be extended to incorporate this fact. Therefore, we also need to specify three neutrino masses and three mixing angles plus a CP-violating phase for the neutrino mixing matrix, bringing the grand total to 26 parameters. Additionally, if neutrinos turn out to be Majorana particles, so that they are their own antiparticles, two more CP-violating phases need to be specified. Notice that at least 20 of the parameters relate to flavour physics.

Many of the ideas for physics beyond the SM that are discussed later have been motivated by attempts to reduce the number of its parameters, or to understand their origins, or at least to make them seem less unnatural, as discussed in subsequent lectures. The SM, however, is not expected to be the final description of the fundamental interactions, but rather an effective low-energy (up to a few TeV) manifestation of a more complete theory.

Some of the outstanding questions in the SM are:

- **How is electroweak symmetry broken?** In other words, how do gauge bosons acquire mass? We have seen that the SM incorporates the Higgs mechanism in the form of a single weak-isospin doublet with a non-zero v.e.v. in order to generate the gauge boson masses, but this is not the only possible way in which the electroweak symmetry can be broken. For instance, there could be more than one Higgs doublet, the Higgs could be a pseudo-Goldstone boson (with a low mass relative to the mass scale of some new interaction) or electroweak symmetry could be broken by a condensate of new particles bound by a new strong interaction. We cover a few of the possibilities in Section 2.

- **How do fermions acquire mass?** Electroweak symmetry breaking is a necessary, but not a sufficient, condition to generate the fermion masses. There also needs to be a mechanism that generates the required Yukawa couplings between the fermions and the (effective) Higgs field. The separation between electroweak symmetry breaking and the generation of fermion masses is made evident in models of dynamical symmetry breaking, such as technicolour (see Section 2), where the breaking is carried out by the formation of a condensate of particles associated with a new interaction, a process which, while breaking electroweak symmetry and giving masses to the gauge bosons, does not necessarily give masses to the fermions. This situation is resolved by adding new interactions which are responsible for generating the fermion masses. Within the SM, there are no predictions for the values of the Yukawa couplings. Moreover, the values required to generate the correct masses for the three charged leptons and the six quarks span six orders of magnitude, which presumably makes the mechanism for the generation of the couplings highly non-trivial.

- **The hierarchy problem.** Why should the Higgs mass remain low, $m_H \lesssim 1$ TeV, in the face of divergent quantum loop corrections? Following [3], the Higgs mass can be expanded in perturbation theory as

$$m_H^2 \left(p^2 \right) = m_{0,H}^2 + \mathcal{C} g^2 \int_{p^2}^{\Lambda^2} dk^2 + \dots, \qquad (14)$$

where $m_{0,H}^2$ is the tree-level (classical) contribution to the Higgs mass squared, g is the coupling constant of the the theory, $\mathcal{C}$ is a model-dependent constant and Λ is the reference scale up to which the SM is assumed to remain valid. The integrals represent contributions at loop level and are apparently quadratically divergent. If there is no new physics, the reference scale is high, like the Planck scale, $\Lambda \sim M_{\text{Pl}} \approx 10^{19}$ GeV or, in Grand Unified Theories (GUTs), $\Lambda \sim M_{GUT} \approx 10^{15} - 10^{16}$ GeV (see Section 4). Clearly, both choices result in large corrections to the Higgs mass. In order for these to be small, there are two alternatives: either the relative magnitudes of the tree-level and loop contributions are finely tuned to yield a net contribution that is small (a feature that is disliked by physicists, but which Nature might have implemented), or there is a new symmetry, like supersymmetry, that protects the Higgs mass, as discussed in Section 3.

- **The vacuum energy problem.** The value of the scalar potential at the v.e.v. $\langle\phi\rangle_0$ of the Higgs boson is

$$V\left(\langle\phi^\dagger\phi\rangle_0\right) = \frac{\mu^2 v^2}{4} < 0. \tag{15}$$

Hence, because the Higgs mass is $m_H^2 = -2\mu^2$, this corresponds to a uniform vacuum energy density

$$\rho_H = -\frac{m_H^2 v^2}{8}. \tag{16}$$

Taking $v = \left(G_F\sqrt{2}\right)^{-1/2} \approx 246$ GeV for the Higgs v.e.v. and using the current experimental lower bound on the Higgs mass [12], $m_H \gtrsim 114.4$ GeV, we have

$$-\rho_H \gtrsim 10^8 \text{ GeV}^4. \tag{17}$$

On the other hand, if the apparent accelerated expansion of the Universe – originally inferred from observations of type 1A supernovae [22] – is attributed to a non-zero cosmological constant corresponding to $\sim 70\%$ of the total energy density of the Universe [12], the required energy density should be

$$\rho_{\text{vac}} \sim 10^{-46} \text{ GeV}^4, \tag{18}$$

which is at least 54 orders of magnitude lower than the corresponding density from the Higgs field, and of the opposite sign! The character of this dark energy remains unexplained [23, 24], and will probably remain so until we have a full quantum theory of gravity.

- **How is flavour symmetry broken?** Part of the flavour problem in the SM is, of course, related to the widely different mass assignments of the fermions ascribed to the Yukawa couplings, which also set the mixing angles between flavour and mass eigenstates. Mixing occurs both in the quark and the lepton sectors, the former being parameterised by the Cabibbo–Kobayashi–Maskawa (CKM) matrix and the latter, by the Maki–Nakagawa–Sakata (MNS) matrix. These are complex rotation matrices, and can each be written in terms of three mixing angles and one CP-violating phase (δ) [12]:

$$V = \begin{pmatrix} c_{12}c_{13} & s_{12}c_{13} & s_{13}e^{-i\delta} \\ -s_{12}c_{23} - c_{12}s_{23}s_{13}e^{i\delta} & c_{12}c_{23} - s_{12}s_{23}s_{13}e^{i\delta} & s_{23}c_{13} \\ s_{12}s_{23} - c_{12}c_{23}s_{13}e^{i\delta} & -c_{12}s_{23} - s_{12}c_{23}s_{13}e^{i\delta} & c_{23}c_{13} \end{pmatrix}, \tag{19}$$

where $c_{ij} \equiv \cos(\theta_{ij})$, $s_{ij} \equiv \sin(\theta_{ij})$. While the off-diagonal elements in the quark sector are rather small (of order 10^{-1} to 10^{-3}), so that there is little mixing between quark families, in the lepton sector the off-diagonal elements (except for $[V_{MNS}]_{e3}$, which is close to zero) are of order 1, so that the mixing between neutrino families is large. The SM does not provide an explanation for this difference.

- **What is the dark matter?** The observation that galaxy rotation curves do not fall off with radial distance from the galactic center can be explained by postulating the existence of a new type of weakly interacting matter, *dark matter*, in the halos of galaxies. Supporting evidence from the cosmic microwave background (CMB) indicates that the dark matter makes up $\sim 25\%$ of the energy density of the Universe [25]. Dark matter is usually thought to be composed of neutral relic particles from the early Universe. Within the SM, neutrinos are the only candidate massive neutral relics. However, they contribute only with a normalised density of $\Omega_\nu \gtrsim 1.2\,(2.2) \times 10^{-3}$ if the mass hierarchy is normal (inverted), or no more than 10% if the lightest mass eigenstate lies around 1 eV, that is, if the hierarchy is degenerate [3]. On top of that, structure formation indicates that dark matter should be cold, i.e., non-relativistic at the time of structure formation, whereas neutrinos would have been relativistic particles. Within the Minimal Supersymmetric extension of the SM (MSSM), the lightest supersymmetric partner, called a *neutralino*, is a popular dark-matter candidate [26].

- **How did the baryon asymmetry of the Universe arise?** The antibaryon density of the Universe is negligible, whilst the baryon-to-photon ratio has been determined, using WMAP data[4] of the CMB [27] to be

$$\eta = \frac{n_b - \overline{n}_b}{n_\gamma} \simeq \frac{n_b}{n_\gamma} = 6.12\,(19) \times 10^{-10}, \qquad (20)$$

where n_b, $\overline{n}_b$, and n_γ are the number densities of baryons, antibaryons and photons, respectively. The fact that the ratio is not zero is intriguing considering that, in a cosmology with an inflationary epoch, conventional thermal equilibrium processes would have yielded equal numbers of particles and antiparticles. In 1967, Sakharov [28] established three necessary conditions (more fully explained in [29]) for the particle–antiparticle asymmetry of the Universe to be generated:

1. violation of the baryon number, B;

2. microscopic C and CP violation;

3. loss of thermal equilibrium.

---

[4]We use here values from the three-year WMAP analysis [27], rather than the five-year analysis [25], in order to be consistent with the values quoted by the Particle Data Group [12] summary tables.

Otherwise, the rate of creation of baryons equals the rate of destruction, and no net asymmetry results. In the perturbative regime, the SM conserves B; however, at the non-perturbative level, B violation is possible through the triangle anomaly. The loss of thermal equilibrium may occur naturally through the expansion of the Universe, and CP violation enters the SM through the complex phase in the CKM matrix [12]. However, the CP violation observed so far, which is described by the Kobayashi–Maskawa mechanism of the SM, is known to be insufficient to explain the observed value of the ratio η, and new physics is needed. One possible solution lies in leptogenesis scenarios, where the baryon asymmetry is a result of a previously existing lepton asymmetry generated by the decays of heavy sterile neutrinos [30].

• **Quantisation of the electric charge**. It is an experimental fact that the charges of all observed particles are simple multiples of a fundamental charge, which we can take to be the electron charge, e. Dirac [31–33] proved that the existence of even a single magnetic monopole (a magnet with only one pole) is sufficient to explain the quantisation of the electric charge, but the particle content of the SM does not include magnetic monopoles. Hence, in the absence of any indication for a magnetic monopole, the explanation of charge quantisation must lie beyond the SM. Indeed, so far there has only been one candidate monopole detection event in a single superconducting loop [34], in 1982, and the monopole interpretation of the event has by now been largely discounted. One expects monopoles to be very massive and non-relativistic at present, in which case time-of-flight measurements in the low-velocity regime ($\beta \equiv v/c \ll 1$) become important. The best current direct upper limit on the supermassive monopole flux comes from cosmic-ray observations [12],

$$\Phi_{1\text{pole}} < 1.0 \times 10^{-15} \text{ cm}^{-2}\text{sr}^{-1}\text{s}^{-1}, \tag{21}$$

for $1.1 \times 10^{-4} < \beta < 0.1$. An alternative route towards charge quantization is via a Grand Unified Theory (GUT) (see Section 4). Such a theory implies the existence of magnetic monopoles that would be so massive that their cosmological density would be suppressed to an unobservably small value by cosmological inflation.

• **How to incorporate gravitation?** One of the most obvious shortcomings of the SM is that it does not incorporate gravitation, which is described on a classical level by general relativity. However, the consistency of our physical theories requires a quantum theory of gravity. The main difficulty in building a quantum field theory of gravity is its non-renormalisability. String theory [35] and loop quantum gravity [36] constitute attempts at building a quantised theory of gravity. If one could answer this question, one would surely also be able to solve the dark energy problem. Conversely, solving the dark energy problem presumably requires a complete quantum theory of gravity.

2 Electroweak Symmetry Breaking beyond the Standard Model

2.1 Theorists Are Getting Cold Feet

After so many years, it seems that we will soon know whether a Higgs boson exists in the way predicted by the SM, or not. Closure at last!

Like the prospect of an imminent hanging, the prospect of imminent Higgs discovery concentrates wonderfully the minds of theorists, and many theorists with cold feet are generating alternative models, as prolifically as monkeys on their laptops. These serve the invaluable purpose of providing benchmarks that can be compared and contrasted with the SM Higgs. Experimentalists should be ready to search for reasonable alternatives, already at the Tevatron and also at the LHC once it is up and running, and they should be on the lookout for tell-tale deviations from the SM predictions if a Higgs boson should appear.

Even within the SM with a single elementary Higgs boson, questions are being asked. As discussed in the previous section, within this framework the experimental data seem to favour a light Higgs boson. However, the interpretation of the precision electroweak data has been challenged. Even if one accepts the data at face value, the SM fit may need to take into account non-renormalisable, higher-dimensional interactions that could conspire to permit a heavier SM Higgs boson. In this section, in addition to these possibilities, we explore several mechanisms of electroweak symmetry breaking beyond the minimal Higgs, i.e., a single elementary $SU(2)$ Higgs doublet whose potential is arranged to have a non-zero v.e.v.

Any successful model of electroweak symmetry breaking must give masses to the matter fermions as well as the weak gauge bosons. This could be achieved using either a single boson, as in the SM, or two of them, as in the Minimal Supersymmetric extension of the Standard Model (MSSM),[5] or by some composite of new fermions with new strong interactions that generate a non-zero v.e.v. as in (extended) technicolour models, or by some Higgsless mechanism.

We do know, however, that the energy scale at which EWSB must occur is $\mathcal{O}(1)$ TeV [37]. This scale is set by the decay constant of the three Goldstone bosons that, through the Higgs mechanism, are transformed into the longitudinal components of the weak gauge bosons:

$$F_\pi = \left(G_F\sqrt{2}\right)^{-1/2} \approx 246 \text{ GeV}. \tag{22}$$

---

[5]We leave the treatment of the Higgs sector within the MSSM for a later section.

If there is any new physics associated to the breaking of electroweak symmetry, it must occur near this energy scale. Another way to see how this energy scale emerges is to consider *s*-wave WW scattering. In the absence of a direct-channel Higgs pole, this amplitude would violate the unitarity limit at an energy scale ~ 1 TeV (8).

It is the scale of 1 TeV, and the typical values of QCD and electroweak cross sections at this energy, $\sigma \simeq 1$ nb $-$ 1fb, that set the energy and luminosity requirements of the LHC: $\sqrt{s} = 14$ TeV and $\mathcal{L} = 10^{34}$ cm^{-2} s^{-1} for pp collisions [12]. This energy scale is to be contrasted with the energy scale of the other unexplained broken symmetry in the SM, namely flavour symmetry, which is completely unknown: it may lie anywhere from 1 TeV up to the Planck scale, $M_P = 1.22 \times 10^{19}$ GeV.

There are some general constraints that any proposed model of electroweak symmetry breaking must satisfy [38]. First, the model must predict a value of the parameter $\rho \equiv m_W^2/m_Z^2 \cos^2 \theta_W$ that agrees with the value $\rho \approx 1$ found experimentally. The desired value $\rho = 1$ is found automatically in models that contain only Higgs doublets and singlets, but would be violated in models with scalar fields in larger $SU(2)$ representations. A second constraint comes from the strict upper limits on flavour-changing neutral currents (FCNCs). These are absent at tree level in the minimal Higgs model, a fact that is in general not true in non-minimal models.

2.2 Interpretation of the Precision Electroweak Data

It is notorious that the two most precise measurements at the Z^0 peak, namely the asymmetries measured with leptons and hadrons, do not agree very well [39].[6] Within the SM, they favour different values of m_H, around 40 and 500 GeV, respectively. Most people think that this discrepancy is just a statistical fluctuation, since the total χ^2 of the global electroweak fit is acceptable ($\chi^2 = 17.3$ for 13 d.o.f., corresponding to a probability of 18% [13]), but it may also reflect the existence of an underestimated systematic error. However, if there were a big error in $A_{FB}^{0,b}$, the preferred value of m_H would be pulled uncomfortably low by the other data, whereas if there was a big error in the interpretation of the leptonic data m_H would be pulled towards much higher values. On the other hand, if we take both pieces of data at face value, perhaps the discrepancy is evidence for new physics at the electroweak scale. In this case there would be no firm basis for the prediction of a light Higgs boson, which is based on an SM fit, and no fit value of m_H could be trusted.

---

[6]Another anomaly is exhibited by the NuTeV data on deep-inelastic $\nu - N$ scattering [40], but this is easier to explain away as due to our lack of understanding of hadronic effects.

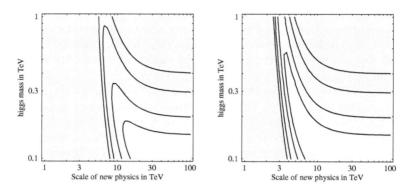

Figure 6. *The 68%, 90%, 99% and 99.9% confidence levels fit for global electroweak fits including two different types of higher-dimensional operators, demonstrating that they might conspire with a relatively heavy Higgs boson to yield an acceptable fit [41].*

2.3 Higher-Dimensional Operators within the SM

The Standard Model should be regarded simply as an effective low-energy theory, to be embedded within some more complete and satisfactory theory. Therefore, one should anticipate that the renormalisable dimension-four interactions of the SM could be supplemented by higher-dimensional operators of the general form

$$\mathcal{L}_{eff} = \mathcal{L}_{SM} + \Sigma_i \frac{c_i}{\Lambda_i^p} \mathcal{O}_i^{4+p}, \tag{23}$$

where Λ_i is a scale at which the supplementary interaction $\mathcal{O}_i^{4+p}$ of dimension $4 + p$ appears to be generated. A global fit to the precision electroweak data suggests that, if the Higgs is indeed light, the coefficients of these additional interactions are small,

$$\Lambda_i > \mathcal{O}(10) \text{ TeV} \tag{24}$$

for $c_i = \pm 1$. It is then a problem to understand the "little hierarchy" between the electroweak scale and Λ_i.

However, conspiracies are in principle possible, which could allow m_H to be large, even if one takes the precision electroweak data at face value [41]. Examples are shown in Fig. 6, where one sees corridors of allowed parameter space extending up to a heavy Higgs mass, if $\Lambda_i \ll 10$ TeV. A theory that predicts a heavy Higgs boson but remains consistent with the precision electroweak data should predict a correlation of the type seen in Fig. 6. At the moment, this may seem unnatural to us, but Nature may know better. In any case, any theory beyond the SM must link the value of m_H and the scales of these higher-dimensional effective operators in some way.

2.4 Little Higgs

One way to address the "little hierarchy problem" and explain the lightness of the Higgs boson (if it is light) is by treating it as a pseudo-Goldstone boson corresponding to a spontaneously broken approximate global symmetry of a new strongly interacting sector at some higher mass scale, the "little Higgs" scenario [42]. Such a theory would work by analogy with the pions in QCD, which have masses far below the generic mass scale of the strong interactions ~ 1 GeV.

If the Higgs is a pseudo-Goldstone boson, its mass is protected from acquiring quadratically divergent loop corrections [43]. This occurs as a result of the particular manner in which the gauge and Yukawa couplings break the global symmetries: more than one coupling must be turned on at a time in order for the symmetry to be broken, a feature known as "collective symmetry breaking" [44, 45]. As a consequence, the quadratic divergences that would normally appear in the SM are cancelled by new particles, sometimes in unexpected ways. For example, the top-quark loop contribution to the Higgs mass-squared has the general form

$$\delta m^2_{H,top}(SM) \sim (115 \text{ GeV})^2 \left(\frac{\Lambda}{400 \text{ GeV}} \right)^2. \tag{25}$$

As illustrated in Fig. 7, in little Higgs models this is cancelled by the loop contribution due to a new heavy top-like quark T with charge $+2/3$ that is a singlet of $SU(2)_L$, leaving a residual logarithmic divergence,

$$\delta m^2_{H,top}(LH) \sim \frac{6G_F m_t^2}{\sqrt{2\pi^2}} m_T^2 \log \frac{\Lambda}{m_T}. \tag{26}$$

Analogously, the quadratic loop divergences associated with the gauge bosons and the Higgs boson of the SM are cancelled by loops of new gauge bosons and Higgs bosons in little Higgs models.

The net result is a spectrum containing a relatively light Higgs boson and other new particles that may be somewhat heavier,

$$M_T < 2 \text{ TeV} \left(\frac{m_H}{200 \text{ GeV}} \right)^2, M_{W'} < 6 \text{ TeV} \left(\frac{m_H}{200 \text{ GeV}} \right)^2, M_{H^{++}} < 10 \text{ TeV}. \tag{27}$$

The extra T quark, in particular, should be accessible to the LHC. In addition, there should be more new strongly interacting physics at some energy scale at or above 10 TeV, to provide the ultra-violet completion of the theory.

2.5 Technicolour

Little Higgs models are particular examples of composite Higgs models, of which the prototypes were technicolour models [46, 47]. In these models, electroweak symmetry is broken dynamically by the introduction of a new

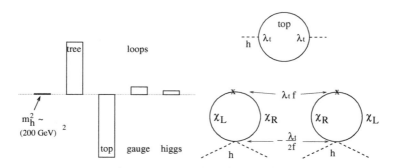

Figure 7. *(a) If the Standard Model Higgs boson weighs around 200 GeV, the top-quark loop contribution to its physical mass (calculated here with a loop momentum cutoff of 10 TeV) must cancel delicately against the tree-level contribution. (b) In "little Higgs" models, the top-quark loop is cancelled by loops containing a heavier charge-2/3 quark [42].*

non-Abelian gauge interaction [48–50] that becomes strong at the TeV scale. The building blocks are massless fermions called technifermions and new force-carrying fields called technigluons. As in the SM, the left-handed components of the technifermions are assigned to electroweak doublets, while the right-handed components form electroweak singlets, and both components carry hypercharge. At $\Lambda_{\rm EW} \sim 1$ TeV the technicolour coupling becomes strong, which leads to the formation of condensates of technifermions with v.e.v.s,

$$\langle \phi \rangle = \langle \overline{f}_L f_R \rangle \equiv v. \tag{28}$$

Because the left-handed technifermions carry electroweak quantum numbers, but the right-handed ones do not, the formation of this technicondensate breaks electroweak symmetry.

The massless technifermions have the chiral symmetry group

$$G_\chi = SU(2N_D)_L \otimes SU(2N_D)_R \supset SU(2)_L \otimes SU(2)_R, \tag{29}$$

where N_D is the number of technifermion doublets. When the condensate forms, this large global symmetry is broken down to

$$S_\chi = SU(2N_D) \supset SU(2)_V, \tag{30}$$

where V refers to the vector combination of left and right currents, and $4N_D^2 - 1$ massless Goldstone bosons appear, with decay constant $F_\pi^{\rm TC}$. Similarly to the Higgs mechanism in the SM, three of these bosons are "eaten" and become the longitudinal components of the $W^\pm$ and Z^0 weak bosons, which acquire masses [37]

$$m_W = \frac{g}{2}\sqrt{N_D}F_\pi^{\rm TC}, \qquad m_Z = \frac{1}{2}\sqrt{g^2 + g'^{\,2}}\sqrt{N_D}F_\pi^{\rm TC} = \frac{m_W}{\cos(\theta_W)}. \tag{31}$$

The scale $\Lambda_{\rm TC}$ at which technicolour interactions become strong is related to the magnitude of electroweak symmetry breaking, namely to the weak scale, by

$$\Lambda_{\rm TC} = \text{few} \times F_\pi^{\rm TC}, \qquad F_\pi^{\rm TC} = F_\pi/\sqrt{N_D}, \qquad (32)$$

where $F_\pi = v \approx 246$ GeV. The breaking of the chiral symmetry in technicolour is reminiscent of chiral symmetry in QCD, which provides a working precedent for the model.[7] Technicolour guarantees $\rho = m_W^2/\left(m_Z^2 \cos^2 \theta_W\right) = 1 + \mathcal{O}\left(\alpha\right)$ through a custodial $SU(2)_R$ flavour symmetry in G_χ [37], which is traceable to the quantum numbers assigned to the technifermions.

Dynamical symmetry breaking addresses the problem of quadratic divergences in the Higgs mass-squared, such as (25), by introducing a composite Higgs boson that "dissolves" at the scale $\Lambda_{\rm TC}$. In this way, it makes loop corrections to the electroweak scale "naturally" small. Moreover, technicolour has a plausible mechanism for stabilizing the weak scale far below the Planck scale. The idea is that technicolour, being an asymptotically free theory, couples weakly at very high energies $\sim 10^{16}$ GeV, and then evolves to become strong at lower energies ~ 1 TeV [46]. However, writing down an explicit GUT scenario based on this scenario has proved elusive.

As described above, the simplest technicolour models could provide masses for the gauge bosons $W^\pm$ and Z^0, but not to the matter fermions. Additions to technicolour could allow for quark and lepton masses by introducing new interaction with technifermions, as in "extended technicolour" models [47, 52]. However, these had severe problems with flavour-changing neutral interactions [53] and a proliferation of relatively light pseudo-Goldstone bosons that have not been seen by experiment [54].

Moreover, a generic problem with technicolour models is presented by the global electroweak fit discussed in the first lecture. The preference within the SM for a relatively light Higgs boson (7) may be translated into constraints on the possible vacuum polarization effects due to generic new physics models. QCD-like technicolour models have many strongly interacting dynamical scalar resonances in the TeV range, e.g., a scalar analogous to the σ meson of QCD that corresponds naively to a relatively heavy Higgs boson, which is disfavoured by the data [55]. Such a model can be reconciled with the electroweak data only if some other effect is postulated to cancel the effects of its large mass. One strategy for evading this problem is offered by "walking technicolour" theories [56], where the coupling strength evolves slowly, i.e., walks. However, the loss of the close analogy with QCD makes it more difficult to calculate so reliably in such models: lattice techniques may come to the rescue here.

---

[7]The condensation phenomenon also occurs in solid-state physics: dynamical symmetry breaking in superconductors is achieved by the formation of Cooper pairs [51], which are condensates of electron pairs with charge $-2e$.

2.6 Interpolating Models

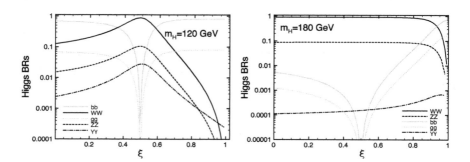

Figure 8. *The dependences of Higgs branching ratios on the parameter ξ (34), for $m_H = 120$ GeV (left) and 180 GeV (right) [57].*

So far, we have examined two extreme scenarios: the orthodox interpretation of the SM in which the Higgs is elementary and relatively light, and hence interacts only weakly, and strongly coupled models exemplified by technicolour. The weakly coupled scenario would require additional TeV-scale particles to stabilize the Higgs mass by cancelling out the quadratic divergences such as (25). A prototype for such models is provided by supersymmetry, as discussed in the next lecture. On the other hand, strongly coupled models such as technicolour introduce many resonances that are required by unitarity and generate important contributions to the oblique radiative corrections, e.g., a vector resonance ρ in W^+W^- scattering would induce

$$\delta\rho \sim \frac{m_W^2}{m_\rho^2}, \tag{33}$$

and the experimental upper limit $|1 - \rho| < 10^{-3}$ at the 95% confidence level imposes $m_\rho > 2.5$ TeV.

One way to interpolate between these two extreme scenarios, and provide a basis for determining how far from the light-SM-Higgs scenario the data permit us to go, is to consider models in which the unitarization of the W^+W^- scattering amplitude is shared between a light Higgs boson with modified couplings and a vector resonance with mass m_ρ and coupling g_ρ, whose relative importance is parameterised by the combination

$$\xi \equiv v\frac{g_\rho}{m_\rho}. \tag{34}$$

The SM is recovered in the limit $\xi \to 0$, but its decay branching ratios may differ considerably as ξ increases towards the strong-coupling limit $\xi = 1$, as seen in Fig. 8. Thus, one signature for such models at the LHC may be the observation of a Higgs boson with couplings that differ from those of the SM.

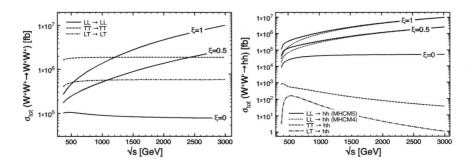

Figure 9. *Left: The cross sections $\sigma(W_T^+ W_T^+ \to W_T^+ W_T^+)$, $\sigma(W_L^+ W_T^+ \to W_T^+ L W_T^+)$ and $\sigma(W_L^+ W_L^+ \to W_L^+ W_L^+)$ as functions of ξ (34). Right: Cross sections for double Higgs production [57].*

Another way to probe such models is to look for effects in $W_L^+ W_L^+$ scattering. Unfortunately, at the LHC the $W^\pm$ bosons that are flashed off from incoming energetic quarks $q \to Wq'$ have predominantly transverse polarizations, so that $\sigma(W_T^+ W_T^+ \to W_T^+ W_T^+) \gg \sigma(W_T^+ W_T^+ \to W_T^+ L W_T^+)$ and $\sigma(W_L^+ W_L^+ \to W_L^+ W_L^+)$ for all $m_{W^+ W^+}$ in the SM, and there is an accidental very small factor [57]:

$$\frac{d\sigma^{LL}/dt}{d\sigma^{TT}/dt} = \frac{1}{2304} \left(\frac{m_{W^+ W^+}}{m_W} \right)^4 \xi^2, \qquad (35)$$

which implies that, even for $\xi = 1$, $\sigma(W_L^+ W_L^+ \to W_L^+ W_L^+) > \sigma(W_T^+ W_T^+ \to W_T^+ W_T^+)$ only for $m_{W^+ W^+} > 1.2$ TeV, which is unlikely to be accessible at the LHC, as seen in Fig. 9. An alternative possibility for the LHC may be double-Higgs production via the reaction $W^+ W^- \to HH$, which may be greatly enhanced as compared with its rate in the SM, as also seen in Fig. 9 – though its observability may be a different matter.

2.7 Higgsless Models and Extra Dimensions

As has already been discussed, if there is nothing like an SM Higgs boson, s-wave WW scattering reaches the unitarity limit at $m_{W^+ W^-} \sim 1$ TeV (9). An immediate reaction might be: who cares? Some non-perturbative strong dynamics will necessarily restore unitarity, even in the absence of a Higgs boson. However, more detailed study in specific models has shown that this strong dynamics is apparently incompatible with the precision data: one needs some perturbative mechanism to break the electroweak symmetry.

How can one break a gauge symmetry? Breaking it explicitly would destroy the renormalisability (calculability) of the gauge theory, whereas breaking the symmetry spontaneously by the v.e.v. of some field everywhere in

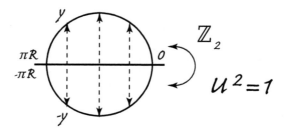

Figure 10. *Compactification on a circle S^1 of radius R with internal coordinate (fifth dimension) y, illustrating the possible orbifolding of this model via the identification S^1/Z_2.*

space does retain the renormalisability (calculability) of the gauge symmetry. But that is the Higgs approach that we are trying to escape: is there another way? The alternative is to break the electroweak symmetry via boundary conditions. This is impossible in conventional $3+1$-dimensional space-time, because it has no boundaries. However, it becomes an option if we postulate finite-size (small) extra space dimensions [58–60].

To see how this works, let us first consider the particle spectrum in the simplest possible model with one extra dimension compactified on a circle S^1 of radius R with internal coordinate (fifth dimension) y, as illustrated in Fig. 10. In this case, the wave function of a boson ϕ at y and $y+2\pi R$ must be identified:

$$\phi(y + 2\pi R) \;=\; \phi(y), \tag{36}$$

so that one can expand the five-dimensional field as follows:

$$\phi(x,y) \;=\; \sum_n \frac{1}{\sqrt{2^{\delta_{n0}}\pi R}} \left(\cos\left(\frac{ny}{R}\right)\phi_n^+(x) + \sin\left(\frac{ny}{R}\right)\phi_n^-(x)\right). \tag{37}$$

The $\phi_n^{\pm}$ are the four-dimensional Kaluza–Klein [61, 62] modes of the field, which appear in four dimensions as particles with masses

$$m_n \;=\; p_y^n \;=\; \frac{n}{R}, \tag{38}$$

and the functions $\cos, \sin(ny/R)$ describe the localizations of these modes along the extra dimension. the lowest-lying mode has a flat wave function ($n=0$), and the excitations have $n>0$.

We now consider what happens if we "fold" the circle by identifying $y \sim -y$. Mathematically, this is the simplest *orbifold* S^1/Z_2, also illustrated in Fig. 10. At the same time as identifying $y \sim -y$, we can also identify the field ϕ up to a sign:

$$\phi(-y) \;=\; U\phi(y) : U^2 \;=\; 1. \tag{39}$$

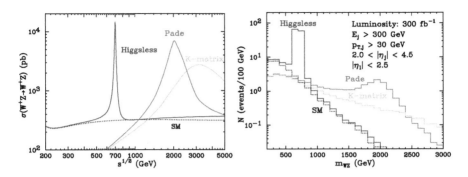

Figure 11. *Left: Calculations of the possible modifications of $\sigma(W^+Z^0 \to W^+Z^0)$. Right: Simulations of the possible numbers of events at the LHC [57].*

This has the effect of projecting out half the Kaluza–Klein wave functions (37). If we choose $U = +1$, we select the even wave functions $\cos(ny/R)$ and hence the Kaluza–Klein modes $\phi_n^+(x)$ whereas, if we choose $U = -1$, we select the odd wave functions $\sin(ny/R)$ and hence the Kaluza–Klein modes $\phi_n^-(x)$. The "even" particles include the massless mode with $n = 0$ whereas all the "odd" particles are massive. The projection U serves to give masses to all the states that are asymmetric.

This mechanism can be extended to break gauge symmetry [58–60]. Let us consider a five-dimensional theory with a gauge field $A_{\mu,5}$, and let us identify it on the orbifold $y \sim -y$ up to a discrete gauge transformation $U : U^2 = 1$:

$$A_\mu = +U A_\mu(y) U^\dagger, \tag{40}$$
$$A_5 = -U A_5(y) U^\dagger. \tag{41}$$

The gauge symmetry group is broken at the end-points of the orbifold $y = 0, \pi R$: the surviving subgroup is the one that commutes with U, and asymmetric particles acquire masses as described above. In this way, one could imagine breaking $SU(2) \otimes U(1) \to U(1)$ with a suitable orbifold construction.

It is a general feature of this construction that a vector resonance should appear in WZ scattering, corresponding to the lowest-lying Kaluza–Klein excitation. The production of such a particle at the LHC has been considered in the context of a Higgsless model, and could well be observable, as seen in Fig. 11.

You might wonder whether this type of vector resonance bears any relation to the vector resonances discussed previously in the context of new strong dynamics. The answer is yes: as was first emphasized in the context of string theory, a strong coupling is equivalent to a new compactified dimension, and there is in general a "holographic" relation between four- and five-dimensional theories, the former being considered as boundaries of the five-dimensional "bulk" theory. These ideas enable the strongly interacting

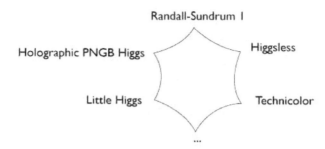

Figure 12. *Relations between different models of electroweak symmetry breaking [64].*

models of electroweak symmetry breaking discussed in this lecture, and many others, to be related through a unified description à la M-theory [63], as seen in Fig. 12 [64]. The alternative is a weakly interacting model of electroweak symmetry breaking, which is favoured, naively, by the indications from precision electroweak data of a light Higgs boson. In the next lecture we discuss supersymmetry, which is the most-developed such alternative.

3   Supersymmetry

We have seen that the Standard Model is a valid description of physical phenomena at energies lower than a few hundreds of GeV. However, there are various reasons to think that supersymmetry might appear at the TeV scale [65–67], and hence play an important role in new discoveries at the LHC, which will explore energies of the order of a TeV. In this lecture we present and discuss supersymmetric models, with a focus on the phenomenological consequences of supersymmetry.

The reasons for its introduction in particle physics are principally physical, and quite diverse in nature, as we now discuss.

• The very special properties of supersymmetric field theories are helpful in addressing the naturalness of a (relatively) light Higgs boson. In the previous lectures we have discussed the existence of enormous radiative corrections to the Higgs mass-squared, m_H^2, which feels the virtual effects of any particle that couples, directly or indirectly to the Higgs field. For example, the correction due to a fermionic loop such as that in Fig. 13(a) yields[8]

$$\Delta m_H^2 = -\frac{y_f^2}{8\pi^2}[2\Lambda^2 + 6m_f^2 \ln(\Lambda/m_f) + ...], \qquad (42)$$

---

[8]For this calculation, we define the Yukawa coupling of the Higgs boson to a fermion as usual, via $y_f H \bar{\psi}\psi$.

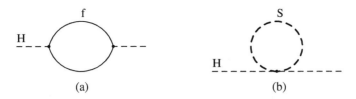

Figure 13. *One-loop quantum corrections to the mass-squared of the Higgs boson due to (a) a fermionic loop, (b) a scalar boson loop.*

where Λ is an ultraviolet cutoff used to represent the scale up to which the SM remains valid, at which new physics appears. We see that the mass of the Higgs diverges quadratically with Λ and, if we suppose that the SM remains valid up to the Planck scale, $M_P \simeq 10^{19}$ GeV, then $\Lambda = M_P$ and this correction is 10^{30} times bigger than the reasonable value of the mass-squared of the Higgs, namely (10^2) GeV)^2! Moreover, there is a similar correction coming from a loop of a scalar field S such as that in Fig. 13(b):

$$\Delta m_H^2 = \frac{\lambda_S}{16\pi^2}[\Lambda^2 - 2m_S^2 \ln(\Lambda/m_S) + ...], \tag{43}$$

where Λ_S is the quartic coupling to the Higgs boson.

Comparing (42) and (43), we see that the divergent contributions terms $\propto \Lambda^2$ are cancelled if, for every fermionic loop of the theory, there is also a scalar loop with $\lambda_S = 2y_f^2$. *We will see later that supersymmetry imposes exactly this relationship!* Thus, supersymmetric field theories have no quadratic divergences, at both the one- and multi-loop levels, which enables a large hierarchy between different physical mass scales to be maintained in a natural way. In addition, other logarithmic corrections to couplings also vanish in a supersymmetric theory [68].

• A second circumstantial hint in favour of supersymmetry is the fact, discussed in the previous lecture, that precision electroweak data prefer a relatively light Higgs boson weighing less than about 150 GeV [13]. This is perfectly consistent with calculations in the MSSM in which the lightest Higgs boson weighs less than about 130 GeV [69].

• A third motivation for supersymmetry is provided by the astrophysical necessity of cold dark matter, which has a density of $\Omega_{CDM}h^2 = 0.1099 \pm 0.0062$ according to the recent measurements of WMAP [25]. This dark matter could be provided by a neutral, weakly interacting particle weighing less than about 1 TeV, such as the lightest supersymmetric particle (LSP) χ [26]. In many supersymmetric models, a conserved quantum number called R parity guarantees that the LSP is stable. As the Universe expanded and cooled, all the particles present at high energies and densities would have annihilated, disintegrated or combined to form baryons, atoms, etc., except for stable weakly interacting particles such as the neutrinos and the LSP. The latter

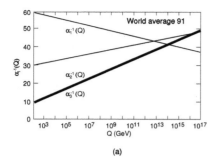
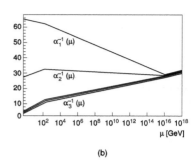

(a)   (b)

Figure 14. *The measurements of the gauge coupling strengths at LEP (a) do not evolve to a unified value if there is no supersymmetry but do (b) if supersymmetry is included [70].*

would be present in the Universe as a relic from the Big Bang, and could have the right density to constitute the majority of the cold dark matter favored by cosmologists.

• Fourthly, let us consider the couplings that characterize each of the fundamental forces. As seen in Fig. 14(a), it has been known for a long time now that if we evolve them with energy according to the renormalisation-group equations of the Standard Model, we find that they never quite become equal at the same scale. However, as seen in Fig. 14(b), when we include supersymmetric particles in the evolution of the couplings, they appear to intersect at exactly the same energy scale (about 2×10^{16} GeV) [70]. Nobody is forced to believe in such a "Grand Unification" on the basis of this possible unification of the couplings, but it is very intriguing that supersymmetry favours unification with high precision.

• Fifthly, supersymmetry seems to be essential for the consistency of string theory [71], although this argument does not really restrict the mass scale at which supersymmetric particles should appear.

• A final hint for supersymmetry may be provided by the anomalous magnetic moment of the muon, $g_\mu - 2$, whose experimental value [72] seems to differ from that calculated in the SM, in a manner that could be explained by contributions from supersymmetric particles. The amount of this discrepancy depends how one calculates the SM contributions to $g_\mu - 2$, in particular that due to low-energy hadronic vacuum polarization, and to a lesser extent that due to light-by-light scattering. The most direct way to calculate the hadronic vacuum polarization contribution is to use low-energy data on $e^+e^- \to$ hadrons: these do not agree perfectly, but may be combined to

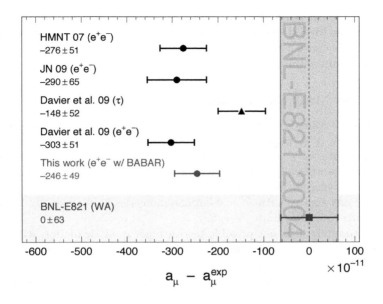

Figure 15. *SM calculations of* $a_\mu \equiv (g_\mu - 2)/2$ *disagree with the experimental measurement [72], particularly if they are based on low-energy* e^+e^- *data.*

yield [73]

$$\delta a_\mu \equiv \delta \left(\frac{g_\mu - 2}{2} \right) = (24.6 \pm 8.0) \times 10^{-10}, \tag{44}$$

a discrepancy of 3.1σ, as illustrated in Fig. 15. Alternatively, and less directly, one may use τ decay data, in which case the discrepancy is reduced to about 2σ. As we have seen, there are several arguments that motivate the study of supersymmetry.[9] Although there are no experimental proofs of its existence, supersymmetry combines so many attractive and useful characteristics that its phenomenology deserves to be studied in detail.

3.1 Low-Energy Supersymmetric Models

In this section we apply the results obtained in the previous section, with the objective of supersymmetrizing the Standard Model while preserving its successful characteristics. The minimal supersymmetric extension of the SM is called the MSSM [66, 67]. We will present its particle content (including the nomenclature of the new particles), we will discuss how the electroweak symmetry may be broken, and we will outline an effective framework for

---

[9]Other extensions of the SM also address some of these issues, though perhaps none do so as naturally as supersymmetry.

describing the breaking of supersymmetry. Later we will present typical predictions of the MSSM. Along the way, we will also mention possible variants of the MSSM, because Nature might very well have chosen a path more complex than this minimal model.

3.1.1   How Many Supersymmetries?

As well as we already mentioned, the number of supersymmetric generators Q_α may be $\mathcal{N} \geq 1$. Supersymmetric theories with $\mathcal{N} \geq 2$ have some characteristic advantages, e.g., they have fewer divergences, which make them very interesting theoretically. Specifically, in the $\mathcal{N} = 2$ case there is only a finite number of divergent Feynman diagrams, and in the $\mathcal{N} = 4$ case there are none, i.e., any theory with $\mathcal{N} = 4$ supersymmetries is intrinsically finite, and it is easy to construct finite $\mathcal{N} = 2$.

Unfortunately, it is not possible to construct realistic models with $\mathcal{N} \geq 2$, because they do not allow the violation of parity that is observed in the weak interactions. This is because a supermultiplet of a theory with $\mathcal{N} \geq 2$ supersymmetries necessarily incorporates both left- and right-handed fermions in the same supermultiplet: applying a supersymmetry charge Q changes the helicity by $1/2$, so applying two charges relates states with helicity $\pm 1/2$, implying that they are in the same representation of the gauge group, and hence have the same interactions. This contradicts experimental observations, which tell us, for example, that the left-handed electron (which forms part of a doublet in the SM) does not have the same interaction with W bosons as the right-handed electron (which is a singlet with zero electroweak isospin that does not feel the $SU(2)$ weak interaction). Models with $\mathcal{N} \geq 2$ cannot describe the physics of the SM particles observed at low energy.

3.1.2   The Particle Content in the MSSM

The supermultiplets in the minimal $\mathcal{N} = 1$ case are:
 • The chiral supermultiplet that includes a fermion of spin $1/2$ and a boson of spin 0,
 • The vector supermultiplet that includes a boson of spin 1 and one fermion of spin $1/2$.

Could we link the particles of the SM in such multiplets, i.e., could we associate quarks and leptons with the bosons W, Z, the photon, and so on? The answer is no, because this would raise problems for the conservation of their quantum numbers. Specifically, the gauge bosons and the fermions do not have the same transformation properties under the SM gauge group, since they possess different quantum numbers, e.g., quarks are triplets of the colour group whereas gauge bosons are either octets (the gluons) or singlets (the other gauge bosons), and leptons carry lepton numbers whereas gauge bosons do not. Simple $\mathcal{N} = 1$ supersymmetry does not modify these quantum numbers, so we cannot associate any gauge boson with a known fermion or vice versa. Therefore, we have to postulate unseen supersymmetric partners for all the known particles. Table 1 lists, for every SM particle, the name, spin and notation for its spartner.

Particle	Spartner	Spin
quarks q	squarks $\tilde{q}$	0
$\rightarrow$ top t	stop $\tilde{t}$	
$\rightarrow$ bottom b	sbottom $\tilde{b}$	
...		
leptons l	sleptons $\tilde{l}$	0
$\rightarrow$ electron e	selectron $\tilde{e}$	
$\rightarrow$ muon μ	smuon $\tilde{\mu}$	
$\rightarrow$ tau τ	stau $\tilde{\tau}$	
$\rightarrow$ neutrinos ν_ℓ	sneutrinos $\tilde{\nu}_\ell$	
gauge bosons	gauginos	1/2
$\rightarrow$ photon γ	photino $\tilde{\gamma}$	
$\rightarrow$ boson Z	Zino $\tilde{Z}$	
$\rightarrow$ boson B	Bino $\tilde{B}$	
$\rightarrow$ boson W	Wino $\tilde{W}$	
$\rightarrow$ gluon g	gluino $\tilde{g}$	
Higgs bosons $H_i^{\pm,0}$	higgsinos $\tilde{H}_i^{\pm,0}$	1/2

Table 1. *Particle Content of the MSSM.*

Before going on to the following sections, we make a few observations. First, we note that the spartners of SM fermions and gauge bosons are of lower spin. *A priori*, one could have considered associating the fermions of the SM with spartners of spin 1, and the gauge bosons with spartners of spin 3/2. However, to introduce a particle of spin 1 would require introducing a new gauge interaction, and hence a non-minimal model. Also, introducing particles of spin > 1 would make the theory non-renormalisable, i.e., it would no longer be possible to absorb the divergences in perturbation theory in a finite number of physical quantities.[10]

Secondly, we recall that in the SM the right-handed fermions have different interactions from the left-handed fermions, e.g., being singlets of $SU(2)$ instead of doublets. In supersymmetry, the left- and right-handed fermions must belong to different supermultiplets, and have distinct spartners, e.g., $q_L \to \tilde{q}_L$ and $q_R \to \tilde{q}_R$. These two squarks are quite different, and we use the chirality index L or R to identify them, even though the concept of handedness does not make physical sense for a scalar particle, whose only helicity is $\lambda = 0$. In general, the $\tilde{f}_L$ and $\tilde{f}_R$ mix, and the physical mass eigenstates are combinations of them. In constructing the Yukawa interactions of the MSSM, it is often convenient to work with superfields that comprise conjugates of the $\tilde{f}_R$ and their scalar spartners: these are left-handed chiral supermultiplets denoted by F^c.

Thirdly, we note that, besides the new spartners, at least two doublets of Higgs bosons are required. To understand why, we recall that, in the study of supersymmetric theories, we introduced the notion of the superpotential. This governs all the possible Yukawa interactions of the matter particles with the Higgs fields. In the SM, if we use a Higgs field h to give masses to the quarks of type "down," *via* Yukawa couplings $q\bar{d}h$, we could use the complex conjugate field h^* to give masses to quarks of type "up," *via* couplings $q\bar{u}h^*$. However, we recall that in a supersymmetric theory the superpotential is an analytic function of the superfields, that cannot depend on their complex conjugates. Therefore, we must use separate Higgs supermultiplets (denoted by capital letters) with opposite hypercharge quantum numbers, and interactions of the forms QD^cH_d and QU^cH_u. Charged leptons may acquire masses through interactions of the form LE^cH_d. We also note that pairs of Higgs superfields are needed in order to cancel the triangle anomalies that would be generated by higgsino fermion loops.

Fourthly, we note that in general the $\tilde{\gamma}, \tilde{Z}, \tilde{W}$ and $\tilde{H}$ mix, and the experimentally observable mass eigenstates are combinations of these gauginos and higgsinos that are generally named neutralinos $\tilde{N}^0_{1,2,3,4}$, which have zero electrical charge, and charginos $\tilde{C}^\pm_{1,2}$,[11] which are electrically charged and mix the $\tilde{W}^\pm$ and the $\tilde{H}^\pm$.

3.1.3 Interactions in the MSSM

The MSSM is the minimal supersymmetric extension of the SM [66,67]. The quarks and the leptons are put together in chiral superfields with their superpartners that have the same charges under $SU(3)_C$, $SU(2)_L$ and $U(1)_Y$. The gauge bosons are placed with their fermionic superpartners in vector superfields. The superpotential of the MSSM is

---

[10]Supergravity does allow a restricted number $\mathcal{N} \leq 8$ of spin-3/2 gravitino partners of the spin-2 graviton to be introduced, but they do not carry conventional gauge interactions.

[11]These are often denoted by $\tilde{\chi}^0_{1,2,3,4}$ and $\tilde{\chi}^\pm_{1,2}$, respectively.

$$\mathcal{W} = \mathcal{Y}_u Q U^c H_u + \mathcal{Y}_d Q D^c H_d + \mathcal{Y}_e L E^c H_d + \mu H_u H_d, \tag{45}$$

where we recall that the Q and L are the superfields containing the left-handed quarks and leptons, respectively, and the U^c, D^c and E^c are the superfields containing the left-handed antiquarks and antileptons, which are the charge conjugates of the right-handed quarks and leptons. Note that, for clarity, we have suppressed the $SU(2)$ indexes. The $\mathcal{Y}$ are 3×3 Yukawa matrices in flavour space, and do not have dimensions. After electroweak symmetry breaking, they give the masses to the quarks and leptons as well as the CKM angles and phases. As already mentioned, two Higgs doublets, H_u and H_d, are needed because of the analytical form of the superpotential.

The $\mu H_u H_d$ term is permitted by the symmetries of the MSSM and is required in order to have a suitable vacuum after electroweak symmetry breaking. The quantity μ has the dimension of a mass, and phenomenology requires it to be of the order of a TeV. The origin of μ is a puzzle: it might be associated to the scale of supersymmetry breaking.

The superpotential (45) determines all the non-gauge interactions of the MSSM, and the form of the effective potential of the theory is given by the gauge and Yukawa interactions.

3.1.4 Soft Supersymmetry Breaking

We have discussed so far the supersymmetric aspects of the MSSM. However, we know that supersymmetry must be broken: the selectron weighs more than the electron, squarks weigh more than quarks, etc. Therefore, we must introduce into the model the breaking of supersymmetry. However, the mechanism and the effective scale of its breaking are still unknown. Hence, we adopt the *ad hoc* strategy of parametrizing the breaking of supersymmetry in terms of effective soft[12] low-energy supersymmetry-breaking terms that are added to the Lagrangian [74]. For a general supersymmetric theory, the form of these soft supersymmetry-breaking terms $\mathcal{L}_{soft}$ in the Lagrangian is:

$$\mathcal{L} \supset \mathcal{L}_{soft} = -\frac{1}{2}(M_\lambda \lambda^a \lambda^a + c.c) - m_{ij}^2 \phi_j^* \phi_i$$
$$+ (\frac{1}{2}b_{ij}\phi_i\phi_j + \frac{1}{6}a_{ijk}\phi_i\phi_j\phi_k + c.c). \tag{46}$$

This breaks supersymmetry explicitly, since only the the the gauginos λ^a and the scalars ϕ_i have mass terms, and the trilinear terms with coefficients a_{ijk} are also not of supersymmetric form. In the case of the MSSM, $\mathcal{L}_{soft}$ takes the following general form in terms of the spartner fields of the MSSM:

$$-\mathcal{L}_{soft} = \frac{1}{2}(M_3 \tilde{g}\tilde{g} + M_2 \tilde{W}\tilde{W} + M_1 \tilde{B}\tilde{B} + c.c)$$

---

[12] Here, the adjective "soft" means that they do not introduce quadratic divergences.

$$+ \quad \tilde{Q}^\dagger m_Q^2 \tilde{Q} + \bar{\tilde{U}}^\dagger m_{\tilde{U}}^2 \bar{\tilde{U}} + \bar{\tilde{D}}^\dagger m_{\tilde{D}}^2 \bar{\tilde{D}} + \bar{\tilde{L}}^\dagger m_{\tilde{L}}^2 \bar{\tilde{L}} + \bar{\tilde{E}}^\dagger m_{\tilde{E}}^2 \bar{\tilde{E}}$$

$$+ \quad (\bar{\tilde{U}}^\dagger a_U \tilde{Q} H_u - \bar{\tilde{D}}^\dagger a_D \tilde{Q} H_d - \bar{\tilde{E}}^\dagger a_E \tilde{L} H_d + \ c.c)$$

$$+ \quad m_{H_u}^2 H_u^* H_u + m_{H_d}^2 H_d^* H_d + (b H_u H_d + \ c.c). \tag{47}$$

The masses M_3, M_2 and M_1 of the gauginos are complex in general, which introduces 6 parameters. The quantities m_Q, m_L and $m_{\bar{u}}$, are the mass matrices of the squarks and sleptons, which are hermitian 3×3 matrices in family space, adding 45 more unknown parameters. The couplings a_U, a_D,..., are also complex 3×3 matrices, characterized by 54 parameters. In addition, the quadratic couplings of the Higgs bosons introduce 4 more parameters, so that the whole $\mathcal{L}_{soft}$ contains a total of 109 unknown parameters, including many that violate CP!

Supersymmetry itself is a very powerful principle, whose implementation introduces only one new parameter (μ) in the MSSM. However, in our present state of ignorance, the breaking of supersymmetry introduces many new parameters. On the other hand, the number of soft parameters can be reduced by postulating symmetries or making supplementary hypotheses. Measuring the parameters of soft supersymmetry breaking would allow us to go beyond the phenomenological parametrization (47), and open the way to testing models of the high-energy dynamics that breaks supersymmetry.

3.1.5 R Parity and Dark Matter

We introduced above the superpotential (47) of the MSSM, which includes only the Yukawa interactions of the SM. However, gauge invariance, Lorentz invariance and analyticity in the SM fields would allow us introduce other terms in the superpotential that do not have any correspondence with the SM, and do not preserve either baryon number and/or lepton number. One way to avoid all such terms is to add to the MSSM a new symmetry called R parity, given by the following combination of baryon number, lepton number and spin S:

$$R = (-1)^{3(B-L)+2S}. \tag{48}$$

This is a multiplicatively conserved quantum number in the SM, since all the SM particles and Higgs bosons have even R parity: $R = +1$. On the other hand, all the sparticles have odd R parity ($R = -1$).

Conservation of R parity would have important phenomenological consequences:

• The sparticles are produced in even numbers (usually two at time), for example: $\bar{p} p \to \tilde{q} \tilde{g} X$, $e^+ e^- \to \tilde{\mu}^+ \tilde{\mu}^-$;

• Each sparticle decays into another sparticle (or into an odd number of them), for example: $\tilde{q} \to q \tilde{g}$, $\tilde{\mu} \to \mu \tilde{\gamma}$;

• The lightest sparticle (LSP) must be stable, since it has $R = -1$. If it is electrically neutral, it can interact only weakly with ordinary matter, and may be a good candidate for the non-baryonic dark matter that is required by cosmology [26].

The dark matter particles should have neither electric charge nor strong interactions, otherwise they would be visible or detectable, e.g., through their binding to ordinary matter to form what would look like anomalous heavy nuclei, which have never been seen. We therefore expect any dark matter particle to have only weak interactions, in which case, if it was produced at a collider such as the LHC, it would carry energy-momentum away invisibly. Accordingly, most LHC searches for supersymmetry focus on events with missing transverse momentum, though searches for signatures of R-violating models are also considered.

The existence of a stable, weakly interacting LSP is a very important prediction of the MSSM, but its nature and its total contribution to the density of dark matter depend on the parameters of the MSSM. One weakly interacting candidate was the lightest sneutrino, but this has already been excluded by direct searches at LEP and by experiments searching directly for dark matter. The remaining candidate particles are the lightest neutralino χ of spin $1/2$, and the gravitino of spin $3/2$. As we discuss later, there are chances to detect a neutralino LSP at the LHC in events with missing energy, or directly as astrophysical dark matter. On the other hand, the interactions of the gravitino are so weak that it could not be directed as astrophysical dark matter, and only indirectly in collider experiments.

3.2 Phenomenology of Supersymmetry

As we have seen, the soft supersymmetry-breaking sector of the MSSM has over a hundred parameters. This renders very difficult the interpretation of experimental constraints and (hopefully) the extraction of the experimental values of these parameters. A simplifying hypothesis is to assume *universality* at a certain scale before renormalisation, leading us to the constrained MSSM (CMSSM):

• The gaugino masses are assumed to be equal at some input GUT or supergravity scale: $M_3 = M_2 = M_1 = m_{1/2}$;

• The scalar masses of squarks and sleptons are assumed to be universal at the same scale: $m_Q^2 = m_{U^c}^2 = \ldots = m_0^2$, as are the soft supersymmetry-breaking contributions to the Higgs masses $m_{H_u}^2 = m_{H_d}^2 = m_0^2$;

• The trilinear couplings are related by a universal coefficient A_0 to the corresponding Yukawa couplings: $a_u = A_0 y_u$, $a_d = A_0 y_d$, $a_e = A_0 y_e$.

Simplifying the MSSM to the CMSSM reduces the number of parameters from over 100 to only 4: $m_{1/2}, m_0, A_0, \tan\beta$ and the sign of μ (the magnitude of μ is fixed by the electroweak vacuum conditions). The CMSSM hypothesis is very practical from a phenomenological point of view, though questionable from a purely theoretical point of view. The CMSSM and the simplification of $\mathcal{L}_{soft}$ are inspired by simple supergravity models where the breaking of supersymmetry is mediated by gravity, though minimal supergravity models actually impose two additional constraints. On the other hand, generic string models often lead to different patterns of soft supersymmetry breaking.

Dropping universality for squarks or sleptons with the same quantum numbers but in different generations would lead to problems with flavour-changing neutral interactions, and Grand Unified Theories relate the soft supersymmetry-breaking masses of squarks and sleptons with different quantum numbers. However, there is no strong theoretical or phenomenological reason to postulate universality for the soft supersymmetry-breaking contributions to the Higgs masses. One may relax this assumption for the Higgs scalar masses-squared, m_H^2, by assuming the same single-parameter non-universal Higgs mass parameter (the NUHM1), or by allowing the non-universal Higgs mass parameters to be different (the NUHM2).

3.3 Renormalisation of the Soft Supersymmetry-Breaking Parameters

In our ignorance of the underlying mechanism of supersymmetry breaking, it is usually assumed that this occurs at some large mass scale far above a TeV, perhaps around the grand unification or Planck scale. They therefore undergo significant renormalisation between this input scale and the electroweak scale. Although quadratic divergences are absent from a softly broken supersymmetric theory, it still has logarithmic divergences that may be treated using the renormalisation group (RG).

At leading order in the RG, which resums the leading one-loop logarithms, the renormalisations of the soft gaugino masses M_a are the same as for the corresponding gauge couplings:

$$Q\frac{dM_a}{dQ} = \beta_a M_a, \tag{49}$$

where β_a is the standard one-loop renormalisation coefficient including supersymmetric particles that is discussed in more detail in the next lecture. As a result of (49), to leading order

$$M_a(Q) = \frac{\alpha_a(Q)}{\alpha_{GUT}} m_{1/2} \tag{50}$$

if the gauge couplings α_a and the gaugino masses are assumed to unify at the same large mass scale M_{GUT}. As a consequence of (50), one expects the gluino to be heavier than the wino: $m_{\tilde{g}}/m_{\tilde{W}} = \alpha_3/\alpha_2$ at leading order.

The soft supersymmetry-breaking scalar masses-squared m_0^2 acquire renormalisations related to the gaugino masses via the gauge couplings, and to the scalar masses and trilinear parameters A_λ via the Yukawa couplings:

$$\frac{Qdm_0^2}{dQ} = \frac{1}{16\pi^2}\left[-g_a^2 M_a^2 + \lambda^2(m_0^2 + A_\lambda^2)\right]. \tag{51}$$

The latter effect is significant for the stop squark, one of the Higgs multiplets, and possibly the other third-generation sfermions if $\tan\beta$ is large. For the other sfermions, at leading order one has

$$m_0^2(Q) = m_0^2 + C m_{1/2}^2, \qquad (52)$$

where the coefficient C depends on the gauge quantum numbers of the corresponding sfermion. Consequently, one expects the squarks to be heavier than the sleptons. Specifically, in the CMSSM one finds at the electroweak scale that:

$$\text{Squarks}: \ m_{\tilde{q}}^2 \ \sim \ m_0^2 + 6 m_{1/2}^2, \qquad (53)$$

$$\text{Left sleptons}: \ m_{\tilde{\ell}_L}^2 \ \sim \ m_0^2 + 0.5 m_{1/2}^2, \qquad (54)$$

$$\text{Right sleptons}: \ m_{\tilde{\ell}_R}^2 \ \sim \ m_0^2 + 0.15 m_{1/2}^2. \qquad (55)$$

The difference between the left and right slepton masses may have implications for cosmology, as we discuss later. A small difference is also expected between the masses of the left and right squarks, but this is relatively less significant numerically.

The CKM mixing between quarks is related in the SM to off-diagonal entries in the Yukawa coupling matrix, and shows up in leading-order charged-current interactions and flavour-changing neutral current (FCNC) interactions induced at the loop level. One would expect additional FCNCs to be induced by similar loop diagrams involving squarks, which would propagate through the RGEs (51) and induce flavour-violating terms in the sfermion mass matrices. However, experiment imposes important upper limits on such additional supersymmetric flavour effects. As already discussed, these would be suppressed (though non-zero) if the soft supersymmetry breaking scalar masses of all sfermions with the same quantum numbers were the same before renormalisation. The hypothesis of Minimal Flavour Violation (MFV) is that flavour mixing of squarks and sleptons is induced only by the CKM mixing in the quark sector and the corresponding MNS mixing in the lepton sector: see the next lecture. The MFV hypothesis requires also that the soft supersymmetry-breaking trilinear parameters A be universal for sfermions with the same quantum numbers: $A_\lambda = A_0 \lambda$. However, the MFV hypothesis does permit the appearance of six additional phases beyond those in the CKM model for quarks: three phases for the different gaugino mass parameters, and three phases for the different A_0 coefficients [75].

Results of typical numerical calculations of these renormalisation effects in the CMSSM are shown in Fig. 16. An important effect illustrated there is that the RGEs may drive $m_{H_u}^2$ negative at some low renormalisation scale Q_N, thanks to the top quark Yukawa coupling appearing in (51).[13] A negative value of $m_{H_u}^2$ would trigger electroweak symmetry breaking at a scale $\sim Q_N$.

---

[13]The effect of the Yukawa coupling is to *increase* m_0^2 as Q increases, i.e., to *decrease* m_0^2 as Q decreases.

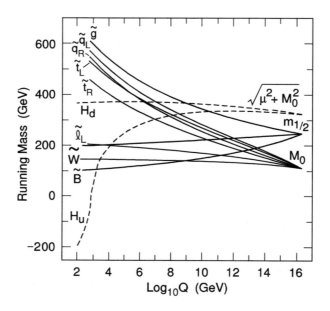

Figure 16. *Calculations of the renormalisation of soft supersymmetry-breaking sparticle masses, assuming universal scalar and gaugino masses* $m_0, m_{1/2}$ *at the GUT scale. Note that strongly interacting sparticles have larger physical masses at low scales, and the* $m_{H_u}^2$ *is driven negative, triggering electroweak symmetry breaking.*

Since the negative value of $m_{H_u}^2$ is due to the logrithmic renormalisation by the top quark Yukawa coupling, electroweak symmetry breaking appears at a scale exponentially smaller than the input GUT or Planck scale:

$$\frac{m_W}{M_{GUT,P}} = \exp\left(-\frac{\mathcal{O}(1)}{\alpha_t}\right) : \alpha_t \equiv \frac{\lambda_t^2}{4\pi}. \tag{56}$$

In this way, it is possible for the electroweak scale to be generated naturally at a scale ~ 100 GeV if the top quark is heavy: $m_t \sim 60$ to 100 GeV, a realization that long predated the discovery of just such a heavy top quark.

3.4 Constraints on the MSSM

Most of the current constraints on possible physics beyond the SM are negative and, specifically, no sparticle has ever been detected. The concordance with the SM predictions means that, in general, one can only set lower limits on the possible masses of supersymmetric particles. However, there are two observational indications of physics beyond the SM that may, in the supersymmetric context, be used for setting *upper* limits of the masses of the supersymmetric particles. As discussed earlier, these two hints for new physics are the anomalous magnetic moment of the muon, $g_\mu - 2$, which seems to disagree with the prediction of the SM (at least if this is calculated using low-energy e^+e^- data as an input), and the density of cold dark matter Ω_{CDM}. However, these discrepancies may be explained either with supersymmetry or with other possible extensions of the SM, so their interpretations require special care. Nevertheless, these may be regarded as additional phenomenological motivations for supersymmetry, in addition to the more theoretical motivations described in the beginning of this section, such as the naturalness of the hierarchy of mass scales in physics, grand unification, string theory, etc. Therefore, in addition to considering the more direct searches for supersymmetry, it is also natural to ask what $g_\mu - 2$ and Ω_{CDM} may imply for the parameters of supersymmetric models. Fig. 17 compiles the impacts of various constraints on supersymmetry, assuming that the soft supersymmetry-breaking contributions $m_{1/2}, m_0$ to the different scalars and gauginos are each universal at the GUT scale (the scenario called the CMSSM), and that the lightest sparticle is the lightest neutralino χ.

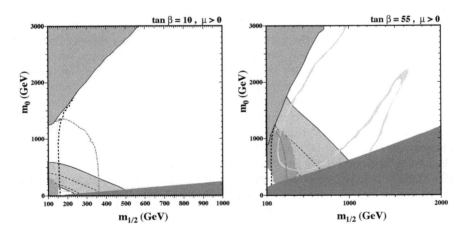

Figure 17. *The CMSSM $(m_{1/2}, m_0)$ planes for (a) $\tan\beta = 10$ and (b) $\tan\beta = 55$, assuming $\mu > 0$, $A_0 = 0$, $m_t = 173.1$ GeV and $m_b(m_b)_{SM}^{\overline{MS}} = 4.25$ GeV. The near-vertical dot-dashed lines are the contours for $m_h = 114$ GeV, and the near-vertical dashed line is the contour $m_{\chi^\pm} = 104$ GeV. Also shown by the dot-dashed curve in the lower left is the region excluded by the LEP bound $m_{\tilde{e}} > 99$ GeV. The medium-shaded region is excluded by $b \to s\gamma$, and the light-shaded area is the cosmologically preferred region. In the dark-shaded region, the LSP is the charged $\tilde{\tau}_1$. The region allowed by the measurement of $g_\mu - 2$ at the 2σ level assuming the e^+e^- calculation of the Standard Model contribution, is shaded and bounded by solid lines, with dashed lines indicating the 1σ ranges (updated from [77]).*

Experiments at LEP and the Tevatron collider, in particular, have made direct searches for supersymmetry using the missing-energy-momentum signature. LEP established lower limits ~ 100 GeV on the masses of many charged sparticles without strong interactions, such as sleptons and charginos. The Tevatron collider has established the best lower limits on the masses of squarks and gluinos, ~ 400 GeV. In view of the greater renormalisation of the squark and gluino masses than for charginos and sleptons, see (50) and (55), these two sets of limits are quite complementary.

Another important constraint is provided by the LEP lower limit on the Higgs mass: $m_H > 114.4$ GeV [17]. This holds in the SM, for the lightest Higgs boson h in the general MSSM for $\tan\beta \lesssim 8$, and almost always in the CMSSM for all $\tan\beta$, at least as long as CP is conserved.[14] Since m_h is sensitive to sparticle masses, particularly $m_{\tilde{t}}$, via loop corrections, the Higgs limit also imposes important constraints on the soft supersymmetry-breaking CMSSM parameters, principally $m_{1/2}$ [77], as seen in Fig. 17.

Important constraints are imposed on the CMSSM parameter space by flavour physics, specifically the agreement with data of the SM prediction for the decay $b \to s\gamma$, as well as the upper limit on the decay $B_s \to \mu^+\mu^-$, which is important at large $\tan\beta$ in particular.

We see in Fig. 17 that narrow strips of the $(m_{1/2}, m_0)$ planes are compatible [77] with the range of the astrophysical cold dark matter density favoured by WMAP and other experiments. However, these strips vary with $\tan\beta$ and A_0. In fact, foliation by these WMAP strips covers large fractions of the $(m_{1/2}, m_0)$ plane as $\tan\beta$ and A_0 are varied. Away from these narrow strips, the relic neutralino density exceeds the WMAP range over most of the $(m_{1/2}, m_0)$ planes shown in Fig. 17. In its left panel, the relic density is reduced into the WMAP range only in the shaded strip at $m_0 \sim 100$ GeV that extends to $m_{1/2} \sim 900$ GeV. This reduction is brought about by coannihilations between the LSP χ (which is mainly a Bino) and sleptons that are only slightly heavier, most notably the lighter stau and the right selectron and smuon, which are significantly lighter than the left sleptons, as discussed earlier. In the right panel of Fig. 17 for $\tan\beta = 50$, this coannihilation strip moves to larger m_0. Also, it is extended to larger $m_{1/2}$, as a result of a reduction in the relic density due to rapid $\chi - \chi$ annihilations through direct-channel heavy Higgs (H, A) states. In addition to these visible WMAP regions, there is in principle another allowed strip at very large values of m_0, called the focus-point region, where the LSP becomes relatively light and acquires a substantial higgsino component, favouring annihilation via W^+W^- final states.

Finally, also shown in the two panels of Fig. 17 are the regions favoured by the supersymmetric interpretation of the discrepancy (44) between the experimental measurement of $g_\mu - 2$ and the value calculated in the SM using

---

[14]The lower bound on the lightest MSSM Higgs boson may be relaxed significantly if CP violation feeds into the MSSM Higgs sector [76].

low-energy e^+e^- data [77]. The favoured regions are displayed as bands corresponding to $\pm 2\sigma$. We see that they can be used to set *upper* limits on the sparticle masses! In particular, $g_\mu - 2$ disfavours the focus-point region, where m_0 is so large that the supersymmetric contribution to $g_\mu - 2$ is negligible, and also the region at large $\tan\beta$ and large $m_{1/2}$ where the neutralinos may annihilate rapidly through direct-channel heavy-Higgs states.

3.5 Frequentist Analysis of the Supersymmetric Parameter Space

In a recent paper [78] the likely range of parameters of the CMSSM and NUHM1 have been estimated using a frequentist approach, by building a χ^2 likelihood function with contributions from the various relevant observables, including precision electroweak physics, $g_\mu - 2$, the lower limit on the lightest Higgs boson mass (taking into taking into account the theoretical uncertainty in the FeynHiggs calculation of M_h [79]), the experimental measurement of BR$(b \rightarrow s\gamma)$(which agrees with the SM), the experimental upper limit on BR$(B_s \rightarrow \mu^+\mu^-)$and Ω_{CDM}. This frequentist analysis used a Markov chain Monte Carlo technique to sample thoroughly the $(m_0, m_{1/2})$ plane up to masses of several TeV, including the focus-point and rapid-annihilation regions, for a wide range of values of A_0 and $\tan\beta$.

We display in Fig. 18 the $\Delta\chi^2$ functions in the $(m_0, m_{1/2})$ planes for the CMSSM (left plot) and for the NUHM1 (right plot). The parameters of the best-fit CMSSM point are $m_0 = 60$ GeV, $m_{1/2} = 310$ GeV, $A_0 = 130$ GeV, $\tan\beta = 11$ and $\mu = 400$ GeV (corresponding nominally to $M_h = 114.2$ GeV and an overall $\chi^2 = 20.6$ for 19 d.o.f. with a probability of 36%), which are very close to the ones previously reported in [80]. The corresponding parameters of the best-fit NUHM1 point are $m_0 = 150$ GeV, $m_{1/2} = 270$ GeV, $A_0 = -1300$ GeV, $\tan\beta = 11$ and $m_{h_1}^2 = m_{h_2}^2 = -1.2 \times 10^6$ GeV^2 or, equivalently, $\mu = 1140$ GeV, yielding $\chi^2 = 18.4$ (corresponding to a similar fit probability to the CMSSM) and $M_h = 120.7$ GeV. The similarities between the best-fit values of m_0, $m_{1/2}$ and $\tan\beta$ in the CMSSM and the NUHM1 suggest that the model frameworks used are reasonably stable: if they had been very different, one might well have wondered what would be the effect of introducing additional parameters, as in the NUHM2 with two non-universality parameters in the Higgs sector.

These best-fit points are both in the coannihilation region of the $(m_0, m_{1/2})$ plane, as can be seen in Fig. 18. The C.L. contours extend to slightly larger values of m_0 in the CMSSM, while they extend to slightly larger values of $m_{1/2}$ in the NUHM1, as was already shown in [80] for the 68% and 95% C.L. contours. However, the qualitative features of the $\Delta\chi^2$ contours are quite similar in the two models, indicating that the preference for small m_0 and $m_{1/2}$ are quite stable and do not depend on details of the Higgs sector. We recall that it was found in [80] that the focus-point region was disfavoured at beyond the 95% C.L. in both the CMSSM and the NUHM1. We see in Fig. 18

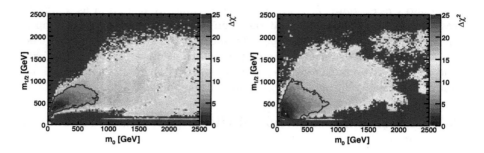

Figure 18. *The $\Delta\chi^2$ functions in the $(m_0, m_{1/2})$ planes for the CMSSM (left plot) and for the NUHM1 (right plot), as found in frequentist analyses of the parameter spaces. We see that the coannihilation regions at low m_0 and $m_{1/2}$ are favoured in both cases [80].*

that this region is disfavoured at the level $\Delta\chi^2 \sim 8$ in the CMSSM and > 9 in the NUHM1.

The favoured values of the particle masses in both models are such that there are good prospects for detecting supersymmetric particles in CMS [81] and ATLAS [82] even in the early phase of the LHC running with reduced centre-of-mass energy and limited luminosity, as seen in Fig. 19. The best-fit points and most of the 68% confidence level regions are within the region of the $(m_0, m_{1/2})$ plane that could be explored with 100 pb^{-1} of data at 14 TeV in the centre of mass, and hence perhaps with 200 fb^{-1} of data at 10 TeV.[15] Almost all the 95% confidence level regions would be accessible to the LHC with 1 fb^{-1} of data at 14 TeV. As seen in Fig. 19, in substantial parts of these regions there are good prospects for detecting $\tilde{q} \to q\ell^+\ell^-\chi$ decays, which are potentially useful for measuring sparticle mass parameters, and the lightest supersymmetric Higgs boson may also be detectable in $\tilde{q}$ decays.

The best-fit spectra in the CMSSM and NUHM1 are shown in Fig 20: they are relatively similar, though the heavier Higgs bosons, the gluinos and the squarks, may be somewhat heavier in the CMSSM, whereas the heavier charginos and neutralinos may be heavier in the NUHM1 [80]. There are considerable uncertainties in these spectra, as seen in Fig. 21 [78]. However, in general there are strong correlations between the different sparticle masses, as exemplified in Fig. 22, though the correlation is weaker, e.g., for the lighter stau and the LSP in the NUHM1.[16]

Finally, a result from this frequentist analysis that also concerns LHC

---

[15]The comparisons are made with experimental simulations for $\tan\beta = 10$ and $A_0 = 0$, whereas the frequentist analysis sampled all values of $\tan\beta$ and A_0. As it happens, the preferred values of $\tan\beta$ in both the CMSSM and the NUHM1 are quite close to 10: the value of A_0 is relatively unimportant for the experimental analysis.

[16]This reflects the possible appearance of rapid direct-channel annihilations also at low $m_{1/2}$ and low $\tan\beta$, allowing an escape from the coannihilation region where $m_\chi \sim m_{\tilde{\tau}_1}$.

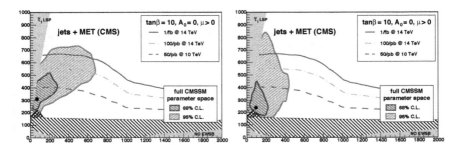

Figure 19. *The $(m_0, m_{1/2})$ planes in the CMSSM (upper) and the NUHM1 (lower) for $\tan\beta = 10$ and $A_0 = 0$. The dark-shaded areas at low m_0 and high $m_{1/2}$ is excluded due to a scalar tau LSP, the light-shaded areas at low $m_{1/2}$ do not exhibit electroweak symmetry breaking. The nearly horizontal line at $m_{1/2} \approx 160$ GeV in the lower panel has $m_{\tilde\chi_1^\pm} = 103$ GeV, and the area below is excluded by LEP searches. Just above this contour at low m_0 in the lower panel is the region that is excluded by trilepton searches at the Tevatron. Shown in each plot is the best-fit point [80], indicated by a star, and the 68 (95)% C.L. contours from the fit as dark grey (light grey) overlays, scanned over all $\tan\beta$ and A_0 values. The plots also show some 5σ discovery contours for CMS [81] with 1 fb^{-1} at 14 TeV, 100 pb^{-1} at 14 TeV and 50 pb^{-1} at 10 TeV centre-of-mass energy [80].*

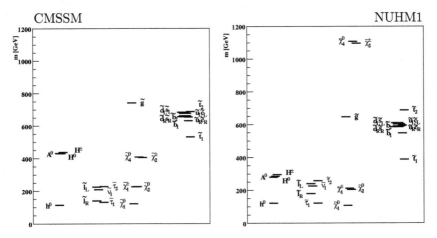

Figure 20. *The spectra at the best-fit points: left – in the CMSSM with $m_{1/2} = 311$ GeV, $m_0 = 63$ GeV, $A_0 = 243$ GeV, $\tan\beta = 11.0$, and right – in the NUHM1 with $m_{1/2} = 265$ GeV, $m_0 = 143$ GeV, $A_0 = -1235$ GeV, $\tan\beta = 10.4$ and $\mu = 1110$ GeV [80].*

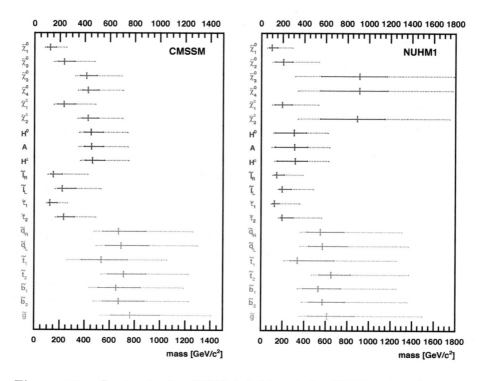

Figure 21. *Spectra in the CMSSM (left) and the NUHM1 (right). The vertical solid lines indicate the best-fit values, the horizontal solid lines are the 68% C.L. ranges, and the horizontal dashed lines are the 95% C.L. ranges for the indicated mass parameters [78].*

physics, but away from the high-energy frontier. We see in Fig. 23 that the branching ratio for $B_s \to \mu^+\mu^-$ may well exceed considerably its value in the SM, particularly at large $\tan\beta$. This is true to some extent in the CMSSM, and even more so in the NUHM1. Particularly in the latter case, this decay might perhaps be accessible to the LHCb experiment during initial LHC running. Therefore, there may be important competition for ATLAS and CMS in their quests to discover supersymmetry!

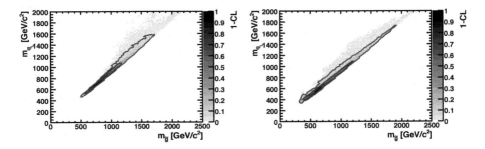

Figure 22. *The correlations between the gluino mass, $m_{\tilde{g}}$, and the masses of the the left-handed partners of the five light squark flavours, $m_{\tilde{q}_L}$ are shown in the CMSSM (left panel) and in the NUHM1 (right panel)* [78].

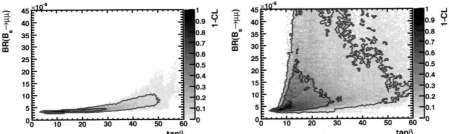

Figure 23. *The correlation between the branching ratio for $B_s \to \mu^+\mu^-$ and $\tan\beta$ in the CMSSM (left panel) and in the NUHM1 (right panel)* [78].

4 Further Beyond: GUTs, String Theory and Extra Dimensions

4.1 Grand Unification

Gauge theories, particularly non-Abelian Yang–Mills theories, are the only suitable framework for describing interactions in particle physics. In the SM, there are three different gauge groups, $SU(3)_C$, $SU(2)_L$ and $U(1)_Y$, and correspondingly there are three different couplings. It is logical to look for a single, more powerful non-Abelian Grand Unified gauge group with a single coupling g_{GUT} that would enable us to unify the three couplings, and might provide interesting relations between the other different SM parameters such as Yukawa couplings and hence fermion masses.[17] As a first approximation, we assume

---

[17]In this chapter, we denote the couplings by g_1 for the $U(1)$ subgroup, g_2 for $SU(2)$ and g_3 for $SU(3)$, which have the appropriate normalisations for Grand Unification: see later.

that the effects of the gravitational interaction are negligible, which is generally true if the Grand Unification scale M_{GUT} is significantly smaller than the Planck mass. As we see later, it turns out that typical estimations, based on extrapolation to very high energies of the known physics of the SM [83], give a Grand Unification scale of the order of 10^{16} GeV, which is about a thousand times smaller than the Planck scale $M_{Pl} = \mathcal{O}(10^{19})$ GeV.

Postulating a single group to describe all the interactions of particle physics also implies new relations between the matter particles themselves, as well as new gauge bosons. Specifically, if the symmetry changes then the representations, and hence the organization of the particles into multiplets, also change. There are some hints for this in low-energy physics, such as charge quantization and the correlation of fractional electrical charges with colour charges, and the cancellation of anomalies between the leptons and the quarks, that also lead us to anticipate an organization simpler than the SM.

Clearly, one must recover the SM at low energy, implying that in these Grand Unified theories (GUTs) one must also study the breaking of the GUT group $G \to SU(3)_C \otimes SU(2)_L \otimes U(1)_Y$.

This section begins with a presentation of the renormalisation group evolution equations of the three SM gauge couplings and studies their possible unification at some GUT scale. Subsequently, some specific examples of GUTs are discussed, notably the prototype based on the group $SU(5)$, which makes possible a simple discussion of many properties of GUTs. This is followed by a short discussion of typical predictions of these models, such as the decay of the proton and the relations between the masses of the quarks and leptons. We finish by discussing some of the advantages, problems and perspectives of GUT models.

4.1.1   The Evolution Equations for Gauge Couplings

The first apparent obstacle to the philosophy of grand unification is the fact that the strong coupling strength $\alpha_3 = g_3^2/4\pi$ is much stronger than the electroweak couplings at present-day energies: $\alpha_3 \gg \alpha_2, \alpha_1$. However, the strong coupling is asymptotically free [9]:

$$\alpha_3(Q) \simeq \frac{12\pi}{(33 - 2N_q)\ln(Q^2/\Lambda_3^2)} + \cdots, \qquad (57)$$

where N_q is the number of quarks, $\Lambda_3 \simeq$ few hundred MeV is an intrinsic scale of the strong interactions, and the dots in (57) represent higher-loop corrections to the leading one-loop behaviour shown. The other SM gauge couplings also exhibit logarithmic violations analogous to (57). For example, the fine-structure constant $\alpha_{em} = 1/137.035999084(51)$ is renormalised to effective value of $\alpha_{em}(m_Z) \sim 1/128$ at the Z mass scale. The renormalisation group evolution for the $SU(2)$ gauge coupling corresponding to (57) is

$$\alpha_2(Q) \simeq \frac{12\pi}{(22 - 2N_q - N_H/2)\ln(Q^2/\Lambda_2^2)} + \cdots, \qquad (58)$$

where we have assumed equal numbers of quarks and leptons, and N_H is the number of Higgs doublets. Taking the inverses of (57) and (58), and then taking their difference, we find

$$\frac{1}{\alpha_3(Q)} - \frac{1}{\alpha_2(Q)} = \left(\frac{11 + N_H/2}{12\pi}\right) \ln\left(\frac{Q^2}{m_X^2}\right) + \ldots \tag{59}$$

Note that we have absorbed the scales Λ_3 and Λ_2 into a single grand unification scale M_X where $\alpha_3 = \alpha_2$.

Evaluating (59) when $Q = \mathcal{O}(M_W)$, where $\alpha_3 \gg \alpha_2 = 0(\alpha_{em})$, we derive the characteristic feature [83]

$$\frac{m_{GUT}}{m_W} = \exp\left(\mathcal{O}\left(\frac{1}{\alpha_{em}}\right)\right), \tag{60}$$

i.e., the grand unification scale is exponentially large. As we see in more detail later, in most GUTs there are new interactions mediated by bosons weighing $\mathcal{O}(m_X)$ that cause protons to decay with a lifetime am_X^4. In order for the proton lifetime to exceed the experimental limit, we need $m_X \gtrsim 10^{14}$ GeV and hence $\alpha_{em} \lesssim 1/120$ in (60) [84]. On the other hand, if the neglect of gravity is to be consistent, we need $m_X \lesssim 10^{19}$ GeV and hence $\alpha_{em} \gtrsim 1/170$ in (60) [84]. The fact that the measured value of the fine-structure constant α_{em} lies in this allowed range may be another hint favouring the GUT philosophy.

Further empirical evidence for grand unification is provided by the prediction it makes for the neutral electroweak mixing angle [83]. Calculating the renormalisation of the electroweak couplings, one finds:

$$\sin^2\theta_W = \frac{\alpha_{em}(m_W)}{\alpha_2(m_W)} \simeq \frac{3}{8}\left[1 - \frac{\alpha_{em}}{4\pi}\frac{110}{9}\ln\frac{m_X^2}{m_W^2}\right], \tag{61}$$

which can be evaluated to yield $\sin^2\theta_W \sim 0.210$ to 0.220, if there are only SM particles with masses $\lesssim m_X$ [83]. This is to be compared with the experimental value $\sin^2\theta_W = 0.23120 \pm 0.00015$ in the $\overline{\text{MS}}$ renormalisation scheme. Considering that $\sin^2\theta_W$ could *a priori* have had any value between 0 and 1, this is an impressive qualitative success. The small discrepancy can be removed by adding some extra particles, such as the supersymmetric particles in the MSSM.

To see this explicitly, we may write

$$\sin^2\theta(m_Z) = \frac{g'^2}{g_2^2 + g'^2} = \frac{3}{5}\frac{g_1^2(m_Z)}{g_2^2(m_Z) + \frac{3}{5}g_1^2(m_Z)}, \tag{62}$$

where g_1 is defined in such a way that its quadratic Casimir coefficient, summed over all the particles in a single generation, is the same as for g_2 and g_3, which is the appropriate normalization within a GUT. Using the one-loop RGEs, we can then write

$$\sin^2\theta(m_Z) = \frac{1}{1 + 8x}\left[3x + \frac{\alpha_{em}(m_Z)}{\alpha_3(m_Z)}\right] = \frac{1}{5}\left(\frac{b_2 - b_3}{b_1 - b_2}\right), \tag{63}$$

where the b_i are the one-loop coefficients in the RGEs for the different SM couplings. Their values in the SM (on the left) and the MSSM (on the right) are:

$$\frac{4}{3}N_G - 11 \;\leftarrow\; b_3 \;\rightarrow\; 2N_G - 9 = -3 \tag{64}$$

$$\frac{1}{6}N_H + \frac{4}{3}N_G - \frac{22}{3} \;\leftarrow\; b_2 \;\rightarrow\; \frac{1}{2}N_H + 2N_G - 6 = +1 \tag{65}$$

$$\frac{1}{10}N_H + \frac{4}{3}N_G \;\leftarrow\; b_1 \;\rightarrow\; \frac{3}{10}N_H + 2N_G = \frac{33}{5} \tag{66}$$

$$\frac{23}{218} = 0.1055 \;\leftarrow\; x \;\rightarrow\; \frac{1}{7}. \tag{67}$$

Experimentally, using $\alpha_{em}(m_Z) = 1/128, \alpha_3 = 0.119 \pm 0.003, \sin^2\theta_W(m_Z) = 0.2315$, we find

$$x = \frac{1}{6.92 \pm 0.07}, \tag{68}$$

in striking agreement with the MSSM prediction in (67)!

Another qualitative success is the prediction of the b quark mass [85, 86]. In many GUTs, such as the minimal $SU(5)$ model discussed shortly, the b quark and the τ lepton have equal Yukawa couplings when renormalised at the GUT scale. The renormalisation group then tells us that

$$\frac{m_b}{m_\tau} \simeq \left[\ln\left(\frac{m_b^2}{m_X^2}\right)\right]^{\frac{12}{33-2N_q}}. \tag{69}$$

Using $m_\tau = 1.78$ GeV, we predict that $m_b \simeq 5$ GeV, in agreement with experiment. Happily, this prediction remains successful if the effects of supersymmetric particles are included in the renormalisation group calculations [87].

4.1.2   Specific GUTs

What groups may be used to construct a GUT [88]?

First, suitable groups must be sufficiently large to include the SM. The latter is of rank 4, i.e., there are four simultaneously diagonalizable symmetry generators:[18] $SU(3)_C$ has two, $SU(2)_L$ one, and $U(1)_Y$ one also. It is striking that all of the diagonal generators are traceless: this is trivial for the non-Abelian groups $SU(3)_C$ and $SU(2)_L$, but non-trival for $U(1)_Y$, and a possible hint that it should be embedded in a non-Abelian GUT group. Therefore, we must first find in the Cartan classification of Lie groups a group of rank higher or equal to four. Secondly, a GUT group must possess complex representations, in order that the matter particles and their antiparticles (described by complex conjugate spinors) could be in inequivalent representations. Thirdly,

---

[18]Each one is associated with a quantum number, a "charge," that may be to label particle states.

we should also keep track of the hypercharges $Y = Q - T_3$. One of the major puzzles of the SM is why

$$\sum_{q,\ell} Q_i = 3Q_u + 3Q_d + Q_e = 0. \tag{70}$$

In the SM, the hypercharge assignments are *a priori* independent of the $SU(3) \times SU(2)_L$ assignments, although constrained by the fact that quantum consistency requires the resulting triangle anomalies to cancel. In a simple GUT group, the relation (70) is automatic: whenever Q is a generator of a simple gauge group, $\sum_R Q = 0$ for particles in any representation R, cf. the values of I_3 in any representation of $SU(2)$.

There are only two groups of rank 4 that have complex representations and hence are suitable *a priori* for GUTs, namely $SU(5)$ and $SU(3) \otimes SU(3)$. However, $SU(3) \otimes SU(3)$ does not allow simultaneously the leptons to have an integer electric charge and the quarks to have a fractional electric charge. Moreover, if one tried to use $SU(3) \times SU(3)$, one would need to embed the electroweak gauge group in the second $SU(3)$ factor. This would be possible only if $\sum_q Q_q = 0 = \sum_\ell Q_\ell$, which is not the case for the known quarks and leptons. Therefore, attention has focussed on $SU(5)$ [88] as the only possible rank 4 GUT group.

The group $SU(5)$ is the simplest GUT group capable of including the SM. Other possible GUT groups have higher rank, and groups that are commonly used are $SO(10)$, the only suitable simple group of rank 5 with complex representations, and the exceptional group E_6 of rank 6. As examples that may help understand the new physics that appears when the symmetry of the SM is enhanced, we are first going to study key aspects of the group $SU(5)$ and then, more briefly, some aspects of the group $SO(10)$.

As in the SM, particles must be arranged in suitable representations of $SU(5)$. This group has a fundamental spinorial representation of dimension 5 and a 2-index antisymmetric spinorial representation of dimension 10. Together they are suitable for accommodating the fermions of a given generation, which consist of $3 \times 2 \times 2 = 12$ quarks + 2 charged leptons + 1 neutrino. To see how this may be done, we first decompose the smallest representations of $SU(5)$ in terms of representations of $SU(3) \otimes SU(2)$:

$$\bar{\mathbf{5}} = (\bar{\mathbf{3}}, \mathbf{1}) + (\mathbf{1}, \mathbf{2}), \tag{71}$$

$$\mathbf{10} = (\bar{\mathbf{3}}, \mathbf{1}) + (\mathbf{3}, \mathbf{2}) + (\mathbf{1}, \mathbf{1}). \tag{72}$$

For example, in (71) the representation $\bar{\mathbf{5}}$ of $SU(5)$ can accommodate a colour antitriplet that is also an $SU(2)$ singlet, and a colour singlet that is also an $SU(2)$ doublet. In addition, it is necessary that the sum of the charges in each of these two multiplets be zero. The only possible combination of first-generation fermions in the SM is:

$$\mathbf{5} : (\psi_i)_L = \begin{pmatrix} \bar{d}_1 \\ \bar{d}_2 \\ \bar{d}_3 \\ e^- \\ -\nu_e \end{pmatrix}_L , \qquad (73)$$

and the rest of the first-generation fermions may be accommodated uniquely, as follows:

$$\mathbf{10} : (\chi^{ij})_L = \frac{1}{\sqrt{2}} \begin{pmatrix} 0 & \bar{u}_3 & -\bar{u}_2 & u_1 & d_1 \\ -\bar{u}_3 & 0 & \bar{u}_1 & u_2 & d_2 \\ u_2 & -\bar{u}_1 & 0 & u_3 & d_3 \\ -u_1 & -u_2 & -u_3 & 0 & e^+ \\ -d_1 & -d_2 & -d_3 & -e^+ & 0 \end{pmatrix}_L , \qquad (74)$$

where we neglect the eventual mixings between the fermions in different generations. We must repeat the previous classification of fermions in $\mathbf{10} + \bar{\mathbf{5}}$ representations for the other two generations: there is no explanation in $SU(5)$ for the presence of three generations.[19]

After discussing the matter fermions, now we discuss the GUT gauge bosons. Groups of type $SU(N)$ have $N^2 - 1$ symmetry generators in an adjoint representation (e.g., $SU(3)_C$ has 8 gluons, $SU(2)$ has 2 W bosons, etc.), so that $SU(5)$ has 24 gauge bosons. Of these 24 gauge bosons, 12 correspond to the SM gluons, $W^\pm$, Z^0 and γ, and 12 are new. Decomposing this 24-dimensional adjoint representation into representations of $SU(3) \otimes SU(2) \otimes U(1)$, we find:

$$\mathbf{24} = \underbrace{(\mathbf{3}, \mathbf{2}, \tfrac{5}{3}) \oplus (\bar{\mathbf{3}}, \mathbf{2}, -\tfrac{5}{3})}_{new\ bosons} \oplus \underbrace{(\mathbf{8}, \mathbf{1}, 0)}_{gluons\ G_a} \oplus \underbrace{(\mathbf{1}, \mathbf{3}, 0)}_{W_i} \oplus \underbrace{(\mathbf{1}, \mathbf{1}, 0)}_{B}. \qquad (75)$$

where the third numbers in the parentheses are the hypercharges of the multiplets. The new bosons, called X and Y, have electric charges 4/3 and 2/3, respectively, carry leptoquark quantum numbers, are coloured and have

---

[19]The pairing of $\bar{\mathbf{5}}$ and $\mathbf{10}$ representations is free of triangle anomalies.

isospin $1/2$.[20] In matrix notation,

$$A = \sum_{a=1}^{24} T_a A^a = \begin{pmatrix} G_i & G_i & G_i & \bar{X} & \bar{Y} \\ G_i & G_i & G_i & \bar{X} & \bar{Y} \\ G_i & G_i & G_i & \bar{X} & \bar{Y} \\ X & X & X & W_i & W_i \\ Y & Y & Y & W_i & W_i \end{pmatrix}, \tag{76}$$

where the T_a are the generators of $SU(5)$ represented by 5×5 matrices (the equivalents for $SU(5)$ of the Pauli matrices of $SU(2)$). The basis is chosen so that $SU(3)_C$ corresponds to the first 3 lines and columns, and $SU(2)_L$ to the last 2 lines. The top-left and bottom-right blocks therefore contain the gluons and W bosons, respectively, and the $U(1)$ boson B (not shown) corresponds to a traceless diagonal generator.

The remaining steps in constructing an $SU(5)$ GUT are the choices of representations for Higgs bosons, first to break $SU(5) \to SU(3) \times SU(2) \times U(1)$ and subsequently to break the electroweak $SU(2) \times U(1)_Y \to U(1)_{em}$. The simplest choice for the first stage is an adjoint **24** of Higgs bosons Φ with a v.e.v.

$$<0|\Phi|0> = \begin{pmatrix} 1 & 0 & 0 & \vdots & 0 & 0 \\ 0 & 1 & 0 & \vdots & 0 & 0 \\ 0 & 0 & 1 & \vdots & 0 & 0 \\ \cdots & \cdots & \cdots & \cdots & \cdots & \cdots \\ 0 & 0 & 0 & \vdots & -\frac{3}{2} & 0 \\ 0 & 0 & 0 & \vdots & 0 & -\frac{3}{2} \end{pmatrix} \times \mathcal{O}(m_{GUT}). \tag{77}$$

It is easy to see that this v.e.v. preserves colour $SU(3)$, which reshuffles the first three rows and columns; weak $SU(2)$, which reshuffles the last two rows and columns; and the hypercharge $U(1)$, which is a diagonal generator. The subsequent breaking of $SU(2) \times U(1)_Y \to U(1)_{em}$ is most economically accomplished by a **5** representation of Higgs bosons H:

$$<0|\phi|0> = (0,0,0,0,1) \times 0(m_W). \tag{78}$$

It is clear that this v.e.v. has an $SU(4)$ symmetry which yields [85] the relation $m_b = m_\tau$ before renormalisation that leads, after renormalisation (69), to a successful prediction for m_b in terms of m_τ. However, the same trick does not work for the first two generations, indicating a need for epicycles in this simplest GUT model.

Making the minimal $SU(5)$ GUT supersymmetric, as motivated by the naturalness of the gauge hierarchy, is not difficult [74]. One must replace the

---

[20]They have direct interactions with quarks and leptons, which we discuss in the next section.

above GUT multiplets by supermultiplets: $\bar{5}$ $\bar{F}$ and 10 T for the matter particles, 24 Φ for the GUT Higgs fields that break $SU(5) \to SU(3) \times SU(2) \times U(1)$. The only complication is that one needs both 5 and $\bar{5}$ Higgs representations H and $\bar{H}$ to break $SU(2) \times U(1)_Y \to U(1)_{em}$, just as two doublets were needed in the MSSM to cancel anomalies and give masses to all the matter fermions. The simplest possible form of the Higgs potential is specified by the superpotential [74]:

$$W = (\mu + \frac{3\lambda}{2}M) + \lambda \bar{H} \Phi H + f(\Phi), \tag{79}$$

where $\mu = \mathcal{O}(1)$ TeV and $M = \mathcal{O}(M_{GUT})$, and $f(\Phi)$ is chosen so that $\partial f / \partial \Phi = 0$ when

$$< 0|\Phi|0 >= M \begin{pmatrix} 1 & 0 & 0 & \vdots & 0 & 0 \\ 0 & 1 & 0 & \vdots & 0 & 0 \\ 0 & 0 & 1 & \vdots & 0 & 0 \\ \cdots & \cdots & \cdots & \cdots & \cdots & \cdots \\ 0 & 0 & 0 & \vdots & -\frac{3}{2} & 0 \\ 0 & 0 & 0 & \vdots & 0 & -\frac{3}{2} \end{pmatrix}. \tag{80}$$

Inserting this into the second term of (79), one finds terms $\lambda M \bar{H}_3 H_3$ and $-3/2 \lambda M \bar{H}_2 H_2$ for the colour-triplet and weak-doublet components of $\bar{H}$ and H, respectively. Combined with the bizarre coefficient of the first term, these lead to terms

$$W \ni (\mu + \frac{5\lambda}{2}M) \bar{H}_3 H_3 + \mu \bar{H}_2 H_2. \tag{81}$$

Thus we have heavy Higgs triplets with masses $\mathcal{O}(M_{GUT})$ and light Higgs doublets with masses $\mathcal{O}(\mu)$. However, this requires fine tuning the coefficient of the first term in W (79) to about 1 part in 10^{13}! In the absence of supersymmetry, such fine tuning would be destroyed by quantum loop corrections [86].

A primary advantage of supersymmetry is that its no-renormalisation theorems guarantee that this fine tuning is *natural*, in the sense that quantum corrections do not destroy it, unlike the situation without supersymmetry. On the other hand, supersymmetry alone does not explain the *origin* of the hierarchy. A second advantage of supersymmetry, as we saw earlier in this section, is that it would make possible a much more precise unification of the gauge couplings. However, a potential snag is that the exchanges of the supersymmetric partners of the heavy Higgs triplets $\bar{H}_3, H_3$ may cause rapid proton decay, as discussed later.

Another possible GUT group that is frequently studied is $SO(10)$ [88,89]. It is a group of rank 5 that contains $SU(5) \otimes U(1)$. The principal advantage of $SO(10)$ over $SU(5)$ is that it possesses a fundamental spinorial representation of dimension 16 that can accommodate all the fermions of one generation, as well as a singlet right-handed neutrino, thanks to its decomposition in terms of $SU(5)$ representations[21]

$$\mathbf{16} = \mathbf{10} \oplus \bar{\mathbf{5}} \oplus \mathbf{1}. \tag{82}$$

The appearance of an $SU(5)$ singlet provides a natural framework for the physics of the neutrinos and the seesaw mechanism.[22] In $SO(10)$ the number of gauge bosons rises to 45, which include 33 additional gauge bosons beyond the SM, and therefore many possible interactions, including additional options for proton decay. In addition, the breaking of $SO(10)$ is more complicated than that of $SU(5)$, because it is done in two steps. One may pass from $SO(10)$ to $SU(5) \otimes U(1)$ or $SU(4) \otimes SU(2)_L \otimes SU(2)_R$, and then to $SU(2) \otimes U(1)$. The Higgs sector is potentially quite extensive, and may include large multiplets of dimensions 10, 16, 45, 54, 120 and 126, depending on the model.

4.2   Towards a Theory of Everything

4.2.1   Problems in Quantum Gravity

One of the most important unfinished tasks for understanding the Universe and the fundamental interactions is the unification of the two great theories of the twentieth century: general relativity and quantum mechanics. To write such a unified Theory of Everything is one of the major challenges to physicists in our century. The solution of the problem of the cosmological constant, for example, will have to find a place in the framework of such a Theory of Everything.

Gravity is a puzzle for conventional quantum theory, in particular because uncontrollable, non-renormalisable infinities appear when one tries to calculate Feynman diagrams that contain loops with gravitons. These correction terms diverge increasingly rapidly as the order of the perturbative calculation increases, essentially because the coupling of gravity has negative mass dimensionality, being $\propto 1/M_P^2$, where $M_P \simeq 1.2 \times 10^{19}$ GeV.

There are also non-perturbative problems in the quantization of gravity, that first appeared in connection with black holes. We recall that a black hole is a non-perturbative solution of the equations of General Relativity, in which the curvature of space-time induced by gravitational forces becomes so strong that no particle can escape the event horizon. The existence of this horizon is linked to the existence of entropy S and a non-zero temperature T of the black hole. From the pioneering work of Bekenstein and Hawking [90] on black hole thermodynamics, we know that the mass of a black hole is proportional to

---

[21]The $SO(10)$ group is anomaly free, so this decomposition explains finally the freedom from anomalies of $SU(5)$ and the SM.

[22]In $SU(5)$, singlet right-handed neutrinos could be added "by hand," in which case they would have no gauge interactions. In the case of $SO(10)$, the gauge interactions of $SO(10)$ do not have any direct influence on accessible neutrino phenomenology, but may provide interesting restrictions on their Yukawa interactions.

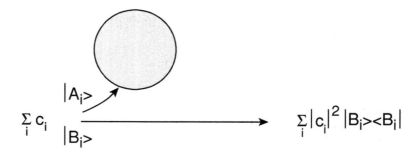

Figure 24. *If a pair of particles $|A\rangle$ $|B\rangle$ is produced near the horizon of a black hole, and one of them ($|A\rangle$, say) falls in, the remaining particle $|B\rangle$ will appear to be in a mixed state, since the state of $|A\rangle$ is unobservable.*

the surface area A of its horizon, which is related in turn to its entropy:

$$S = \frac{1}{4}\,A \tag{83}$$

The appearance of non-zero entropy means that the quantum description of a black hole must involve mixed states. The intuition underlying this feature is that information can be lost through the event horizon. To see how this may happen, consider, for example, a pure quantum-mechanical pair state $|A,B\rangle \equiv \sum_i c_i |A_i\rangle |B_i\rangle$ prepared near the horizon, and what happens if one of the particles, say A, falls through the horizon while B escapes, as seen in Fig. 24. In this case, all the information about the component $|A_i\rangle$ of the wave function is lost, so that

$$\sum_i c_i |A_i B_i\rangle \rightarrow \sum_i |c_i|^2 |B_i\rangle\langle B_i| \tag{84}$$

and B emerges in a mixed state, as in Hawking's original treatment of the black hole radiation that bears his name [90]. The problem is that conventional quantum mechanics does not permit the evolution of a pure initial state into a mixed final state.

For a discussion of these and other open problems in quantum black hole physics, see [91]. Many theorists consider that these problems point to a fundamental conflict between the proudest achievements of early twentieth-century physics, namely quantum mechanics and general relativity. One or the other should be modified, and perhaps both. Since quantum mechanics is sacred to field theorists, most particle physicists prefer to modify General Relativity by elevating it to string theory, as we now discuss.

4.2.2 Introduction to String Theory

As was just mentioned, one of the major issues of quantum gravity is that it has an infinite number of infinities. These divergences can be traced to the absence of a short-distance cut-off in conventional field theories, where the particles are points. The problem is that one can in principle approach infinitely near a point particle, giving rise to interactions of infinite strength:

$$\int^{\Lambda \to \infty} d^4 k \left(\frac{1}{k^2} \right) \leftrightarrow \int_{1/\Lambda \to 0} d^4 x \left(\frac{1}{x^6} \right) \sim \Lambda^2 \to \infty. \tag{85}$$

Such divergences can be avoided or removed if one replaces point particles by extended objects. The simplest possibility is to extend in just one dimension, leading to a theory of strings. In such a theory, instead of point particles moving along one-dimensional world lines, one has strings moving over two-dimensional world sheets. Historically, closed loops of string have been the most popular, and the corresponding world sheet would be tubes. The "wiring diagrams" generated by the Feynman rules of conventional point-like particle theories become "plumbing circuits" generated by the junctions and connections of these tubes of closed string. One could imagine generalizing this idea to higher-dimensional extended objects such as membranes describing world volumes, etc., and we return later to this option.

Among consistent string theories, one may enumerate the following. The *Bosonic String* exists in 26 dimensions, but this is not even its worst problem! It contains no fermionic matter degrees of freedom, and the flat-space vacuum is intrinsically unstable. *Superstrings* exist in 10 dimensions, have fermionic matter and also a stable flat-space vacuum. On the other hand, the ten-dimensional theory is left-right symmetric, and the incorporation of parity violation in four dimensions is not trivial. The *Heterotic String* was originally formulated in 10 dimensions, with parity violation already incorporated, since the left and right movers were treated differently. This theory also has a stable vacuum, but still suffers from the disadvantage of having too many dimensions. *Four-Dimensional Heterotic Strings* may be obtained either by compactifying the six surplus dimensions: $10 = 4$ plus 6 compact dimensions with size $R \sim 1/m_P$, or by direct construction in four dimensions, replacing the missing dimensions by other internal degrees of freedom such as fermions or group manifolds or ...? In this way it was possible to incorporate a GUT-like gauge group or even something resembling the SM.

What are the general features of such string models? First, they predict there are no more than 10 dimensions, which agrees with the observed number of 4! Secondly, they suggest that the rank of the four-dimensional gauge group should not be very large, in agreement with the rank 4 of the SM![23] Thirdly, the simplest four-dimensional string models do not accommodate large matter representations, such as an **8** of SU(3) or a **3** of SU(2), again in agreement with the known representation structure of the SM! Fourthly, simple string models predict fairly successfully the mass of the top quark, from the requirement that the theory make sense at all energies up to the Planck mass. Fifthly, string theory makes a fairly successful prediction for the gauge unification

---

[23] However, the number of gauge symmetries may be enhanced by non-perturbative effects.

scale in terms of m_P. If the intrinsic string coupling g_s is weak, one predicts

$$M_{GUT} = O(g) \times \frac{m_P}{\sqrt{8\pi}} \simeq \text{few} \times 10^{17}\text{GeV}, \qquad (86)$$

where g is the gauge coupling, which is $\mathcal{O}(20)$ higher than the value calculated on the basis of LEP measurement of the gauge couplings. Nevertheless, it would be nice to obtain closer agreement, and this provides the major motivation for considering strongly coupled string theory, which corresponds to a large internal dimension $l > m_{GUT}^{-1}$, as we discuss next.

4.2.3 M Theory

As was already said, the bosonic string model has many more disadvantages than other models. It has 26 dimensions, does not contain fermions, and has an unstable vacuum! Consequently, physicists focused on superstring models, of which five types exist:

- Type IIA, that reduces at low energy to a non-chiral $N = 2$ supergravity in $d = 10$ dimensions;

- Type IIB, that reduces at low energy to a chiral $N = 2$ supergravity in $d = 10$ dimensions;

- The Heterotic $E(8) \times E(8)$ theory, that reduces at low energy to an $N = 1$ supergravity in $d = 10$, connected to a Yang–Mills gauge theory with an $E(8) \times E(8)$ gauge group;

- The Heterotic theory $SO(32)$, that reduces at low energy to an $N = 1$ supergravity in $d = 10$, connected to a Yang–Mills gauge theory with an $SO(32)$ gauge group;

- Type I, that contains simultaneously opened and closed strings, and that reduces at low energy to an $N = 1$ supergravity in $d = 10$ connected to a Yang–Mills gauge theory with an $SO(32)$ gauge group.

These theories all look different. For example, the Type I theory is the only one that contains simultaneously open and closed strings, whereas the others contain only closed strings. In addition, the low-energy gauge structures of the five theories are different. It seems then, that we have five distinct theories that may describe gravity at the quantum level. How may we understand this? Is it possible that there is a link between the different theories?

Current developments involve going beyond string to consider higher-dimensional extended objects, such as generalized membranes with various numbers of internal dimensions. These can be regarded as solitons (non-perturbative classical solutions) of string theory [92], with masses

$$m \propto \frac{1}{g_s}, \qquad (87)$$

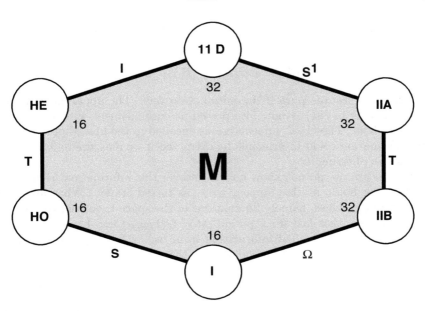

Figure 25. *The different limits of the M theory are joined by different duality relations. The numbers* 16 *and* 32 *are the numbers of spinor components in the theory.*

somewhat analogously to monopoles in gauge theory. It is evident from (87) that such membrane-solitons become light in the limit of strong string coupling: $g_s \to \infty$.

It was observed some time ago that there should be a strong-coupling/weak-coupling duality between elementary excitations and monopoles in supersymmetric gauge theories. These ideas were confirmed in a spectacular solution of $\mathcal{N} = 2$ supersymmetric gauge theory in four dimensions [93]. Similarly, it was shown that there are analogous dualities in string theory [94], whereby solitons in some strongly coupled string theory are equivalent to light string states in some other weakly coupled string theory. Indeed, it appears that all string theories are related by such dualities. A peculiarity of this discovery is that the string coupling strength g_s is related to an extra dimension, in such a way that its size $R \to \infty$ as $g_s \to \infty$. This then leads to the idea of an underlying 11-dimensional framework called M theory [63] that reduces to the different string theories in different strong-/weak-coupling linits, and reduces to 11-dimensional supergravity in the low-energy limit (see Fig. 25).

A particular class of string solitons called D-branes offers a promising approach to the black hole information paradox mentioned previously. According to this picture, black holes are viewed as solitonic balls of string, and their entropy simply counts the number of internal string states. These are in principle countable, so string theory may provide an accounting system for the information contained in black holes. Within this framework, the previously paradoxical process (84) becomes

$$|A, B\rangle + |BH\rangle \rightarrow |B'\rangle + |BH'\rangle \qquad (88)$$

and the final state is pure if the initial state was. The apparent entropy of the final state in (84) is now interpreted as entanglement with the state of the black hole. The "lost" information is encoded in the black-hole state, and this information could in principle be extracted if we measured all properties of this ball of string [95].

In practice, we do not know how to recover this information from macroscopic black holes, so they appear to us as mixed states. What about microscopic black holes, namely fluctuations in the space-time background with $\Delta E = O(m_P)$, that last for a period $\Delta t = O(1/m_P)$ and have a size $\Delta x = O(1/m_P)$? Do these steal information from us, or do they give it back to us when they decay? Most people think there is no microscopic leakage of information in this way, but not all of us [96] are convinced. The neutral kaon system is among the most sensitive experimental areas for testing this speculative possibility.

How large might be the extra dimension in M theory? Remember that the naïve string unification scale (86) is about 20 times larger than m_{GUT} as inferred from LEP data. If one wants to maintain consistency of LEP data with supersymmetric GUTs, it seems that the extra dimension may be relatively large, with size $L_{11} \gg 1/m_{GUT} \simeq 1/10^{16}$ GeV $\gg 1/m_P$ [97]. This may be traced to the fact that the gravitational interaction strength, although growing rapidly as a power of energy

$$\sigma_G \sim E^2/m_P^4, \qquad (89)$$

is still much smaller than the gauge coupling strength at $E = m_{GUT}$. However, if an extra space-time dimension appears at an energy $E < m_{GUT}$, the gravitational interaction strength grows faster, as indicated in Fig. 26. Unification with gravity around 10^{16} GeV then becomes possible, *if* the gauge couplings do not also acquire a similar higher-dimensional kick. Thus we are led to the startling capacitor-plate framework for fundamental physics shown in Fig. 27.

Each capacitor plate is *a priori* 10-dimensional, and the bulk space between them is *a priori* 11-dimensional. Six dimensions are compactified on a scale $L_6 \sim 1/m_{GUT}$, leaving a theory which is effectively five-dimensional in the bulk and four-dimensional on the walls. Conventional gauge interactions and observable matter particles are hypothesized to live on one capacitor plate, and there are other hidden gauge interactions and matter particles living on the other plate. The fifth dimension has a characteristic size which is estimated to be $\mathcal{O}(10^{12} \text{ to } 10^{13} \text{ GeV})^{-1}$. Physics at smaller energies (large distances) looks effectively four dimensional, whereas gravitational physics at larger energies (smaller distances) looks five dimensional, and the strength of

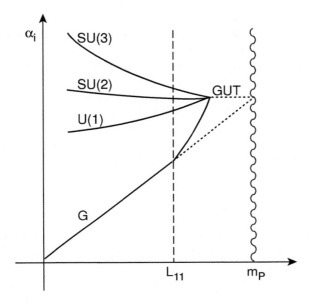

Figure 26. *Sketch of the possible evolution of the gauge couplings and the gravitational coupling G: if there is a large fifth dimension with size $\gg m_{GUT}^{-1}$, G may be unified with the gauge couplings at the GUT scale [97].*

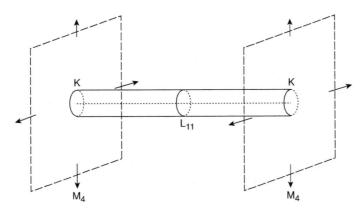

Figure 27. *The capacitor-plate scenario favoured in 11-dimensional M theory. The eleventh dimension has a size $L_{11} \gg M_{GUT}^{-1}$, whereas dimensions 5, ..., 10 are compactified on a small manifold K with characteristic size $\sim M_{GUT}^{-1}$. The remaining four dimensions form (approximately) a flat Minkowski space M_4 [97].*

the gravitational coupling rises rapidly to unify with the gauge couplings. Supersymmetry breaking is expected to originate on the hidden capacitor plate in this scenario, and to be transmitted to the observable wall by gravitational-strength interactions in the bulk.

The phenomenological richness of this speculative M-theory approach is only beginning to be explored, and it remains to be seen whether it offers a realistic phenomenological description. However, it does embody all the available theoretical wisdom as well as offering the prospect of unifying all the observable gauge interactions with gravity at a single effective scale $\sim m_{GUT}$, including the interactions of the SM. As such, it constitutes our best contemporary guess about the Theory of Everything within and beyond the SM.

4.3   Extra Dimensions

We have seen that string theories suggest that there may be extra unseen dimensions of space, but this speculation did not originate with string theorists. The idea of extra dimensions was first developed by Kaluza [61] and Klein [62]. They noticed that gravitational and electromagnetic interactions, being so alike in many ways, could be descendants of a common ancestor. Indeed, if we formulate a theory with extra spatial dimensions, is possible to unify gravity and electromagnetism. In the same way, non-Abelian gauge fields can be unified with Einstein's gravity in more complicated models with extra dimensions. Thus, the first reason why extra dimensions were studied was to unify the gravitational and gauge interactions. These initial discussions concerned gravitation at the classical level. If you want to quantize gravity, you would be well advised to look at the best available candidate, namely string or M theory, which, as we have seen, can be formulated consistently in a space with six or seven extra dimensions. From this point of view, the quantization of gravitational interactions becomes a second reason for extra dimensions.

In all the scenarios considered above, the extra were very small, close to the Planck size or perhaps somewhat larger, but undetectable in conceivable experiments.

However, it was suggested by Antoniadis [98] that an extra dimension might be a good way to break supersymmetry, in which case its size would be $\sim 1/$ TeV, in which case it might have some observable manifestations at the LHC.

Another suggestion, discussed in Section 2, was the possibility that boundary conditions in an extra dimension might be used to break the electroweak gauge symmetry. In this case also, the size of the extra dimension should be $\sim 1/$ TeV, and potentially detectable at the LHC [58–60].

Arkani-Hamed, Dimopoulos and Dvali (ADD) [99] went even further, observing that the Higgs mass hierarchy problem might be addressed in models with large extra dimensions, if they were of a millimetre or micron in size.

Because the extra dimensions are so large in the ADD framework, their effects might be measurable even in low-energy table-top experiments. These models can be embedded in string theory framework, as discussed in [100]. The main ingredients of the simplest ADD scenario are [101]:

- The particles of the SM live on a 3-brane, while gravity spreads to all $4+N$ dimensions;

- There is a new fundamental scale of gravity in extra dimensions, M_*, which together with the ultraviolet completion scale of the SM is around a few TeV or so, thus eliminating the Higgs mass hierarchy problem;

- N extra dimensions are compactified.

If we define in this context the four-dimensional Planck mass

$$M_{Pl}^2 = M_*^{2+N}(2\pi L)^N, \tag{90}$$

and postulate that the quantum gravity scale $M_* \sim$ TeV, we can estimate the size of the extra dimensions to be

$$L \sim 10^{-17+30/N} \text{ cm}. \tag{91}$$

For one extra dimension, $N = 1$, we obtain $L \sim 10^{13}$ cm, which is excluded within the ADD framework, because gravity would have become higher-dimensional at distances $\sim 10^{13}$ cm. On the other hand, for $N = 2$ we get $L \sim 10^{-2}$ cm. This case is very interesting, because it predicts a modification of the four-dimensional laws of gravity at submillimetre distances, which has become the subject of active experimental studies [101]. For larger N, the value of L should decrease but, even for $N = 6$, L is very large compared to $1/M_P$.

Randall and Sundrum (RS) went much further still [102], showing that a model with an *infinite* warped extra dimension could provide an attractive way to reformulate the hierarchy problem. In this scenario, four-dimensional gravity on a brane is obtained through the phenomenon of localization of gravity. The brane is embedded in a five-dimension bulk space with negative cosmological constant. In this case we find a relation between the four-dimensional Planck mass and M_*,

$$M_{Pl}^2 = M_*^3(2L). \tag{92}$$

This is similar to the relation between the fundamental scale M_*, the size L of the extra dimension and the Planck mass M_P in the ADD model with one extra dimension (90). This similarity is based on the fact that in both theories the effective size of the extra dimension that is felt by the zero-mode graviton is finite and $\sim L$.

So, are extra dimensions very small, small, large or infinite, and how do we tell? There are several ways to search for extra dimensions in experiments at the TeV scale at the LHC.

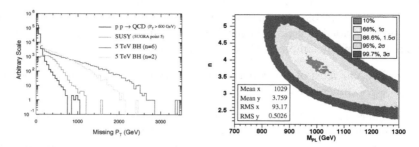

Figure 28. *Left: A comparison of the missing transverse momentum spectra in the SM, in a typical supersymmetric model, and in two black hole scenarios. Right: The results of a fit to the number of extra dimensions n and the higher-dimensional Planck mass M_{PL} on the basis of simulated black hole production at the LHC, taken from [104].*

Typical examples in theories with TeV-scale extra dimensions are the appearance of Kaluza–Klein excitations, corresponding to particle wave functions that wrap themselves around the extra dimension. These show up as resonances that can appear in cross sections at specific energies related to the compactification scale. These Kaluza–Klein excitations occur in "towers" that can be understood by analogy with a quantum-mechanical particle in a potential well. Its energy is quantized due to the boundary conditions at the walls of the well. In our case, the supplementary dimension plays the role of the wall of the well.

In models with very large extra dimensions, there are many Kaluza–Klein excitations of the graviton, which may be detectable via missing-energy events.

Another speculative possibility is the creation of a microscopic black hole [103]. Any concentration of energy or mass m will be transformed into a black hole if it is squeezed below its Schwarzschild radius: G/m. The larger the mass, the easier it can be squeezed below its Schwarzschild radius. Moreover, as we have seen, extra dimensions can increase the value of G. Hence, if there are a few extra dimensions of sufficient size, it is conceivable that collisions in the LHC might squeeze a pair of partons below their combined Schwarzschild radius, and hence create a microscopic black hole. These should evaporate rapidly, since Hawking radiation implies that the black hole loses energy at a rate inversely proportional to its mass. Studies performed by the CMS [81] and ATLAS [82] collaborations have demonstrated that such Hawking radiation would be visible in LHC via energetic jets, leptons and photons, as well as missing energy carried away by neutrinos. See Fig. 28 for some results for simulated black hole production at the LHC [104].

4.4   And Now for Something Completely Different?

In 1982, Prime Minister Thatcher of the United Kingdom visited CERN. I was placed in the receiving line, and introduced as a theoretical physicist. "So what do theoretical physicists *do*?" she boomed. I replied, "We think of things for the experimentalists to look for, and we hope they find something different." Mrs Thatcher was not sure about this, and asked, "Wouldn't it be better if they found what you had predicted?" My response was, "In that case, we would not be learning anything new." In the same spirit, let us hope that new experiments, particularly at the LHC, will soon reveal new physics beyond the Standard Model. Perhaps it will look something like the possibilities discussed in these lectures, but let us hope that it will take us beyond the beyonds imagined by theorists.

References

[1] P. Q. Hung and C. Quigg, Science **210** (1980) 1205.

[2] S. Weinberg, Int. J. Mod. Phys. A **23** (2008) 1627.

[3] C. Quigg, Ann. Rev. Nucl. Part. Sci. **59** (2009) 505 [arXiv:0905.3187 [hep-ph]].

[4] F. J. Hasert *et al.* [Gargamelle Collaboration], Phys. Lett. B **46** (1973) 121; Phys. Lett. B **46** (1973) 138.

[5] G. Arnison *et al.* [UA1 Collaboration], Phys. Lett. B **122** (1983) 103.

[6] M. Banner *et al.* [UA2 Collaboration], Phys. Lett. B **122** (1983) 476.

[7] P. Bagnaia *et al.* [UA2 Collaboration], Phys. Lett. B **129** (1983) 130.

[8] C. Rubbia, Rev. Mod. Phys. **57** (1985) 699.

[9] J. R. Ellis, arXiv:hep-ph/9812235.

[10] J. R. Ellis, arXiv:hep-ph/0203114.

[11] J. Welzel, D. Gherson and J. R. Ellis, arXiv:hep-ph/0506163.

[12] C. Amsler *et al.* [Particle Data Group], Phys. Lett. B **667** (2008) 1.

[13] ALEPH, CDF, D0, DELPHI, L3, OPAL and SLD Collaborations, LEP and Tevatron Electroweak Working Groups, SLD Electroweak and Heavy Flavour Groups, arXiv:0911.2604.

[14] H. Flacher, M. Goebel, J. Haller, A. Hocker, K. Moenig and J. Stelzer, Eur. Phys. J. C **60** (2009) 543 [arXiv:0811.0009 [hep-ph]].

[15] J. R. Ellis and G. L. Fogli, Phys. Lett. B **231** (1989) 189.

[16] J. R. Ellis, G. L. Fogli and E. Lisi, Phys. Lett. B **274** (1992) 456.

[17] S. Schael *et al.* [ALEPH, DELPHI, L3, OPAL Collaborations and LEP Working Group for Higgs boson searches], Eur. Phys. J. C **47** (2006) 547 [arXiv:hep-ex/0602042].

[18] B. W. Lee, C. Quigg and H. B. Thacker, Phys. Rev. Lett. **38** (1977) 883; Phys. Rev. D **16** (1977) 1519.

[19] Tevatron New Phenomena & Higgs Working Group, arXiv:0911.3930; http://tevnphwg.fnal.gov/results/SM_Higgs_Fall_09/.

[20] J. Ellis, J. R. Espinosa, G. F. Giudice, A. Hoecker and A. Riotto, Phys. Lett. B **679** (2009) 369 [arXiv:0906.0954 [hep-ph]].

[21] J. R. Ellis and D. Ross, Phys. Lett. B **506** (2001) 331 [arXiv:hep-ph/0012067].

[22] A. G. Riess *et al.* [Supernova Search Team Collaboration], Astron. J. **116** (1998) 1009 [arXiv:astro-ph/9805201]; S. Perlmutter *et al.* [Supernova Cosmology Project Collaboration], Astrophys. J. **517** (1999) 565 [arXiv:astro-ph/9812133].

[23] A. Dobado and A. L. Maroto, Astrophys. Space Sci. **320** (2009) 167 [arXiv:0802.1873 [astro-ph]].

[24] A. Harvey, Eur. J. Phys. **30** (2009) 877.

[25] J. Dunkley *et al.* [WMAP Collaboration], Astrophys. J. Suppl. **180** (2009) 306 [arXiv:0803.0586 [astro-ph]]; E. Komatsu *et al.* [WMAP Collaboration], Astrophys. J. Suppl. **180** (2009) 330 [arXiv:0803.0547 [astro-ph]].

[26] J. Ellis, J.S. Hagelin, D.V. Nanopoulos, K.A. Olive and M. Srednicki, Nucl. Phys. B **238** (1984) 453; see also H. Goldberg, Phys. Rev. Lett. **50** (1983) 1419.

[27] D. N. Spergel *et al.* [WMAP Collaboration], Astrophys. J. Suppl. **170** (2007) 377 [arXiv:astro-ph/0603449].

[28] A. D. Sakharov, Pisma Zh. Eksp. Teor. Fiz. **5** (1967) 32 [JETP Lett. **5**, 24 (1967 SOPUA,34,392-393.1991 UFNAA,161,61-64.1991)].

[29] J. M. Cline, arXiv:hep-ph/0609145.

[30] A. Pilaftsis, J. Phys. Conf. Ser. **171** (2009 012017) [arXiv:0904.1182 [hep-ph]].

[31] P. A. M. Dirac, Proc. Roy. Soc. Lond. A **133** (1931) 60.

[32] P. A. M. Dirac, Phys. Rev. **74** (1948) 817.

[33] P. A. M. Dirac, in *Coral Gables 1976, Proceedings, New Pathways in High-Energy Physics, Vol. I*, New York 1976, 1-14.

[34] B. Cabrera, Phys. Rev. Lett. **48** (1982) 1378.

[35] J. H. Schwarz and N. Seiberg, Rev. Mod. Phys. **71** (1999) S112 [arXiv:hep-th/9803179].

[36] A. Ashtekar, Nuovo Cim. **122B** (2007) 135 [arXiv:gr-qc/0702030].

[37] K. D. Lane, arXiv:hep-ph/9401324.

[38] J. F. Gunion, H. E. Haber, G. Kane, and S. Dawson, *The Higgs Hunter's Guide*, Perseus Publishing, 1990.

[39] M. S. Chanowitz, Phys. Rev. D **66** (2002) 073002 [arXiv:hep-ph/0207123].

[40] G. P. Zeller *et al.* [NuTeV Collaboration], Phys. Rev. Lett. **88** (2002) 091802 [Erratum ibid. **90** (2003) 239902] [arXiv:hep-ex/0110059].

[41] R. Barbieri and A. Strumia, arXiv:hep-ph/0007265.

[42] For a review of little Higgs models, see M. Schmaltz and D. Tucker-Smith, Ann. Rev. Nucl. Part. Sci. **55** (2005) 229 [arXiv:hep-ph/0502182].

[43] N. Arkani-Hamed, A. G. Cohen and H. Georgi, Phys. Lett. B **513** (2001) 232 [arXiv:hep-ph/0105239].

[44] H. C. Cheng and I. Low, JHEP **0408** (2004) 061 [arXiv:hep-ph/0405243].

[45] M. Perelstein, Prog. Part. Nucl. Phys. **58** (2007) 247 [arXiv:hep-ph/0512128].

[46] E. Farhi and L. Susskind, Phys. Rept. **74** (1981) 277.

[47] C. T. Hill and E. H. Simmons, Phys. Rept. **381** (2003) 235 [Erratum ibid. **390** (2004) 553] [arXiv:hep-ph/0203079].

[48] S. Weinberg, Phys. Rev. D **19** (1979) 1277.

[49] L. Susskind, Phys. Rev. D **20** (1979) 2619.

[50] A. Martin, arXiv:0812.1841 [hep-ph].

[51] J. Bardeen, L. N. Cooper and J. R. Schrieffer, Phys. Rev. **108** (1957) 1175.

[52] T. Appelquist, M. Piai and R. Shrock, Phys. Rev. D **69** (2004) 015002 [arXiv:hep-ph/0308061].

[53] S. Dimopoulos and J. R. Ellis, Nucl. Phys. B **182** (1982) 505.

[54] J. R. Ellis, M. K. Gaillard, D. V. Nanopoulos and P. Sikivie, Nucl. Phys. B **182** (1981) 529.

[55] J. R. Ellis, G. L. Fogli and E. Lisi, Phys. Lett. B **343** (1995) 282.

[56] M. T. Frandsen, arXiv:0710.4333 [hep-ph].

[57] R. Contino, C. Grojean, M. Moretti, F. Piccinini and R. Rattazzi, JHEP **1005** (2010) 089 [arXiv:1002.1011 [hep-ph]].

[58] S. K. Rai, Int. J. Mod. Phys. A **23** (2008) 823 [arXiv:hep-ph/0510339].

[59] R. Barbieri, G. Marandella and M. Papucci, Phys. Rev. D **66** (2002) 095003 [arXiv:hep-ph/0205280].

[60] J. F. Gunion and B. Grzadkowski, arXiv:hep-ph/0004058.

[61] Th. Kaluza, Sitzungsber. Preuss. Akad. Wiss. Phys. Math. Klasse (1921) 996; Reprinted with an English translation in *Modern Kaluza-Klein Theories*, eds. T. Appelquist, A. Chodos and P.G.O. Freund (Addison-Wesley, Menlo Park, 1987).

[62] O. Klein, Z.F. Physik, **37** (1926) 895; Reprinted with an English translation in *Modern Kaluza–Klein Theories*, eds. T. Appelquist, A. Chodos and P.G.O. Freund (Addison-Wesley, Menlo Park, 1987).

[63] For a review, see: Miao Li, hep-th/9811019.

[64] H. C. Cheng, arXiv:0710.3407 [hep-ph].

[65] P. Fayet and S. Ferrara, Phys. Rept. **32** (1977) 249.

[66] H. P. Nilles, Phys. Rept. **110** (1984) 1.

[67] H. E. Haber and G. L. Kane, Phys. Rept. **117** (1985) 75.

[68] S. P. Martin, *A Supersymmetry Primer*, arXiv:hep-ph/9709356.

[69] Y. Okada, M. Yamaguchi and T. Yanagida, Prog. Theor. Phys. **85** (1991) 1; J. R. Ellis, G. Ridolfi and F. Zwirner, Phys. Lett. B **257** (1991) 83; H. E. Haber and R. Hempfling, Phys. Rev. Lett. **66** (1991) 1815.

[70] J. Ellis, S. Kelley and D.V. Nanopoulos, Phys. Lett. B **260** (1991) 131; U. Amaldi, W. de Boer and H. Furstenau, Phys. Lett. B **B260** (1991) 447; P. Langacker and M. Luo, Phys. Rev. D **44** (1991) 817; C. Giunti, C. W. Kim and U. W. Lee, Mod. Phys. Lett. A **6** (1991) 1745.

[71] M. B. Green, J. H. Schwarz and E. Witten, *Superstring Theory* (Cambridge Univ. Press, 1987).

[72] H. N. Brown *et al.* [Muon g-2 Collaboration], Phys. Rev. Lett. **86** (2001) 2227 [arXiv:hep-ex/0102017].

[73] M. Davier, A. Hoecker, B. Malaescu, C. Z. Yuan and Z. Zhang, Eur. Phys. J. **C66** (2010) 1 [arXiv:0908.4300 [hep-ph]].

[74] S. Dimopoulos and H. Georgi, Nucl. Phys. B **193** (1981) 150.

[75] J. R. Ellis, J. S. Lee and A. Pilaftsis, Phys. Rev. D **76** (2007) 115011 [arXiv:0708.2079 [hep-ph]].

[76] M. Carena, J. R. Ellis, A. Pilaftsis and C. E. Wagner, Nucl. Phys. B **586** (2000) 92 [arXiv:hep-ph/0003180]; Phys. Lett. B **495** (2000) 155 [arXiv:hep-ph/0009212]; and references therein.

[77] J. R. Ellis, K. A. Olive, Y. Santoso and V. C. Spanos, Phys. Lett. B **565** (2003) 176 [arXiv:hep-ph/0303043].

[78] O. Buchmueller *et al.*, Eur. Phys. J. C **64**, (2009) 391 [arXiv:0907.5568 [hep-ph]].

[79] T. Hahn, S. Heinemeyer, W. Hollik, H. Rzehak and G. Weiglein, Comput. Phys. Commun. **180** (2009) 1426.

[80] O. Buchmueller *et al.*, JHEP **0809** (2008) 117 [arXiv:0808.4128 [hep-ph]].

[81] G. L. Bayatian *et al.* [CMS Collaboration], J. Phys. G **34** (2007) 995.

[82] G. Aad *et al.* [The ATLAS Collaboration], arXiv:0901.0512 [hep-ex].

[83] H. Georgi, H. Quinn and S. Weinberg, Phys. Rev. Lett. **33** (1974) 451.

[84] J. Ellis and D. V. Nanopoulos, Nature **292** (1981) 436.

[85] M. Chanowitz, J. Ellis and M. K. Gaillard, Nucl. Phys. B **128** (1977) 506.

[86] A. J. Buras, J. Ellis, M. K. Gaillard and D. V. Nanopoulos, Nucl. Phys. B **135** (1978) 66.

[87] D. V. Nanopoulos and D. A. Ross, Phys. Lett. B **118** (1982) 99.

[88] H. Georgi and S.L. Glashow, Phys. Rev. Lett. **32** (1974) 438.

[89] H. Fritzsch and P. Minkowski, Annals Phys. **93** (1975) 193.

[90] J. Bekenstein, Phys. Rev. D **12** (1975) 3077;
S. Hawking, Comm. Math. Phys. **43** (1975) 199.

[91] A. Strominger, Nucl. Phys. Proc. Suppl. **192-193** (2009) 119 [arXiv:0906.1313 [hep-th]].

[92] J. Polchinski, Phys. Rev. Lett. **75** (1995) 4724 [arXiv:hep-th/9510017].

[93] N. Seiberg and E. Witten, Nucl. Phys. B **426** (1994) 19 [Erratum ibid. B **430** (1994) 485] [arXiv:hep-th/9407087].

[94] C. M. Hull and P. K. Townsend, Nucl. Phys. B **438** (1995) 109 [arXiv:hep-th/9410167].

[95] For a recent take on this, see S. B. Giddings, Class. Quant. Grav. **28** (2011) 025002 [arXiv:0911.3395 [hep-th]].

[96] J. Ellis, N. E. Mavromatos and D. V. Nanopoulos, Mod. Phys. Lett. **A10** (1995) 425 and references therein.

[97] P. Horava and E. Witten, Nucl. Phys. B **460** (1996) 506 and Nucl. Phys. B **475** (1996) 94;
P. Horava, Phys. Rev. D **54** (1996) 7561.

[98] I. Antoniadis, Phys. Lett. B **246** (1990) 377.

[99] N. Arkani-Hamed, S. Dimopoulos and G. R. Dvali, Phys. Lett. B **429** (1998) 263 [arXiv:hep-ph/9803315]; N. Arkani-Hamed, S. Dimopoulos and G. R. Dvali, Phys. Rev. D **59** (1999) 086004 [arXiv:hep-ph/9807344].

[100] I. Antoniadis, N. Arkani-Hamed, S. Dimopoulos and G. R. Dvali, Phys. Lett. B **436** (1998) 257 [arXiv:hep-ph/9804398].

[101] G. Gabadadze, arXiv:hep-ph/0308112.

[102] L. Randall and R. Sundrum, Phys. Rev. Lett. **83** (1999) 4690 [arXiv:hep-th/9906064].

[103] S. B. Giddings and S. D. Thomas, Phys. Rev. D **65** (2002) 056010 [arXiv:hep-ph/0106219];
S. Dimopoulos and G. L. Landsberg, Phys. Rev. Lett. **87** (2001) 161602 [arXiv:hep-ph/0106295].

[104] C. M. Harris, M. J. Palmer, M. A. Parker, P. Richardson, A. Sabetfakhri and B. R. Webber, JHEP **0505** (2005) 053 [arXiv:hep-ph/0411022].

Section II: The Large Hadron Collider

The LHC Accelerator: Performance and Technology Challenges

Philippe Lebrun

CERN, Geneva, Switzerland

1 Introduction

The Large Hadron Collider (LHC), now operating for physics at CERN, the European Laboratory for Particle Physics in Geneva (Switzerland), is the largest scientific instrument in the world. It accelerates intense proton and ion beams up to energies of 7 TeV (protons) or 2.759 TeV/nucleon (Pb ions), and brings them into collision at the heart of four large detectors located around its 26.7 km circumference (Figure 1). The twin accelerators are composed of eight arcs, 2987 m long, connected via 528-m-long straight sections. At the transition between arcs and straight sections, dispersion suppressors cancel horizontal dispersion arising in the arcs and help match the optics. In order to guide and focus its very rigid beams, the LHC makes use of several thousand high-field superconducting magnets, operating in superfluid helium below 2 K. Although well above the preceding state-of-the-art in terms of beam energy and luminosity, the LHC is built on the knowledge and experience gained at previous high-energy accelerators: physics of hadron colliders was developed at the CERN ISR, the SPS antiproton collider and the TeVatron at Fermilab, while the TeVatron, HERA at DESY, RHIC at Brookhaven National Laboratory and CEBAF at Thomas Jefferson National Laboratory pioneered the technology of superconducting accelerators and the use of superfluid helium cooling.

In the following we will try and demonstrate how the performance requirements of the LHC drive its technological choices, and conversely how advanced technology permits unprecedented performance within affordable boundary conditions and constraints. The most important such constraint was the re-use of the large tunnel which had previously housed the LEP electron-positron

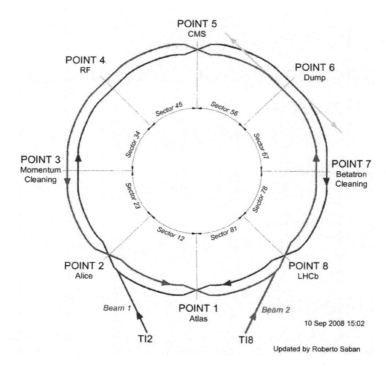

Figure 1. *Schematic layout of the LHC.*

collider, as well as of the CERN complex of particle accelerators used in the LHC injector chain.

The comprehensive reference document for the LHC main ring is the Design Report [1], also available on-line from the CERN web site, while reference [2] gives a good introduction to its accelerator physics and technology challenges. Table 1 lists the main design parameters of the machine.

2 Beam Energy

The design energy of LHC beams for physics is 7 TeV, thus allowing 14 TeV center-of-mass collisions: this is about seven times the highest energy previously available at the TeVatron.

The momentum p of a particle of charge e, bending field B and radius of curvature R in a circular accelerator are related by

$$p = eBR. \tag{1}$$

Circumference	26.7	km
Beam energy at collision	7	TeV
Beam energy at injection	0.45	TeV
Dipole field at 7 TeV	8.33	T
Luminosity	10^{34}	$cm^{-2}.s^{-1}$
Beam current	0.58	A
Protons per bunch	1.15×10^{11}	
Number of bunches per beam	2808	
Nominal bunch spacing	24.95	ns
Normalized emittance	3.75	μm.rad
Total crossing angle	285	μrad
Energy loss per turn	6.7	keV
Critical synchrotron energy	43.1	eV
Radiated power per beam	3.9	kW
Stored energy per beam	362	MJ
Stored energy in magnet circuits	11	MJ
Operating temperature	1.9	K

Table 1. *The main design parameters of the LHC (proton collider).*

Or, introducing numerical values,

$$p[\text{GeV/c}] \approx 0.3B[\text{T}]R[\text{m}]. \tag{2}$$

Hence, in order to reach 7 TeV/c momentum in the existing tunnel, with a radius of curvature of 2804 m, the LHC needs bending magnets with a field of 8.33 T. This is well above the saturation of iron, which sets the practical field limit in conventional magnets, and thus requires superconducting magnets.

Superconductors [3] are characterised by their "critical surface" in the temperature, magnetic field, current density space: the material remains in the superconducting state as long as its working point is below the critical surface. Conversely, superconductivity can be destroyed by increasing temperature, magnetic field or transport current density, or a combination of these. An electromagnet is a device which aims at producing magnetic field using current, therefore an appropriate representation of the working space is the current density versus magnetic field plane. In this plane, the critical surface projects itself as a set of critical lines, one for each operating temperature. At any temperature, the current-carrying capacity of the superconductor decreases with the magnetic field (Fig. 2).

The most common technical superconductors available today are alloys

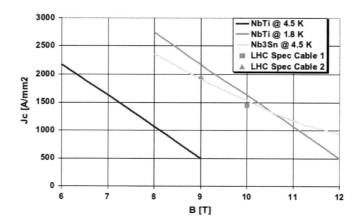

Figure 2. *Current density of technical superconductors and LHC design points.*

of Nb and Ti with about equal content of both elements. While adequate for building magnets operating around 5 T in normal boiling helium at 4.2 K, they lose most of their current-carrying capacity when one approaches 8 to 10 T. A solution, studied as a potential alternative in the early developments for the LHC, could have been to use other, more exotic superconductors such as Nb$_3$Sn, which conserve current-carrying capacity at high fields in normal liquid helium. These materials are however brittle and thus difficult to use in magnet windings. Moreover, they are expensive and were neither fully developed nor available in industrial quantities at the time the LHC project was launched.

The approach retained for the LHC was thus to rely on the well-proven, industrially available and affordable Nb-Ti alloy, and to improve its current-carrying capacity at high magnetic fields by operating it at lower temperature than 4.2 K. Operating at 1.8 K in superfluid helium shifts the critical curve by about 3 T to the right, thus rendering the material adequate for building magnets operating close to 10 T (Fig. 2). An important consequence, further discussed in Section 7, is the need to cool the magnets at lower temperature and design a more demanding cryogenic system operating with superfluid helium.

3   Luminosity

The luminosity of a collider is, for a given reaction, the event rate produced per unit of nuclear cross section. It is commonly expressed in $\mathrm{cm}^{-2}\mathrm{s}^{-1}$. For colliding bunched beams of equal characteristics, the luminosity is given by

$$L = \frac{N_b^2 n_b f_{\text{rev}} \gamma}{4\pi\varepsilon_n \beta^*} F, \tag{3}$$

where n_b is the number of bunches in each beam, N_b is the bunch population, f_{rev} is the revolution frequency, γ is the relativistic factor, ε_n and β^* are the normalized emittance and the beta function at the crossing point, respectively, and F is a geometrical factor linked to the crossing angle θ_c,

$$F = \frac{1}{\sqrt{1 + (\theta_c \sigma_z / 2\sigma^*)^2}}. \tag{4}$$

To maximize the luminosity for a given collider at any beam energy, one should therefore increase the bunch number and bunch population (and hence the circulating beam current), reduce the beam emittance and beta function at the collision point, and cross the beams at small angle. This can be done within limits set by different physical phenomena, which we will briefly discuss now.

The energy stored in the circulating beams is

$$E_{\text{beam}} = m_0 c^2 \gamma N_b n_b. \tag{5}$$

For the LHC nominal conditions, i.e., 2808 bunches of 1.1×10^{11} protons of 7 TeV in each beam, the stored energy amounts to 362 MJ, i.e., equivalent to some 80 kg of TNT. Limiting the beam stored energy is therefore an important design and operational issue. The luminosity formula can be rewritten as

$$L = \frac{N_b f_{\text{rev}}}{4\pi m_0 c^2 \varepsilon_n \beta^*} F.E_{\text{beam}} = \frac{1}{4\pi m_0 c^2} f_{\text{rev}} \frac{N_b}{\varepsilon_n} \frac{F}{\beta^*} E_{\text{beam}}. \tag{6}$$

For a given beam stored energy, increasing the luminosity can only be achieved by increasing the brilliance of the beams N_b/ε_n, that is controlled by the injector chain, or by crossing the beams at small angle and reducing the beta function at the collision point.

Another limit comes from the beam–beam effect, i.e., the perturbation of the particle trajectories in one beam by the electromagnetic field produced by the other. The beam-beam effect has two components, the "head-on" effect at the bunch crossings, and the "long-range" effect stemming from the small beam separation on either side of the crossing at small angle. These two components excite betatron resonances and generate tune spread, with the risk of the machine tune crossing low-order resonances and thus losing the beams.

The intensity of the beam–beam tune shift is characterized by the beam–beam parameter, ξ, given by

$$\xi = \frac{k N_b r_p}{4\pi\varepsilon_n} F < \xi_{\max}, \tag{7}$$

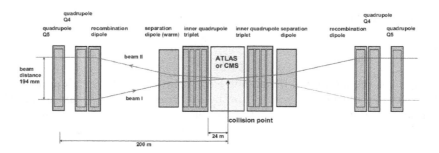

Figure 3. *Layout of high-luminosity collision region.*

where k is the number of crossing points around the machine, and r_p is the classical proton radius. A maximum value for ξ of 0.015 including all effects was derived empirically from experience of operating previous hadron colliders at the beam–beam limit.

The luminosity formula can now be rewritten showing explicitly the beam-beam parameter

$$L = \frac{\xi N_b n_b f_{\mathrm{rev}} \gamma}{k r_p \beta^*} \tag{8}$$

This yields the strategy for maximising luminosity, i.e., operating at the beam-beam limit, increasing the number of bunches and the bunch population within the limits given by beam stored energy (and beam losses), and decreasing the beta function at the collision point. In nominal conditions, each beam in the LHC will have 2808 bunches, each containing 1.15×10^{11} protons. Around the collision points (Fig. 3), beam separation/recombination dipoles permit adjusting the crossing angle at a typical value of 285 μrad, a compromise between loss of luminosity through the geometrical parameter F, the minimum spacing of the beams required to limit the long-range beam-beam effect and the aperture of the single vacuum chamber which has to accommodate both beams in the central part of the insertion. The vertical beta function, which oscillates between 30 m and 180 m in the arcs, is reduced to 0.55 m at the high-luminosity collision points by triplets of strong super-conducting quadrupole magnets installed on either side of the detector [4].

In the course of colliding-beam operation, the luminosity will decay with time due to the degradation of beam intensity and emittance, through several processes: intra-beam scattering, i.e., multiple Coulomb scattering between particles in the same bunch, nuclear scattering of particles by residual gas molecules, and the particle collisions proper. The luminosity decay can be characterized by a time constant τ_L,

$$\frac{1}{\tau_L} = \frac{1}{\tau_{\mathrm{IBS}}} + \frac{2}{\tau_{\mathrm{gas}}} + \frac{1}{\tau_{\mathrm{nuclear}}}, \tag{9}$$

where τ_{IBS}, τ_{gas} and τ_{nuclear} characterise the effects of intra-beam scattering, beam-gas scattering, and nuclear reactions in the collisions, respectively. τ_{nuclear} is inversely proportional to luminosity, and has an initial value of 45 h at nominal conditions. τ_{IBS} is about 80 h. In order not to degrade luminosity lifetime, τ_{gas} must be large with respect to the other characteristic times: the design value is 100 h, which sets the requirement on the quality of the beam vacuum (discussed in Section 8). With the above values, the total luminosity lifetime is about 15 h, which defines the typical operating pattern for the LHC in colliding beam operation: one or two fills per day, to be optimized by monitoring the luminosity decay.

4   Synchrotron Radiation and Electron Cloud

The LHC is the first proton machine in which synchrotron radiation shows tangible effects. While this phenomenon produces the beneficial effect of emittance damping, it also brings detrimental consequences to the operation of the accelerator.

The total synchrotron power radiated by the beam is proportional to the fourth power of the relativistic factor, and inversely proportional to the bending radius

$$P_{\text{syn}} = \frac{Z_0 e^2 c \gamma^4}{3R} N_b n_b f_{\text{rev}}, \tag{10}$$

where Z_0 is the impedance of free space. Note that the group, $N_b \, n_b \, f_{\text{rev}}$, is proportional to the circulating beam current. Each nominal LHC beam radiates 3.9 kW; while restoring this loss is not a big issue for the RF acceleration system, the power is radiated in a cryogenic environment, with 0.22 W/m in the arcs, a heat load comparable in value to the heat in-leak of the cryostats, which would thus strongly impact the sizing of the cryogenic refrigeration system if it were absorbed at the 1.9 K temperature level of the superconducting magnets.

The spectrum of the emitted radiation can be characterized by its median, the "critical" photon energy

$$u_c = \frac{3}{2} \hbar c \frac{\gamma^3}{R}. \tag{11}$$

For the LHC arcs, the critical energy is 43.1 eV; the photons are emitted in the ultra-violet range. They can be easily intercepted before they fall onto the 1.9 K level and deposit their energy with maximum thermodynamic impact. This is the primary role of the beam screen, a non-magnetic stainless steel tube lined with high-conductivity copper and cooled at 5 to 20 K by circulation of supercritical helium through capillaries, fitted into the magnet cold bore (Fig. 4).

Photons emitted by synchrotron radiation, hitting the surface of the beam screen, will extract electrons from the wall by photo-electric effect. Some of these electrons can be resonantly accelerated by the electrical field produced

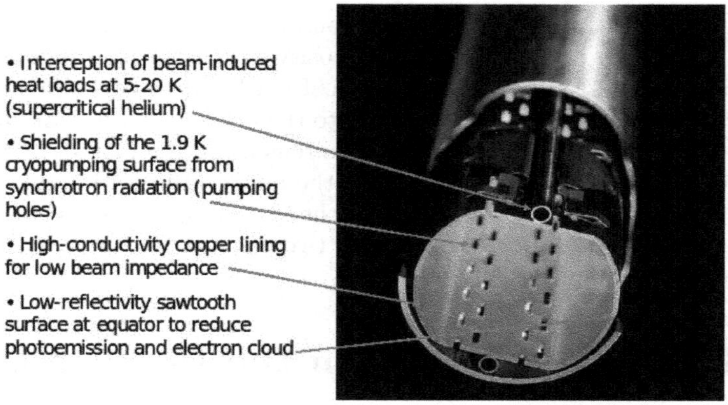

- Interception of beam-induced heat loads at 5-20 K (supercritical helium)
- Shielding of the 1.9 K cryopumping surface from synchrotron radiation (pumping holes)
- High-conductivity copper lining for low beam impedance
- Low-reflectivity sawtooth surface at equator to reduce photoemission and electron cloud

Figure 4. *The LHC beam screen.*

by the successive proton bunches, and in turn, they can extract secondary electrons from the walls, some of which will also be resonantly accelerated. This leads to the build-up of an electron cloud in the beam chamber, leading to beam instabilities and deposition of energy [5]. The intensity of the electron cloud is governed by the photon flux, which can be reduced by lowering the optical reflectivity of the wall—transverse saw-tooth indentations on the inner surface of the beam screen equator intercept photons at near normal incidence—the bunch repetition rate and the secondary electron emission yield. The latter depends on the wall material and surface coverage: it can be gradually improved by the synchrotron photons bombarding the wall, producing a "beam scrubbing" effect. In this fashion, the secondary electron emission yield can be reduced to 1.3, a value below which the development of the electron cloud is practically inhibited.

5   Field Quality and Dynamic Aperture

In superconducting magnets, the magnetic field is produced directly by the current in the windings (Biot–Savart law), and the field quality determined by the positioning precision of the conductors and not by the geometry of the iron yoke. As a consequence, the field quality cannot be as good as with "iron-dominated" magnets, in which the surfaces of the pole pieces coincide with equipotentials (as long as they are not saturated). The useable "good field" region is therefore substantially smaller than the geometrical aperture of the magnet. The variation of the magnetic field with position in the magnet aperture is conventionally represented by the components b_n and a_n of a Fourier series expansion, describing the field in a sum of multipole components of increasing order n

$$B_y + iB_x = B_1 \sum_{n=1}^{\infty} (b_n + ia_n) \left(\frac{x + iy}{r_{\text{ref}}} \right)^{n-1}. \tag{12}$$

The b_n and a_n are, respectively, called the normal and skew harmonics of the field. Tolerances on these harmonics are derived from beam physics criteria [6]: overall, a total relative field error smaller than 10^{-3} is required in the "good field" region of the magnets.

This leads to define the "dynamic aperture" of the accelerator as the aperture inside which the particle orbits are asymptotically stable. The dynamic aperture can be particularly critical at injection, when the beam has a bigger transverse size. In practice, the dynamic aperture is estimated by computer tracking of many sample particles around virtual models of the accelerator, with distributed random and systematic imperfections [7]. Computing resources set practical limits to the number of turns over which the sample particles are tracked: Fig. 5 shows simulation results over 10^6 turns, which only corresponds to about 100 seconds of accelerator operation. The results of such tracking simulations must then be corrected for many sources of uncertainties, from rounding errors in numerical calculations to residual ripple in power converters, and extrapolated to higher number of turns in order to obtain a safe value for dynamic aperture. Starting from the 10 to $12\,\sigma$ given by the simulations in Fig. 5, the remaining effective dynamic aperture after correction is estimated to only $6\,\sigma$. Dynamic aperture simulations are then used to define maximum allowable values for systematic and random variations of each field multipole. The variations which cannot be guaranteed by the reproducibility of magnet construction and powering need to be corrected by sets of correcting magnets, strategically distributed in the accelerator lattice.

Figure 6 shows a schematic layout of one cell in the LHC arcs, with a length of 106.9 m, a pattern which reproduces itself 23 times in each arc of the machine. At the ends of the main dipoles, sextupole and decapole/octupole correctors locally correct their multipole errors. In each half-cell, a short straight section houses the main focusing or defocusing quadrupole, a combined sextupole for correcting chromaticity and dipole for orbit correction, as well as tuning or skew quadrupoles, or an octupole for Landau damping of coherent beam oscillations. Apart from the individually powered orbit correction dipoles, all these magnets are grouped in families and powered in series. Three main powering circuits extend over the length of each arc, for the main dipoles and the main quadrupoles, focusing and defocusing.

6   Superconducting Magnets

There are more than 8000 superconducting magnets of all types in the LHC. In the following, we will, however, only discuss the 1232 twin-aperture main dipoles which occupy most of the machine circumference and constitute the

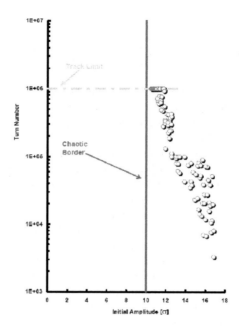

Figure 5. *Typical survival plot of dynamic aperture simulation.*

Schematic layout of one LHC cell (23 periods per arc)

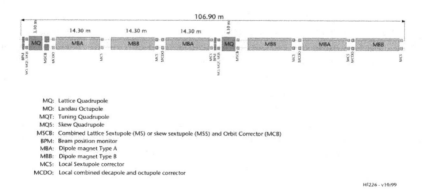

MQ: Lattice Quadrupole
MO: Landau Octupole
MQT: Tuning Quadrupole
MQS: Skew Quadrupole
MSCB: Combined Lattice Sextupole (MS) or skew sextupole (MSS) and Orbit Corrector (MCB)
BPM: Beam position monitor
MBA: Dipole magnet Type A
MBB: Dipole magnet Type B
MCS: Local Sextupole corrector
MCDO: Local combined decapole and octupole corrector

HF226 - v10/99

Figure 6. *Schematic layout of one LHC cell.*

largest part of the capital investment and the most challenging technical objects in the project [8].

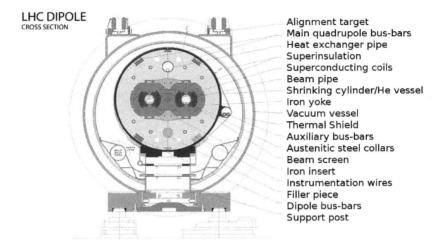

LHC DIPOLE
CROSS SECTION

Alignment target
Main quadrupole bus-bars
Heat exchanger pipe
Superinsulation
Superconducting coils
Beam pipe
Shrinking cylinder/He vessel
Iron yoke
Vacuum vessel
Thermal Shield
Auxiliary bus-bars
Austenitic steel collars
Beam screen
Iron insert
Instrumentation wires
Filler piece
Dipole bus-bars
Support post

Figure 7. *Transverse cross section of LHC dipole in its cryostat.*

The LHC needs two magnetic channels of equal and opposite field in order to guide the two counter-rotating particle beams in the arcs of the machine, before and after they are brought together at the collision points. An efficient and space-saving technical solution is to design twin-aperture dipole magnets sharing the same magnetic and mechanical structure, housed in a single cryostat (Fig. 7).

The current distribution which produces a uniform magnetic field is that of two intersecting ellipses with uniform current density, as shown in a simple argument by I. Rabi [9]. As this shape is not easy to realize by winding practical conductors, it is usually approximated by an annular layer of conductor with its current density varying as the cosine of the polar angle, hence the name of $\cos\theta$ for this magnet configuration. Finally, the $\cos\theta$ average current distribution is obtained by diluting blocks of conductor carrying fixed current density, by means of adequate spacers placed at strategic angular locations.

The dipole field produced by a single-block annular layer of thickness w, extending up to a polar angle of 60°, and carrying constant average current density j_{tech}, is

$$B = \frac{\mu_0\sqrt{3}}{\pi} j_{\text{tech}} w. \tag{13}$$

As j_{tech}, resulting from superconductor properties and coil filling factor, is a decreasing function of the field B, the width w of the annular layer may become prohibitive in high-field magnets, thus making it necessary to design coils with multiple nested layers. The LHC dipole coils are made of two layers wound from a cable with a width of 15.1 mm each. Although complicating design and construction of the magnets, this opens the way to use different cables for the two layers and thus to optimize the working points of each layer in the current density versus magnetic field plane: this technique, known as "current grading," permits the outer cable, which sees a lower magnetic field, to operate at higher current density while preserving equal distance to the critical curve, and therefore equal operating safety margin (Fig. 2).

Before being wound, the superconducting cable must be electrically insulated with an insulation system of adequate dielectric performance, resisting high forces and permeable to liquid helium. These conflicting requirements can be met by a double wrap of polyimide tapes, 50 % overlapped, covered by a final wrap of adhesive polyimide in "barber-pole" fashion, thus providing cooling channels in the gaps between successive turns.

Once the coil configuration is established, the next issue to be solved is that of the containment of the electromagnetic forces, exerted normal to the conductors and the magnetic field. These forces, essentially transverse over the length of the magnet, also have axial components at the coil ends. Their modulus is very high: it is easy to calculate that a single conductor, seeing a field of $10\,T$ and carrying a typical current of $10\,kA$, is submitted to a force of $10^5\,N$ per metre length. Thus the horizontal component acting transversely on the long sides of an LHC dipole coil at nominal field amounts to $3.6\,MN/m$. While the azimuthal component of the force can be transferred through the annular coil ("roman arch" principle), the radial component must be reacted by the magnet structure, in order to prevent deformations which could lead to field errors and to energy dissipation inducing resistive transitions ("quenches"). This is the role of the collars, precision-punched out of non-magnetic austenitic stainless steel sheet and keyed together, holding the coil assemblies and maintaining their geometry under pre-stress.

The coil-collar assembly is then inserted into a magnetic steel yoke, which closes most of the field lines within the magnet, thus improving its "transfer function" (the ratio of field produced to current in the windings) and reducing the stray field. At nominal excitation, the yoke contributes to about 12% of the total field in the LHC dipoles. By construction, the yoke is split and its two halves held together in an austenitic stainless steel shrinking cylinder, also acting as the outer wall of the magnet helium vessel. Thanks to shrinkage of the cylinder upon welding and differential contractions happening during cool-down, the gap in the yoke closes, and the two halves remain in contact during magnet excitation, thus rigidifying further the structure.

After manufacturing, the magnets are assembled into their cryostats (see Section 7), connected to test benches, cooled down and reception tested in normal operating conditions at $1.9\,K$. Based on several sorting criteria, geometrical and magnetic characteristics measured during reception tests the cryo-magnets are allocated to positions in the ring and installed there. One then proceeds to connect them in series, electrically and hydraulically. Electrical connections must be made with limited heating of the cable in order to preserve its superconducting properties. They must reliably show a very low residual resistance, which is achieved by brazing using induction heating in an automated machine recording all parameters of the operation, and periodically producing witness samples taken to the cryogenic laboratory for testing. Cryogenic pipe connections are made by TIG (Tungsten Inert Gas) welding using automated orbital machines, in order to guarantee quality and reproducibility. Overall, there are some 65,000 electrical connections, and 40,000 cryogenic pipe connections in the machine.

Superconducting magnets are metastable systems. While they inductively store a large magnetic energy (6.9 MJ for a 15 m LHC dipole, 1.1 GJ for a series-powered 3.3 km dipole string in every LHC sector), they operate, by design, close to the critical line. The temperature margin of the superconductor in an LHC dipole is only about 1.5 K, which, combined with the low specific heats of solid materials at cryogenic temperatures, yields specific quench energy of only about 10 mJ/cm^3. Superconducting magnet circuits therefore need active protection in case of resistive transitions.

Once a transition, or "quench," is detected in the winding of a magnet (appearance of a resistive voltage), heaters thermally coupled to the coils are powered by fast discharge of capacitor banks in order to spread the transition over the maximum volume of conductor, thus limiting the "hot spot" temperature resulting from Joule heating. At the same time, the power supply is switched off, and extraction resistors inserted in series with the magnets. In addition, the current in the circuit must bypass the quenched magnet by means of an automatic switching device, a cold high-current diode in parallel to its terminals. By the combined actions of this protection system, the magnets can withstand resistive transitions that, either intrinsic or provoked by beam losses, may happen during operation of the accelerator.

An overview of the diverse aspects of superconducting magnet design and construction is given in Ref. [10].

7   Cryogenics

The LHC cryogenic system is among the most powerful and complex ever built. It contains some 120 t of helium, mostly in the superfluid state inside the magnets [11].

The prime reason for superfluid helium cooling of the LHC superconducting magnets is the lower temperature of operation, which improves the current-carrying capacity of the Nb-Ti superconductor at high field and thus enables production of fields in the 8 to 10 T range. Operating at 1.9 K instead of the usual 4.2 K, however, has some drawbacks: it decreases the specific heat of the superconductor and other coil materials, making them more sensitive to thermal perturbations, and increases the thermodynamic cost of refrigeration, a particular concern in a large cryogenic system. It is therefore necessary to make full use of the peculiar transport properties of superfluid helium to counteract these effects [12].

The very low viscosity of superfluid helium allows it to permeate the coils where its high relative specific heat—2000 times that of the conductor per

LHC magnet string cooling scheme

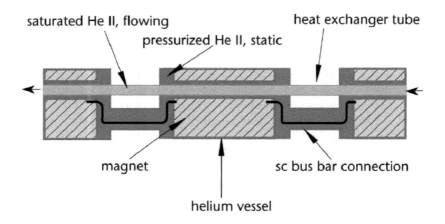

Figure 8. *Principle of LHC magnet cooling scheme.*

unit volume—locally buffers thermal perturbations, while its excellent thermal conductivity—peaking at 1000 times that of high-conductivity copper as long as the heat flux remains moderate—can evacuate the heat loads by conduction to the magnet bath outside the confined geometry of windings. Superfluid helium, combined with an appropriate design of the magnet coils and insulation, is therefore an essential ingredient of magnet stability.

The thermal conductivity of superfluid helium however remains finite and insufficient to transport heat over long distances, such as the length of a cryogenic sector of the LHC, 3.3 km from the furthest magnet to the refrigerator serving the sector. Moreover, limiting the thermodynamic penalty leads to allocation of only 0.1 K for heat transport, i.e., producing refrigeration at 1.8 K to keep all magnets at temperatures below or equal to 1.9 K. This is achieved by a cooling scheme combining the advantages of static pressurized superfluid helium in the magnets, with flowing saturated superfluid helium in a heat exchanger tube running over lengths of 109 m, along each magnet string (Fig. 8). In this fashion, the magnets operate in a quasi-isothermal bath of single-phase, high thermal conductivity liquid, while the latent heat of vaporisation of two-phase helium flowing in the tube absorbs the heat with very low temperature difference.

The low temperature and large size of the LHC magnet system impose a tight management of the heat loads. A guiding design principle is to try and intercept heat loads at the highest possible temperature level, in order to limit the residual load at 1.9 K. In the magnet cryostat (Fig. 7), the magnet, surrounded by an actively cooled shield at 50–70 K and wrapped with multilayer reflective insulation, is supported by low-conduction posts made of non-metallic glass-fiber/epoxy composite, with two levels of heat interception (at 50 K and 5 K). The beam-induced heat loads, e.g., synchrotron radiation, are intercepted by the beam screens installed in the magnet apertures, cooled by forced flow of supercritical helium between 5 K and 20 K. Construction of the cryostat favours industrial solutions, robust against variability of series production, to maintain thermal performance within tight limits.

Refrigeration power is provided by eight large cryogenic helium refrigerators of unit power equivalent to 18 kW at 4.5 K located around the LHC circumference. Their coefficient of performance, i.e., the ratio of work supplied to the refrigerator to cooling duty, reaches 230, corresponding to an efficiency of 28.5 % with respect to the Carnot cycle. They are complemented at their cold end by eight cold boxes at 2.4 kW and 1.8 K, using a combination of cold hydrodynamic and room-temperature volumetric compressors to produce the pressure ratio of 80 from the 16 mbar saturation pressure of helium at 1.8 K, up to atmospheric pressure. The measured coefficient of performance at 1.8 K is 930.

8 Beam Vacuum

With 54 km of beam pipes, the beam vacuum system of the LHC is unprecedented in size. A general description can be found in Ref. [13]. A recent overview of accelerator vacuum issues is presented in Ref. [14].

The beam-vacuum lifetime τ_{gas}, dominated by nuclear scattering of protons on the residual gas, must be greater than 100 h in order to limit beam decay and keep energy deposition by scattered protons to below 30 mW/m. This sets the maximum values of acceptable molecular densities for the different gas species present in the beam pipes,

$$\frac{1}{\tau_{gas}} = -\frac{1}{N}\frac{dN}{dt} = \nu \sum_i \sigma_i n_i, \tag{14}$$

where ν is the proton velocity, n_i the molecular density and σ_i the nuclear scattering cross section of species i, and the sum is over all species present. Since nuclear cross sections are in the range of hundreds of mbarn, this requires molecular gas densities to remain in the few $10^{14}\,\mathrm{m}^{-3}$. The partial pressure is related to the molecular gas density through

$$P_i = n_i kT. \tag{15}$$

The partial pressures must therefore remain in the few 10^{-9} to 10^{-8} mbar. In the absence of beam, this is easily achieved, for all species except helium, by condensation cryo-pumping onto the cold surface of the beam screen. In the presence of beam, beam-induced desorption of molecules trapped on the wall surface creates a dynamic situation, leading to a degradation of the vacuum.

The pumping holes in the beam screen (Fig. 4) help restore a low dynamic pressure, by allowing desorbed molecules to migrate to the cold bore that provides a surface at 1.9 K sheltered by the beam screen where they will be re-condensed.

For hydrogen, condensation cryo-pumping at 5 K is not sufficient to maintain a low enough residual pressure, and must be supplemented by cryosorption onto a mesh of carbon fiber attached to the outside of the beam screen. Finally, in the few kilometre lengths of beam pipe at room temperature, pumping is achieved *in situ* by a non-evaporable getter coating of the inner surface of the vacuum chambers, acting as a distributed pump.

9 Conclusion

This short presentation is aimed at addressing the main accelerator challenges specific to the LHC and explaining how they determine the choices made for the key technologies of superconducting magnets, cryogenics and vacuum. It is by nature incomplete, and many other accelerator systems, not even mentioned here, are essential to the operation of the machine. The interested reader is referred to the LHC Design Report [1] for a complete description.

Acknowledgement

The author is indebted to Lyn Evans for many discussions and invaluable help in the preparation of these lectures.

References

[1] O. Brüning *et al.*, (editors), LHC Design Report, Vol. 1, The LHC Main Ring, CERN-2004-003 (2004) http://lhc.web.cern.ch/lhc/LHC-DesignReport.html

[2] L. Evans, Proc. PAC 1999 New York, JACoW (1999) 21.

[3] P. Schmüser, Proc. CAS Superconductivity and Cryogenics for Accelerators and Detectors Erice, CERN-2004-008 (2004) 1.

[4] O. Brüning, W. Herr & R. Ostojic, LHC Project Report 315, CERN (1999).

[5] F. Zimmermann, Proc. ECLOUD'02, CERN-2002-001 (2002) 47.

[6] S. Fartoukh & O. Brüning, LHC Project Report 501, CERN (2001).

[7] M. Böge *et al.*, Proc. PAC 1997 Vancouver, JACoW (1997) 1356.

[8] L. Rossi, Trans. Applied Superconductivity **17** (2007) 1005.

[9] I. I. Rabi, Rev. Sci. Instr. **5** (1934) 78.

[10] M. N. Wilson, Superconducting magnets, Clarendon Press Oxford (1983).

[11] Ph. Lebrun, IEEE Trans. Applied Superconductivity **10** (2000) 1500.

[12] Ph. Lebrun & L. Tavian, Proc. CAS Superconductivity and Cryogenics for Accelerators and Detectors Erice, CERN-2004-008 (2004) 375.

[13] J. M. Jiménez, Vacuum 84 (2009) 2.

[14] Proc. CAS Vacuum in Accelerators Platja d'Aro, CERN-2007-003 (2007).

LHC Detectors and Early Physics

Günther Dissertori

ETH Zurich, Switzerland

1 Introduction

At the time of writing these proceedings (December 2009), we have just witnessed the successful re-start (after the sudden stop of operations in September 2008) of the world's most powerful particle accelerator ever built. CERN's Large Hadron Collider (LHC) has provided first collisions at injection energy (450 GeV per beam) and at the world's record energy of 1.18 TeV/beam. The centre-of-mass energy was increased to 7 TeV for the runs in 2010 and 2011.

The unprecedented energies and luminosities will give particle physicists the possibility to explore the TeV energy range for the first time and hopefully discover new phenomena, which go beyond the very successful Standard Model (SM). Among the most prominent new physics scenarios are the appearance of one (or several) Higgs bosons, of supersymmetric particles and of signatures for the existence of extra spatial dimensions.

In this review I will try to sketch the basic criteria and boundary conditions which have guided the design of the LHC detectors. The discussion will concentrate on the so-called general-purpose experiments, ATLAS and CMS. After an overview of the detector's characteristics and performance, I will elaborate on the expected measurements of hard processes, with emphasis on jet and vector boson production, i.e., tests of Quantum Chromodynamics (QCD) and Electroweak Physics.

There are many excellent reviews on the topics addressed here. Without making an attempt to be comprehensive, I would like to mention the various articles which can be found in a recent book edited by Kane and Pierce [1]. Extensive material has been compiled by the ATLAS and CMS collaborations in their Technical Design Reports on Physics performance [2,3], and detailed descriptions of the detectors are given in [4,5]. Finally, an excellent overview of the two detector's performance and their comparison can be found in [6].

2   Design Criteria for the Detectors

When designing large, *general-purpose* detector systems such as ATLAS and
CMS, which have to operate in collider mode at an accelerator such as the
LHC, obviously many aspects, constraints, boundary conditions, etc. have to
be taken into account. First of all, the expected physics and the accelerator's
parameters drive the design in a significant way. The overall layout of the
experiment is strongly determined by the choice of the magnet system(s).
Finally, the tracking, calorimeter, muon and data-acquisition systems are built
with the goal of optimal performance under the imposed boundary conditions
and with the available technologies. In the following I will address these
aspects and show how they have led to the experiments as they are built
today. Here the aim is not to go into every detail of the (technical) choices
of ATLAS and CMS, but rather to discuss, via some simple calculations, the
basic parameters and their role in the development of the detectors.

Before beginning the discussion, it is necessary to define the most relevant
and often used kinematic variables and relations. The transverse momentum p_T is defined as the component of a particle's (or jet's) momentum $\vec{p}$
perpendicular to the beam line, i.e., $p_T = p \sin\theta$, where $p = |\vec{p}|$ and θ is
the angle w.r.t. the beam line. When talking about energy (E) deposits
in calorimeters (or jets built out of them), the transverse energy is introduced, $E_T = E \sin\theta$. If the energy deposits are defined as vectors (by using
their directions w.r.t. the interaction point), their negative vector sum gives
the Missing Transverse Energy (MET, $\not{E}_T$). At hadron colliders the *rapidity* $y = 0.5 \ln\left[(E + p_L)/(E - p_L)\right]$ turns out to be a well-suited kinematic
variable, because differences in rapidity are invariant under boosts along
the beam direction. Here p_L denotes the momentum component along the
beam line. For massless particles the rapidity is equal to the *pseudorapidity* $\eta = -\ln(\tan(\theta/2))$.[1] Indeed, the detector elements at hadron colliders
are typically segmented in (pseudo-)rapidity intervals. Finally, the azimuthal
angle around the beam direction is usually denoted as ϕ or Φ.

With these definitions at hand, we can now look at the basic kinematic
relations relevant for the discussion of inelastic proton-proton scattering. Figure 1 shows the basic Feynman diagram for the description of inelastic hadron-
hadron scattering. Two partons (quarks or gluons), which carry a fraction $x_{1,2}$
of the incoming hadron's momenta $p_{1,2}$, interact at a momentum transfer (or
invariant mass) scale of $\hat{s} = x_1 x_2 s$, where s is the squared centre-of-mass energy of the incoming hadrons. In the case of a heavy particle (or resonance)
of mass M being produced in the interaction of the two partons, we need
$\hat{s} = M^2$. The energy and longitudinal momentum of the newly produced
state are given by $E = (x_1 + x_2)\sqrt{s}/2$ and $p_L = (x_1 - x_2)\sqrt{s}/2$ (we neglect
parton masses). By inserting this into the definition for the rapidity we obtain
the relations

---

[1] Thus, polar angles of $\theta = 1, 10, 90°$ correspond to $\eta = 5, 2.4, 0$.

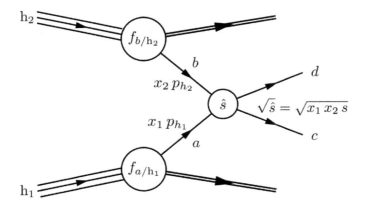

Figure 1. *Basic Feynman diagram for the description of inelastic hadron–hadron scattering.*

$$e^y = \sqrt{\frac{x_1}{x_2}} \quad , \quad x_{1,2} = \frac{M}{\sqrt{s}} e^{\pm y}, \tag{1}$$

where y is the rapidity of the particle with mass M. If M is very large (of the order of the total centre-of-mass energy), both momentum fractions have to be very large (and not too dissimilar), leading to $e^y \rightarrow 1$, i.e., such states are produced at so-called central rapidity ($y \rightarrow 0$). On the other hand, if the momentum fractions x_i are very different, e.g., one of them is much smaller than the other, then the produced final state will have a strong boost and appear at large (positive or negative) rapidities. As a consequence, the decay products of heavy states (or particles from reactions at very large momentum transfer) will tend to appear at smaller rapidities (and thus "more centrally," i.e., at larger polar angles) than those from the bulk of softer interactions.

2.1   Expected Physics

The main aspects of the physics to be expected can be understood by studying Fig. 2. There we find the cross sections for various processes as a function of the proton-proton centre-of-mass energy. The total inelastic cross section, of the order of 60–70 mb, is completely dominated by soft collisions at low momentum transfer. Typically this bulk of interactions is called *minimum bias events*, since they will be registered with a minimal set of trigger conditions. Obviously, since these are the processes with the highest cross section (probability), they will show up first and require only a minimal selection. Depending on the exact geometrical acceptance and the detailed definition of these minimal trigger settings, also single- and double-diffractive events will contribute here, whereas the measurement of purely elastic proton-proton scattering re-

proton-(anti)proton cross sections

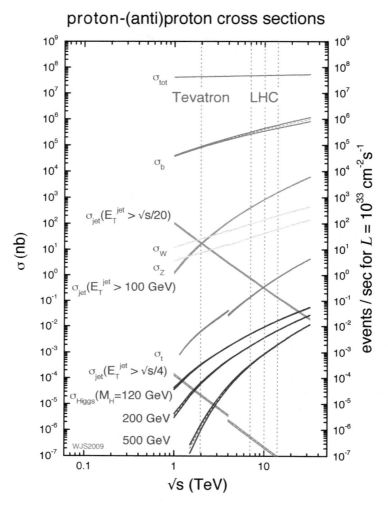

Figure 2. *Cross sections for various processes in proton-(anti)proton collisions as a function of centre-of-mass energy, √s (from [7]).*

lies on dedicated detectors in the far-forward part of the interaction region. It is worth noting that also hard-scattering events at large momentum transfer will pass the minimum bias trigger, and thus, in principle, are part of the minimum bias event sample. However, since their rate is so much smaller, the sample is basically populated by relatively soft collisions.

The average transverse momentum of charged particles produced in minimum bias events at 14 TeV centre-of-mass energy will be around 600 MeV and the most likely transverse momentum in the 200–300 MeV range (the exact values strongly depend on the model parameters, which are not well known so far, or on the extrapolations of fits to lower-energy data). From pure phase–space considerations, $d^3p/(2E) = \pi/2\,dp_T^2\,dy$, we expect a basically flat distribution of charged particles in rapidity, up to the kinematic boundaries of around 5 units in rapidity where there is a sharp drop. Whereas earlier measurements (as well as the very first LHC measurements at 900 GeV centre-of-mass energy, [8]) give slightly more than three charged particles per unit rapidity for the central plateau, at the higher LHC energies we expect up to six charged and 2–3 neutral particles per unit rapidity, uniformly distributed in ϕ. These numbers allow us to make the following conclusions:

- Roughly integrating the charged particle densities over the whole acceptance, we obtain $(2 \times y_{\max}) \times (6 + 3) \approx 90$ particles per minimum bias event, for a maximum rapidity coverage of $y_{\max} = 5$, and a total transverse momentum of $\approx 90 \times 0.6 = 54$ GeV. If several minimum bias events occur at the same beam crossing, these numbers have to be multiplied accordingly. Thus it is important to have detectors with excellent coverage (hermeticity) up to rapidities of around 5, in order to collect the largest fraction of these particles and thus to avoid spurious measurements of $\not{E}_T$;

- It will be important to actually measure these numbers and their energy dependence as quickly as possible, in order to tune the Monte Carlo predictions, which later will be used to model the contributions of such soft events to the energy flow in overlapping soft and hard-scattering proton-proton collisions;

- In a strong solenoidal magnetic field, many of these soft charged particles will become curling tracks ("loopers"). If we would like to avoid too many of the soft particles to reach the calorimeters, thus reducing the occupancy there, we can use the simple equation $p[\text{GeV}] = 0.3\,R[\text{m}]\,B[\text{T}]$ to derive a minimal radius $L \leq 2R = p/(0.15\,B) \approx 1$ m of a cylindrical tracking detector inside a calorimeter system. Here R denotes the bending radius of charged particles in a magnetic field B. We have used $p \approx 0.6$ GeV and $B = 4$ T. Stronger fields and/or larger tracking systems will keep the calorimeters cleaner from these soft-event contributions, but induce more loopers, thus higher occupancy and more difficult pattern/track recognition in the tracker.

The next most-likely events to appear will be collisions containing two or more jets. Depending on the minimal threshold in E_T, applied for triggering on such jets, the cross sections can grow up to hundreds of μb. More interestingly, even the cross section for the production of jets above several hundreds of GeV are in the tens-of-nb range and thus much larger than at the Tevatron. Therefore LHC will quickly extend the jet E_T range into uncharted territory and become sensitive to high-mass di-jet resonances, new contact interactions and/or quark compositeness. However, achieving such physics goals quickly will require an excellent understanding of jet reconstruction. This vast topic includes the usage of modern jet algorithms (for a recent overview we refer to [9]), the data-driven determination of the jet energy scale, the combination of calorimeter and tracker information for obtaining an optimal jet energy scale and resolution, the understanding of pile-up energy and underlying event contributions, etc. The resulting detector requirements are an excellent (granular) calorimeter system up to large rapidities, combined with a high-performance tracking detector if the overall (charged and neutral) particle-flow reconstruction is used as input to the jet finding.

Once the integrated luminosities allow sensitivity to processes with cross sections in the nb range, we start to explore the domain of electroweak physics, starting with the production of W bosons (plus anything), followed by Z and then top-pair production. The latter is up to 100 times larger than at the Tevatron, which justifies calling the LHC a top-factory. Since the cross sections for QCD jet production are many orders of magnitude larger than for these electroweak processes, it is obviously hopeless to look for W and Z decays into jets. Vector boson production is rather triggered on by their leptonic decays. This explains why a major effort went into the design of detectors with excellent lepton (electron, muon, tau) reconstruction capabilities. Further arguments for this can be found by looking at a list of benchmark processes, which were identified in the early days of the LHC planning to be well suited for the study of electroweak symmetry breaking, namely $pp \to W^+W^- \to \mu^+\nu_\mu\mu^-\bar{\nu}_\mu$, $pp \to ZZ \to \mu^+\mu^-\mu^+\mu^-$, $pp \to ZZ \to \mu^+\mu^-\nu_\mu\bar{\nu}_\mu$, $pp \to H \to \gamma\gamma$, $pp \to H$ jet jet (Vector Boson Fusion, VBF), as well as new physics such as $pp \to Z' \to \mu^+\mu^-$. All these processes have cross sections (times branching ratios) of order 1–100 fb, which immediately determines the necessary integrated luminosities for obtaining sensitivity to them. Again, this requires excellent lepton, photon and jet reconstruction. Since here the leptons (photons) are produced in decays of heavy objects, they will be triggered on by their relatively large transverse momentum (typically above 20 GeV). Furthermore, as we have seen above, they tend to be produced at central rapidities, thus the relevant sub-detectors, such as the electromagnetic calorimeter or the muon system, have to be optimized for and cover only a restricted (pseudo)-rapidity range, typically up to $\eta = 2.5$. Altogether the above considerations lead us to the following additional detector requirements:

- The detectors must be capable of triggering on and identifying extremely rare events, with cross sections some 10^{-14} times the total cross section. This requires an online rejection of $\sim 10^7$. Nevertheless, the overall event rates are very large and impose strong requirements on data handling and storage, with a yearly yield of $\sim 10^9$ events of a few MByte each;

- Leptons and photons should be well triggered on and measured with high resolution. This determines the performance of the electromagnetic calorimetry and the muon systems. In particular, an efficient electron reconstruction together with a jet fake rate[2] of about 10^{-5} has to be achieved, in order to cope with the enormous jet background. Furthermore, an excellent energy resolution of the electromagnetic calorimeter will ease the identification of resonances over a very large background, such as $H \to \gamma\gamma$;

- Many of the electroweak processes, such as top production, as well as new physics scenarios such as supersymmetric Higgs decays, involve the production of b-quarks and taus. This asks for silicon strip and pixel detectors, which give efficient b-tagging and tau identification. Here, jet rejection factors of at least ~ 100 are needed, for a b-tagging efficiency of $\sim 50\%$.

2.2 The LHC Parameters

Evidently, the LHC machine parameters play a fundamental role in the design of the experiments. The ultimate centre-of-mass energy of 14 TeV will be reached in several steps, currently the most likely scenario is $0.9 \to 2.36 \to 7 \to 10 \to 14$ TeV. The maximal energy is basically fixed by the radius of the LEP tunnel and the available superconducting magnet technologies. During the early planning phases it became a clear design goal that the lower energy compared to the SSC in the US (20 TeV/beam) had to be compensated by a much higher luminosity. Previously we have seen that processes related to the electroweak symmetry breaking have cross sections of order 1–100 fb. If we assume a canonical running time per year of $T = 10^7$ sec, we will need an instantaneous luminosity of $\mathcal{L} = 10^{34}$ cm^{-1}sec^{-1} $= 10^{-5}$ fb^{-1}sec^{-1} in order to accumulate $N = (\mathcal{L} \cdot T)\,\sigma \sim 100$ events per year for a process with cross section $\sigma = 1$ fb. Luminosities in this range can only be achieved by large bunch intensities ($\sim 10^{11}$ protons/bunch), a large number of bunches or equivalently a small bunch spacing (25 ns) and a small beam size at the interaction regions ($\mathcal{O}(15\text{–}20)\mu$m). It is worth noting that a simple multiplication of instantaneous peak luminosity and running time gives a too optimistic estimate for the integrated luminosity. Decreasing luminosities during a fill, the time needed for the filling and acceleration cycles, machine commissioning and development, as well as other down-time periods can be accounted for by a heuristic efficiency factor of roughly 0.2.

At such high luminosities of $\mathcal{O}(10^{34}$ cm^{-1}sec$^{-1})$, due to the very large total cross section, we expect the enormous rate of inelastic events of about $R = \sigma_{\text{inel}}\,\mathcal{L} \approx \mathcal{O}(100)$ mb $\times$ $(10^7$ mb^{-1}/sec$) = 10^9$ events per second. This allows us to calculate the number of inelastic events per bunch crossing, namely 10^9/sec $\times$ 25×10^{-9} sec $= 25$ events, which will pile-up on top of a possibly interesting high-p_T scattering event. These pile-up events will be soft proton-proton interactions, simply because it is the most likely thing to happen.

---

[2]This is the probability that a jet is misidentified as an electron by the reconstruction algorithms.

Previously we have estimated that about 90 particles are produced in such a soft collision, with a total p_T of order 50 GeV. Multiplying this by the number of pile-up events, we therefore expect more than 2000 particles carrying a total p_T of more than 1 TeV per bunch crossing. Good coverage and hermeticity become an even stronger requirement. At the same time, it is clear that dealing with these pile-up events in the detectors (and individual detector channels), as well as with the induced radiation levels, will lead to strong boundary conditions for the experiment's design. Indeed, at the LHC design luminosity we expect ionising doses of $\sim 2 \times 10^6$ Gy / $r_\perp^2$ /year, where $r_\perp$ is the transverse distance (in cm) to the beam. Damage can also be caused by photons created in electromagnetic showers and by the very high neutron fluences, in particular in the forward regions (up to 10^{17} n/cm^2 over 10 years of LHC running).

The high bunch-crossing frequency, combined with the large-sized detectors, imposes a further technical challenge, related to the timing of the trigger and readout. Interactions occur every 25 ns. During such an interval, the produced particles travel a distance of roughly 7.5 m. This is to be compared to a typical LHC detector radius exceeding 10 m, and overall half-length beyond 20 m. Thus, the particles created in one or two previous crossings have not yet left the detector when the next ones are produced at the interaction point. Furthermore, one has to consider that the electronic signals from the detectors travel some 5 m during a 25-ns interval, and typical cable lengths are of order 100 m.

Altogether, these challenging conditions ask for highly granular and radiation-hard detectors, combined with fast readout (20- to 50-ns response time). High granularity helps to minimize pile-up effects in a single detector element. However, detectors with many channels (e.g., about 100 million pixels, 200 000 cells in an electromagnetic calorimeter) represent a strong cost factor.

2.3   The Choice of the Magnet System

The layout of the magnet system is among the most important of all design choices, since it fixes many other parameters of the experiment. Therefore it also has to come very early on in the development. Of course, the main purpose of the magnetic field(s) is to bend charged particles and thus to determine their momentum and their charge sign. Furthermore, strong fields help to keep most of the soft particles (cf. Section 2.1) within cylindrical regions of small radius (see Fig. 3, right), thus reducing the occupancy and pile-up effects in the calorimeters. Concerning the momentum measurement, let's recall the most relevant formulas. Typically the momentum of a charged particle track is not determined directly from the bending radius, but from the sagitta s of a track's segment within a detector region of length L (see Fig. 3, left). Within the approximation $r \gg L/2$ we find $s = L^2/(8r)$. Then the momentum p and its relative uncertainty δp are given by

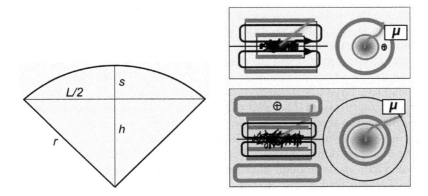

Figure 3. *Left: Definition of the sagitta s, as used in momentum measurements based on the curvature of charged particle tracks in magnetic fields; Right: Two possible magnetic field configurations for large particle detectors. On top a single large solenoidal field (CMS), on the bottom a smaller solenoid combined with a toroidal system (ATLAS). The field lines are in black, the coil windings are drawn in blue and the red line indicates a charged particle track. The black cross indicates a vector field orientation perpendicular to the drawing plane (from [10]).*

$$p = \frac{0.3\,L^2\,B}{8\,s} \quad \Rightarrow \quad \frac{\delta p}{p} = \frac{\delta s}{s} = \frac{8}{0.3}\frac{1}{L^2\,B}\,p\,\delta s, \tag{2}$$

where B is the magnetic field strength (expressed in teslas). We see that the measurement error can be minimized by maximizing the product $L^2\,B$, i.e., it is best to have large tracking systems (large lever arm) and strong fields. Obviously, both parameters will drive the overall cost of the experiment, in particular L. It is also worth noting that the total bending power is proportional to $\int B\,dl_\perp$, where $l_\perp$ is the particle's path perpendicular to the magnetic field. With these basic relations in hand, we can discuss the choices made by ATLAS and CMS (Fig. 3, right). A detailed comparison and technical parameters can be found, e.g., in table 3 of Ref. [6].

CMS has opted for a single magnet system based on a large, high-field superconducting solenoid and an instrumented iron return yoke. This has led to a simple and compact overall design (determining by the way also part of the experiment's name), giving excellent momentum resolution when combined with a powerful inner tracking system. The radius R of the solenoid was its main cost driver, and $R \sim 3$ m turned out to be affordable and technically doable. A magnetic field of 4 T was realizable, whereas 3.5 T would still deliver good physics performance. The current operating field of CMS is 3.8 T. A single solenoid has the disadvantage of limiting the momentum resolution in the forward direction, at large rapidities (remember $\int B\,dl_\perp$), and choosing a large winding radius implies making the solenoid also very long in order to cover the largest possible rapidity. The instrumented return yoke (iron) limits the momentum resolution at low p because of multiple scattering. This also impacts the trigger rates and the choice of the lower trigger thresholds. This is because the steeply falling momentum distributions, folded with a bad resolution, lead to a large feedthrough of soft particles beyond the trigger thresholds and thus to saturation. This is of particular concern for the stand-alone muon triggering at the foreseen large Super-LHC (SLHC) rates, since the multiple scattering effect cannot be overcome by installing additional and/or more precise muon tracking stations. A further important choice was to make the solenoid radius large enough in order to place the complete calorimetry (both electromagnetic and hadronic) inside the coil. Whereas this has the advantage of minimizing the amount of dead material in front of the calorimeter and therefore not compromising its intrinsic resolution, it implies that less than 2 m of radial extension are left for placing all absorbers, active calorimeter materials and readout, given that the tracker has a radius somewhat larger than 1 m. The implications are discussed in Section 2.5.

ATLAS is characterized by a two-magnet system, with a solenoid (smaller radius than in CMS) in the inner part and a large air-core toroid surrounding the calorimeters. In view of an optimal performance with muons, a toroid has the advantage of very large L^2B and good bending power also in the forward direction. The problems related to multiple scattering and its impact on muon stand-alone triggering in a high-rate environment, as discussed above, can be minimized by an air-core toroid, which has no return yoke. It also keeps the calorimeters free of field and leaves large enough space for them. However, it determines the extremely large overall size of ATLAS and leads to a rather complex structure and complex magnetic field configurations. The latter is of particular concern in view of precise particle tracking, which requires excellent knowledge of the field map. Many coils would give a more uniform field, but obviously increase the cost considerably. Indeed, the original proposal of 12 coils had to be reduced to 8 coils because of this. Furthermore, the large magnet structures require very large muon chambers, which then have to be aligned at precisions up to 30 μm, a formidable task. Because of the closed field lines around the calorimeters, a toroidal system has to be complemented by a solenoid which provides a field for inner tracking close to the interaction point. Since this solenoid is placed in front of the liquid-argon electromagnetic calorimeter, a lot of design work went into optimizing the materials and integrating the vacuum vessel for the coil with the cryostat for the barrel liquid-argon calorimeter, in order to minimize the dead material in front of the electromagnetic calorimeter. Also, the ATLAS solenoid is shorter than the one in CMS, which impacts somewhat the inner tracking performance at large rapidities because of the reduced field uniformity.

2.4 Tracking and Muon Systems

The basic requirements of a tracking system are:

- Allow for a robust and redundant pattern recognition, which is necessary for an efficient and precise reconstruction of all charged particles with momentum above $\sim$0.5 GeV, up to pseudorapidities of $\sim$2.5;

- Provide high-level triggering capabilities for electrons, taus and b-jets;

- Allow for an efficient and precise reconstruction of secondary vertices and impact parameters, which is of paramount importance for final states involving heavy flavours, in particular b-quarks;

- Complement and improve the electron reconstruction and triggering performance of the electromagnetic calorimeter by matching isolated tracks to calorimeter clusters;

- Provide some particle identification power, such as electron/pion separation, by a measurement of the specific ionization (dE/dx) or some other techniques, such as transition radiation.

When designing a tracking system, we have to deal with a fundamental problem of "conflicting interests": many tracking layers will provide many hits for a robust track reconstruction. However, many channels will also require a large number of supports (cables, cooling, support structures, etc.), which leads to a considerable amount of (dead) material in front of the calorimeters. This obviously jeopardizes the intrinsic resolution of the electromagnetic calorimeter, leads to a large fraction of photon conversions inside the tracker before reaching the calorimeter and causes multiple scattering of low-momentum particles. Indeed, the issue of the so-called material budget of the tracking systems has led to an "unfortunate similarity" between ATLAS and CMS, since in both cases the total material has increased by a factor of $\sim$2–2.5 from their approval in 1994 to now. The material distributions reach peak levels of $\sim$1.5–2 radiation lengths at rapidities around 1.5 (the transition region between barrel and endcap tracking stations is a preferred cable and support routing area). The consequences are that electrons lose between 25 and 70% of their energy by bremsstrahlung and that 20 to 60% of the photons convert into e^+e^- pairs before reaching the electromagnetic calorimeter. There exist algorithms for the recovery of energy loss by bremsstrahlung and for finding conversions, but obviously the overall reconstruction of electrons and photons is hampered by this.

Concerning the intrinsic resolution requirements of the tracking detectors, we can look again at Eq. (2) and plug in some typical numbers, such as $L = 1$ m and $B = 4$ T. If we aim for a relative momentum resolution of 1% at $p = 100$ GeV, we have to measure the sagitta with a precision of $\delta s = 15$ μm. Therefore the individual hit reconstruction precision should be in this ballpark and definitely not much worse than $\mathcal{O}(50\text{--}100)$ μm. Pattern recognition,

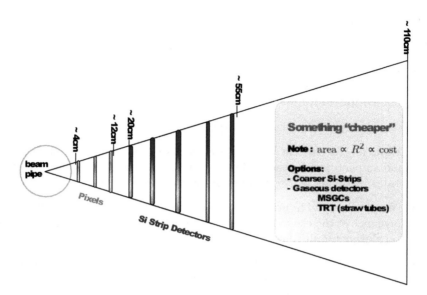

Figure 4. *Very generic layout of a possible tracking system for an LHC experiment.*

impact parameter resolution and the large soft-particle flux at small $r_\perp$ (cf. Section 2.2) ask for the smallest cell sizes and possibly three-dimensional hit reconstruction at the smallest radii. This optimizes the single-hit resolution and minimizes the occupancy and thus fake hit assignment. When going to larger radii, the requirements become less stringent, since the particle flux falls like $r_\perp^2$ and multiple scattering in the inner layers puts a natural limit on the achievable momentum resolution and thus on the necessary cell size (or pitch). On the other hand, the detector area grows with $r_\perp^2$, and with it the number of channels and ultimately the costs. A careful optimization of all these ingredients has led the two collaborations to go for tracker designs with a basic structure as depicted in Fig. 4. Extremely high-performance pixel and silicon strip detectors in the innermost regions, with unprecedented numbers of channels and overall areas, are complemented with a straw-tube tracker (ATLAS) or a larger-pitch silicon strip system (CMS) at larger radii. The straw-tube tracker gives a very large number of hits and in addition, via transition radiation layers, also provides an electron/pion separation. For the all-silicon choice of CMS it is worth noting that the outermost tracking layer has a similarly fine pitch like a few layers at intermediate radii. This provides better accuracy at the end of the lever arm than in ATLAS, hence a more than a factor of 2 better momentum resolution at $\eta = 0$. The CMS pixel detector is engineered in a manner which allows quick and easy installation and removal, a non-negligible feature in view of many years of LHC running and the necessary shut-down periods in between. A comparison of the final tracking systems of ATLAS and CMS can be found in tables 4 and 6, and of their performance in table 7 of Ref. [6]. In terms of momentum resolution the CMS tracker turns out to be superior to that of ATLAS (1.5% at $p = 100$ GeV and central rapidity, compared to 3.8%), in particular at larger rapidities. This is because of the stronger magnetic field and its better uniformity over the whole tracker region. On the other hand, the vertexing and b-tagging performances are similar, and the impact of the material and the larger B-field seem to be visible in slightly lower reconstruction efficiencies (80–85%) for low-momentum pions and electrons in CMS, compared to ATLAS (84–90%).

The tracking systems described above play an important role also for the momentum measurements of muons. However, when designing the muon systems, further requirements have to be taken into consideration:

- Muon momenta up to 1 TeV should be reconstructed at a precision of 10%, over a wide rapidity range;

- The mass of high-mass di-muon resonances, such as a hypothetical Z' with mass of $\mathcal{O}(1\,\text{TeV})$, should be reconstructed at a precision of 1%;

- Muon identification has to be performed in a very dense environment, and if needed the systems should be capable of triggering on and measuring muons above $p \sim 5$ GeV in stand-alone mode;

- Muon identification and triggering should be based on robustness and redundancy, with radiation-hard detectors and various readout speeds. This can be achieved by combining different technologies for the muon chambers.

The issue of stand-alone muon reconstruction has already been addressed above. Whereas the interaction point as additional constraint helps to achieve this in CMS, the multiple scattering poses some limitations, which are avoided by air-core toroids in ATLAS. The effect of multiple scattering on the momentum resolution can be modeled by

$$\frac{\delta p_{\text{ms}}}{p} \approx \frac{52\ 10^{-3}}{\beta\,B\,\sqrt{L\,x_0}}. \tag{3}$$

If we choose $\beta \sim 1$ for the particle's velocity, $B = 2$ T, $L = 2$ m and a radiation length of $x_0 = 0.14$ m, as is the case in the iron return yoke of CMS, we find a relative uncertainty of 5%, which places an absolute limit on the achievable resolution. Further issues are the unprecedented challenges to be faced by the alignment systems and the punch-through of pions from the calorimeters.

The ATLAS muon spectrometer excels by its stand-alone reconstruction and triggering capabilities and its large coverage ($|\eta| < 2.7$) in open geometry. At the same time, the complicated geometry and field configurations lead to large fluctuations in acceptance and performance over the full potential $\eta - \phi$

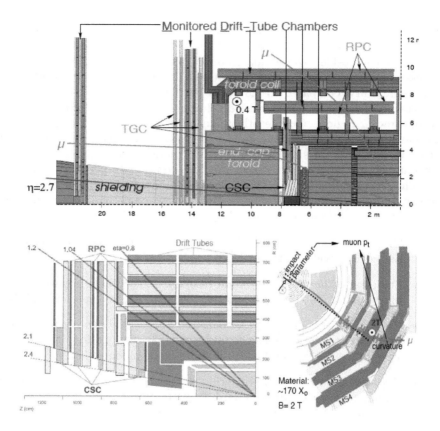

Figure 5. *Basic layouts of the muons systems in ATLAS (top) and CMS (bottom).*

area. The CMS muon system allows for a superior momentum resolution in the central detector regions, when combining the information from the inner tracker with the muon chambers. This overall excellent resolution is degraded in the forward regions ($|\eta| > 2$), where the solenoid bending power becomes insufficient. In addition, the limitations on stand-alone triggering under high-rate conditions have already been discussed. Again, a detailed comparison of the systems can be found in [6] (tables 11 and 12), and their basic layouts are depicted in Fig. 5.

2.5   Calorimetry

Calorimeters play a central role in the reconstruction and (transverse) energy measurement of electrons, photons and hadronic jets. Whereas the best energy/momentum resolution for low- and medium-p_T particles is obtained with spectrometers, at very high p_T the calorimeters take over, as can be seen from

$$\text{Calorimeter: } \frac{\delta E}{E} \propto \frac{1}{\sqrt{E}} \qquad \text{Spectrometer: } \frac{\delta p}{p} \propto p. \qquad (4)$$

This is also one of the reasons why calorimeters provide essential information for triggering on high-p_T objects. At LHC they have to absorb particles and jets with energies up to the TeV region. This has the following implications: The position of the maximum of a shower, which develops in a calorimeter, grows like $\ln E$, namely $x_{\max} \propto x_0 \ln(E/E_c)$, where x_0 and E_c are the radiation length and critical energy of an absorber material (see, e.g., [11]). In order to well contain an electromagnetic shower of order 1 TeV we need an absorber thickness of $\sim 25x_0$. Similarly, for containing a hadronic jet of 1 TeV we need roughly $11\lambda_0$. Here λ_0 is the effective interaction length of the system. Now let's take some concrete examples, such as lead-tungstate (PbWO$_4$) for the electromagnetic calorimeter (ECAL). We have $(x_0)_{\text{PbWO}_4} = 0.89$ cm, which implies that for this ECAL, leaving also some space for the readout electronics, we have to foresee at least some 50–60 cm of radial space. On the other hand, if we choose iron as the absorber material for the hadronic calorimeter (HCAL), we find $(\lambda_0)_{\text{Fe}} = 16.8$ cm and thus require about 180 cm of radial thickness in order to fully contain 1 TeV jets. However, if we remember the constraints in the CMS design, mostly given by the coil diameter and the tracker size, we observe that $R_{\text{coil}} - R_{\text{tracker}} - \text{ECAL}(+\text{electronics}) \sim 1$ m (cf. Fig. 6). Thus, instead of 11 interaction lengths there is only space for 6, or $7\lambda_0$ when counting also the ECAL material in front of HCAL. Indeed, a detailed analysis shows that the CMS coverage, for $|\eta| \lesssim 1$, is smaller than the qualitative requirement of $11\lambda_0$. In order to remedy this situation, a so-called *tail-catcher* (or HO=HCAL-Outer) is installed externally to the coil, just before the first muon stations. In the case of ATLAS no such problem exists. Because of the different magnet system, as discussed above, there is enough space left between the central solenoid and the external toroids to place electromagnetic and hadronic calorimeters of sufficient absorption lengths.

Further issues to be considered when choosing a calorimeter system are:

- Homogeneous vs. sampling calorimeter: whereas the former typically has the advantage of providing a better resolution (especially the stochastic term), a sampling calorimeter offers the possibility to measure also the longitudinal shower development at different sampling depths;

- The very-forward calorimeters (at pseudorapidities up to 5) can either be put at some larger distance in order to reduce the radiation load, or be kept as close as possible to the other calorimeter parts, thus giving better uniformity of the rapidity coverage. A careful choice of radiation-hard materials has to be made in this case;

- When choosing the projective calorimeter tower sizes the relevant parameters are the Molière radius and the expected and acceptable occupancy. For example, a very simple solution for an HCAL segmentation

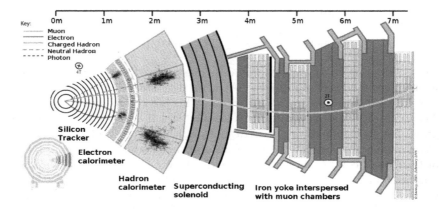

Figure 6. *Transverse slice through CMS, from [12].*

could be $\Delta\eta \times (\Delta\Phi/2\pi) = 0.1 \times 0.1$, which for a complete rapidity coverage of $2\,\eta_{max} = 10$ would lead to about 10 000 towers with the corresponding number of readout channels.

ATLAS and CMS have made some quite distinct choices in calorimeter technologies and layouts. Detailed comparisons and the main design parameters can be found in tables 8, 9 and 10 of Ref. [6]. A rough comparison is also given in Table 1 below.

2.6 Data Acquisition (DAQ)

When discussing the main LHC parameters in Section 2.2, we have already encountered some of the relevant numbers, which determine the design of the multi-level Trigger/DAQ architectures. The online requirements can be roughly summarized by a collision rate of 40 MHz, an event size of ~ 1 Mbyte, a Level-1 Trigger input of 40 MHz, a Level-2 (or High-Level) Trigger input of 100 kHz, a mass storage rate of ~ 100 Hz, thus an overall event rejection power of $\sim 10^7$ with a system dead-time not exceeding the per-cent level. Further DAQ design issues are a data network bandwidth (for the event builder) of $\sim$ TByte/sec, a computing power needed for the High-Level Trigger (HLT) of ~ 10 Tflop, corresponding to about 10 000 computing cores and a local storage need of ~ 300 TByte. The systems have to be robust, i.e., the operational efficiency should be independent of detector noise and machine conditions. In order to estimate the trigger efficiencies in a purely data-driven way (without Monte Carlo simulations), multiple overlapping triggers have to be carefully designed. Finally, also triggers and data streams for alignment and calibration have to be provided. A guiding principle in order to meet this formidable task was to minimize custom design as much as possible, and rather exploit the fast developments in data communication and computing technologies. Indeed, at

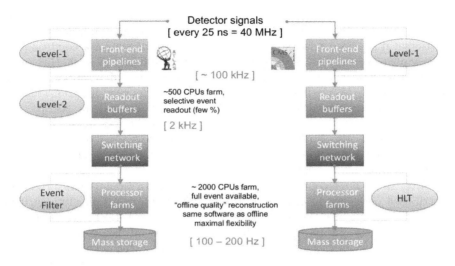

Figure 7. *Overview of the trigger and readout schemes in ATLAS and CMS, from [13].*

the early times some technologies, such as network switches, were not yet in the performance range required. However, the decision to count on and extrapolate the technology advances has paid off in the end.

The basic trigger schemes are depicted in Fig. 7. Both experiments have chosen similar approaches, with some differences however, in particular regarding the steps from the first level to the high-level trigger. Obviously, in both cases one starts with a channel data sampling at 40 MHz. The Level-1 trigger has to select events at a rate of 100 kHz, with decisions based on the identification of (relatively) high p_T electrons, muons, jets, as well as $\not{E}_T$. This is achieved by local pattern recognition and by energy estimates from prompt macro-granular information. The time budget for taking a decision is 3.2 μsec, i.e., 128 bunch crossings. This budget includes the time needed to transfer the signals to the central logic and back, as well as the time needed by the logic (implemented on custom-designed boards) itself. During this time, until a Level-1 "Accept" or "Reject" arrives, the signals are stored in a pipeline (readout buffers on the front-end boards) of at least 3 μsec length. All this requires a high-precision ($\sim$ 100 ps) timing, trigger and control distribution system.

The next level ("Level-2") differs somewhat between the experiments. Whereas in ATLAS it is a dedicated trigger stage, implemented on a farm of CPUs, in CMS it can be regarded as a first stage of a more general HLT processing. Indeed, in CMS the whole Level-1 output bandwidth has to be passed through a switching network, into a large farm of CPUs, where the events are processed in a parallel manner, i.e., one event per processor. At this stage the full event information is available and a reconstruction of almost "offline quality" is possible. This has the great advantage of full flexibility, in particular if the need arises to react to special machine/detector conditions or to new physics scenarios. The rate reduction from 100 kHz to ~ 100 Hz can be achieved with about 10000 cores and a time budget of 100 msec/event. The price to pay for this flexible and scale-free model is the need for an extremely high-performance network switch for the data distribution. In ATLAS the switch requirements are less demanding because of the intermediate Level-2 stage. However, this implies that only a very selective event readout (a few % of the whole event information) is processed here, on the basis of so-called regions of interest as identified at Level-1.

After an HLT "Accept," the events are fed into data streams which then are distributed onto several primary datasets. The latter are typically identified by a set of trigger bits. These datasets are transferred to the Tier-1 centres for a first reprocessing step and then further separated into smaller (secondary) datasets or special data skims, for user analysis at the Tier-2 and Tier-3 computing centres. All this is based on the LHC Computing Grid concept, explained in a dedicated set of lectures [14].

3   Overview of ATLAS and CMS

All the issues, requirements and boundary conditions, as sketched in the sections above, have been taken into account for the final designs of the two large general-purpose LHC experiments, ATLAS and CMS. A number of their most characteristic features have already been hinted at before, and therefore will not be repeated here. Their general layouts are drawn in Fig. 8, and a rough comparison of their most important components and their performance is given in Table 1. However, this table is only intended to give a quick and rough overview. For a much more detailed comparison, in particular regarding the expected and thus far observed performance, the relevant literature should be consulted, such as [6] or the set of publications and presentations, which have resulted from the commissioning with cosmic rays and first beams. Here it is also worth noting that the performance in terms of jet energy and $\not{E}_T$ resolution, as expected from pure calorimeter resolutions, can be considerably improved, especially at low p_T. This is possible via so-called *Particle Flow* approaches, where the information from the calorimeters is combined in an optimal way with the input from the other sub-detector systems, in particular the tracking. Such an algorithm attempts to reconstruct all particles in an event and thus extract the maximal information from the available data.

There is one more design feature, which has not been discussed before and which distinguishes the two experiments quite considerably. CMS is strongly characterized by a very modular design, which has been a guiding principle from its conception onwards. It was clear that it was not possible to build 13-m-long muon chambers, which would cover the whole barrel. This led to

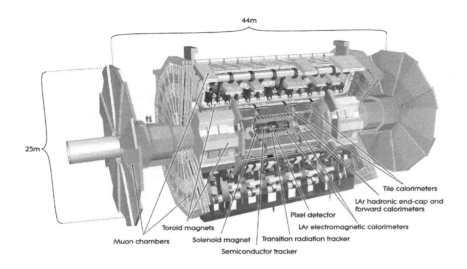

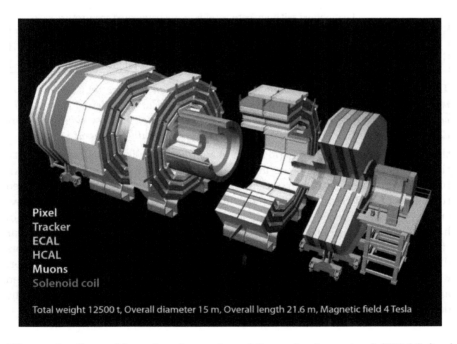

Figure 8. *General layout and overview of the main elements of ATLAS (top) and CMS (bottom).*

	ATLAS A Toroidal LHC ApparatuS	**CMS** Compact Muon Solenoid
Magnet(s)	Air-core toroids + solenoid in inner cavity four magnets calorimeters in field-free region	Solenoid only one magnet calorimeters inside field
Tracker	Pixels and Si-strips Pid: TRT + dE/dx $B = 2$ T $\sigma_{p_T}/p_T \sim 5 \times 10^{-4} p_T \oplus 0.01$	Pixels and Si-strips Pid: dE/dx $B = 3.8$ T $\sigma_{p_T}/p_T \sim 1.5 \times 10^{-4} p_T \oplus 0.005$
Electromagnetic Calorimetry	Lead-liquid argon sampling $\sigma_E/E \sim 10\%/\sqrt{E} \oplus 0.007$ longitudinal segmentation	Lead-tungstate crystals homogenous $\sigma_E/E \sim 3\%/\sqrt{E} \oplus 0.5\%$ no longitudinal segmentation
Hadronic Calorimetry	Fe-scint. + Cu-liquid argon $\sigma_E/E \sim 50\%/\sqrt{E} \oplus 0.03$ $\gtrsim 11\lambda_0$ longit. segmented readout	Brass-scint. $\sigma_E/E \sim 100\%/\sqrt{E} \oplus 0.05$ $\gtrsim 7\lambda_0$ + tail catcher single (full) depth in readout
Muon System	combined with air-core toroids $\sigma_{p_T}/p_T \sim 2\%$ (at 50 GeV) $\sim 10\%$ (at 1 TeV) in stand-alone mode	instrumented iron return yoke $\sigma_{p_T}/p_T \sim 1\%$ (at 50 GeV) $\sim 10\%$ (at 1 TeV) when combined with tracker

Table 1. *A simple comparison of the main design choices and performance numbers for the ATLAS and CMS detectors. Pid=Particle identification. The calorimeter energy resolutions are for the barrel parts. For considerably more detailed information and discussions Refs. [4–6] should be consulted.*

the idea of separating the barrel into five completely independent wheels, with the central one supporting the magnet coil. This obviously offers great flexibility in construction and maintenance, in particular since it was anticipated to construct and test the experiment at the surface, in parallel with the cavern excavation. Indeed, CMS was lowered into the cavern by a sequence of heavy-lifting operations of its individual elements. In the cavern the modular structure has the further advantage of easy access to sub-parts of the detectors during shutdown periods. The most dramatic example is the pixel detector, which can be removed and re-installed relatively quickly. ATLAS has been assembled inside its cavern, and access to some of the inner elements of the detector poses a greater challenge and requires more time than in CMS.

3.1   From Construction to First Collisions

The numerous quality checks during the construction and beam tests, let us conclude that the detectors as built should give a good starting-point performance. However, commissioning of the detectors with cosmic rays, beam-

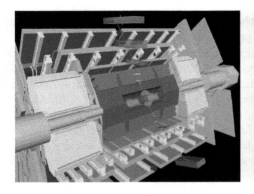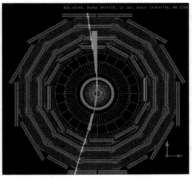

Figure 9. *Event displays of cosmic rays in ATLAS (left) and CMS (right).*

splash events and first collisions has been and still is an invaluable tool to prepare the experiments for the high-energy operations. The beam-splash events, which were available already in 2008 during the very first LHC injection tests, turned out to be extremely useful to time-in many subsystems and to align and intercalibrate some of the subdetectors. Such splash events were observed when the 450 GeV injection beam, with about 10^9 protons in a single bunch, struck collimators some 150 m upstream of the experiments, giving $\sim 10^5$ muons traversing the detectors, mostly horizontal. Halo muons were also observed once the beam started passing through the detectors. These very high-energy particles give almost straight tracks across the systems, and thus can be used for alignment studies.

Before the first LHC start, and between the incident on Sept. 19, 2008 and the re-start in Nov. 2009, all the LHC experiments made extensive use of cosmic rays. Besides setting up the online operations, bringing all subsystems into a unified readout and training the data-taking procedures as far as possible, the cosmic runs gave an astonishing number of commissioning, calibration and even some physics results. Each experiment collected several hundred million cosmic events (cf. Fig. 9), producing up to ~ 1 PByte of raw data. Track reconstruction in the muon and tracker systems, in stand-alone mode or combined, could be exercised and its performance, such as the momentum resolution, measured. The modeling of the magnetic field maps could be verified and corrected where necessary. Energy deposits by muons in the calorimeters were registered and compared to predictions. All this has helped the experiments to approach the first collision period in an extremely well-prepared fashion, with often more than 99% of the subdetector channels fully functional and well understood. Obviously, with the first collisions in hand, the trigger and data acquisition systems were finally timed in, the data coherence checked, subsystems synchronized and reconstruction algorithms debugged and calibrated.

Indeed, the speed, at which the experiments analyzed the first collision data at the end of 2009, as well as the quality of the results and the agreement of the data with the Monte Carlo predictions were a surprise to many.

Nevertheless, considerable effort will have to be invested to obtain the ultimate calibration and alignment precision. The electromagnetic and hadronic calorimeters will be calibrated with physics events. For example, the initial crystal inter-calibration precision of about 4% for the CMS ECAL will be improved to about 2% by using the ϕ-symmetry of the energy deposition in minimum-bias and jet events. Later the ultimate precision ($\approx 0.5\%$) and the absolute calibration will be obtained using $Z \rightarrow e^+e^-$ decays and the E/p measurements for isolated electrons, such as in $W \rightarrow e\nu$ decays [15]. The latter requires a well-understood tracking system. The uniformity of the hadronic calorimeters can be checked with single pions and jets. In order to obtain the jet energy scale (JES) to a few per-cent precision or better, physics processes such as $\gamma + \text{jet}$, $Z(\rightarrow \ell\ell) + \text{jet}$ or $W \rightarrow 2$ jets in top pair events will be analyzed. Finally, the tracker and muon system alignment will be carried out with generic tracks, isolated muons or $Z \rightarrow \mu^+\mu^-$ decays. Regarding all these calibration and alignment efforts, the ultimate statistical precision should be achieved very quickly in many cases. Then systematic effects have to be faced.

4 Measurements of Hard Processes

Before entering the discovery regime, considerable efforts will be invested in the measurements of SM processes. They will serve as a proof for a working detector (a necessary requirement before any claim of discovery is made). Indeed, some of the SM processes are also excellent tools to calibrate parts of the detector. However, such measurements are also interesting in their own right. We will be able to challenge the SM predictions at unprecedented energy and momentum transfer scales, by measuring cross sections and event features for minimum-bias events, jet production, W and Z production with their leptonic decays, as well as top quark production. This will allow to check the validity of the Monte Carlo generators, both at the highest-energy scales and at small momentum transfers, such as in models for the omnipresent underlying event. The parton distribution functions (pdfs) can be further constrained or measured for the first time in kinematic ranges not accessible at HERA. Important tools for pdf studies will be jet + photon production or Drell-Yan processes. Finally, SM processes such as W/Z+jets, multi-jet and top pair production will be important backgrounds to a large number of searches for new physics and therefore have to be understood in detail.

The very early goals to be pursued by the experiments, once the first data are on tape, are three-fold : (1) it will be of utmost importance to commission and calibrate the detectors *in situ*, with physics processes as outlined below. The trigger performance has to be understood in a possibly unbiased manner, by analyzing the trigger rates of minimum-bias events, jet events for various thresholds, single and di-lepton as well as single and di-photon events. (2) It will be necessary to measure the main SM processes and (3) prepare the road for possible discoveries. It is instructive to recall the event statistics collected for different types of processes. For an integrated luminosity of 100 pb^{-1} per experiment, we expect about 10^6 $W \to e\nu$ events on tape, a factor of 10 less $Z \to e^+e^-$ and some 10^4 $t\bar{t} \to \mu + X$ events. If a trigger bandwidth of about 10% is assumed for QCD jets with transverse momentum $p_T > 150$ GeV, $b\bar{b} \to \mu + X$ and minimum-bias events, we will write about 10^5 events to tape, for each of these channels. This means that the statistical uncertainties will be negligible relatively quickly, for most of the physics cases. The analysis results will be dominated by systematic uncertainties, be it the detailed understanding of the detector response, theoretical uncertainties or the uncertainty from the luminosity measurements.

The anticipated detector performance leads to the following estimates for the reconstruction precision of the most important physics objects: isolated electrons and photons can be reconstructed with a relative energy resolution characterized by a stochastic term of a few per-cent and an aimed-for 0.5% constant term. Typically isolation requirements are defined by putting a cone around the electron/photon and counting the additional electromagnetic and hadronic energy and/or track transverse momentum within this cone. The optimal cone size in $\eta - \phi$ space depends on the particular analysis and event topology. For typical acceptance cuts, such as a transverse momentum above 10–20 GeV and $|\eta| < 2.5$, electrons and photons can be expected to be reconstructed with excellent angular resolution, high efficiency ($\geq$ 80–90%) and small backgrounds. Again, the precise values depend very much on the final state topology and the corresponding tightness of the selection cuts. Most importantly, the systematic uncertainty on the reconstruction efficiency should be controllable at the 1–2% level, using *in situ* measurements such as $Z \to e^+e^-$ decays, with one of the electrons serving as tag lepton and the other one as probe object for which the efficiency is determined.

Isolated muons, with similar acceptance cuts as mentioned above for electrons, should be reconstructed with a relative transverse momentum resolution of 1– 5% and excellent angular resolution up to several hundreds of GeV. Again, a systematic uncertainty on the reconstruction efficiency of 1–2% appears to be achievable.

Hadronic jets will be reconstructed up to pseudorapidities of 4.5–5, with good angular resolution. The energy resolution depends rather strongly on the specific calorimeter performance. For example, in the case of ATLAS (CMS) a stochastic term of the order of 50–60% (100–150%) is to be expected when energy deposits in projective calorimeter towers are used for the jet clustering procedure. However, as mentioned above, important improvements in the CMS jet energy resolution are expected from new approaches such as particle flow algorithms. Well above the trigger thresholds jets will be reconstructed with very high efficiency; the challenge is the understanding of the efficiency turn-on curves. In contrast to leptons, for jets the experimental systematic uncertainties are much more sizeable and difficult to control. A more detailed discussion follows below. A further important question is the lowest p_T threshold above which jets can be reconstructed reliably. Contrary to the naïve expectation that only high-p_T objects (around 100 GeV and higher) are relevant, it turns out that many physics channels require jets to be reconstructed with rather low transverse momentum of ~ 20–30 GeV. One reason for this is the importance of jet veto requirements in searches for new physics, such as in the $H \rightarrow WW^* \rightarrow 2\ell 2\nu$ channel, where a jet veto is necessary to reduce the top background. The experimental difficulties related to the understanding of the low-p_T jet response,[3] the thresholds due to noise suppression, the impact of the underlying event and additional pile-up events and ultimately the knowledge of the JES lead to the conclusion that it will be extremely challenging to reliably reconstruct jets below a p_T of 30 GeV. In addition, the theoretical predictions are challenged by very low-p_T effects. Here fixed-order calculations may have to be supplemented by resummations of large logarithms.

Finally, the missing transverse energy will be a very important "indirect" observable, which is constructed from measurements of other quantities, such as all calorimeter energy deposits. Many searches for new physics, such as supersymmetry, rely very much on this observable. However, it turns out that it is also an extremely difficult quantity to measure, since it is sensitive to almost every detail of the detector performance. Here it is even more difficult to give estimates of the expected systematic uncertainties. Also, the reconstruction performance depends very much on the details of the particular final state, such as the number of jets and/or leptons in the event, the existence of "true" missing energy, e.g., from neutrinos, the number of pile-up events and in general the overall transverse energy deposited in the detector. The first data are of paramount importance for a timely understanding of this quantity.

In the following I will concentrate on some of the early measurements to be performed on the first few hundred pb^{-1} up to 1 fb^{-1} of integrated luminosity. Many reviews exist on this topic, such as Refs. [1, 16, 17] to mention only a few. However, before entering the discussion of physics measurements, it is worth recalling some recent developments in the area of jet algorithms, which will play an important role in almost all of the LHC analyses.

---

[3]The jet response is defined as the ratio of the reconstructed and the "true" jet momentum.

4.1 Jet Algorithms

In hard interactions, final-state partons and hadrons appear predominantly in collimated bunches. These bunches are generically called *jets*. To a first approximation, a jet can be thought of as a hard parton that has undergone soft and collinear showering and then hadronization. Jets are used both for testing our understanding and predictions of high-energy QCD processes and also for identifying the hard partonic structure of decays of massive particles like top quarks. In order to map observed hadrons onto a set of jets one uses a *jet definition*. Good jet definitions are infrared and collinear safe, simple to use in theoretical and experimental contexts, applicable to any type of inputs (parton or hadron momenta, charged particle tracks and/or energy deposits in the detectors) and lead to jets that are not too sensitive to non-perturbative effects. An extensive treatment of the topic of jet definitions is given in [18] (for e^+e^- collisions) and [9, 19] (for pp or $p\bar{p}$ collisions). Here I will briefly discuss the two main classes: cone algorithms, extensively used in hadron colliders, and sequential recombination algorithms, more widespread in e^+e^- and ep colliders.

Very generically, most (iterative) cone algorithms start with some seed particle i, sum the momenta of all particles j within a cone of opening-angle R, typically defined in terms of (pseudo)-rapidity and azimuthal angle. They then take the direction of this sum as a new seed, repeat until the cone is stable and call the contents of the resulting stable cone a jet if its transverse momentum is above some threshold $p_{T,\min}$. The parameters R and $p_{T,\min}$ should be chosen according to the needs of a given analysis.

There are many variants of cone algorithm, and they differ in the set of seeds they use and the manner in which they ensure a one-to-one mapping of particles to jets, given that two stable cones may share particles ("overlap"). The use of seed particles is a problem w.r.t. infrared and collinear safety, and seeded algorithms are generally not compatible with higher-order (or sometimes even leading-order) QCD calculations, especially in multi-jet contexts, as well as potentially subject to large non-perturbative corrections and instabilities. Seeded algorithms (JetCLU, MidPoint and various other experiment-specific iterative cone algorithms) are therefore to be deprecated. A modern alternative is to use a seedless variant, SISCone [20].

Sequential recombination algorithms at hadron colliders (and in DIS) are characterized by a distance $d_{ij} = \min(k_{t,i}^{2p}, k_{t,j}^{2p}) \Delta_{ij}^2 / R^2$ between all pairs of particles i, j, where Δ_{ij} is their distance in the rapidity–azimuthal plane, $k_{t,i}$ is the transverse momentum w.r.t. the incoming beams and R is a free parameter. They also involve a "beam" distance $d_{iB} = k_{t,i}^{2p}$. One identifies the smallest of all the d_{ij} and d_{iB} and if it is a d_{ij} then i and j are merged into a new pseudo-particle (with some prescription, a recombination scheme, for the definition of the merged four-momentum). If the smallest distance is d_{iB} then i is removed from the list of particles for the next iteration and called a jet. As with cone algorithms one usually considers only jets above some transverse-momentum threshold $p_{T,\text{min}}$. The parameter p determines the kind of algorithm: $p = 1$ corresponds to the (inclusive-)k_t algorithm [21–23], $p = 0$ defines the *Cambridge-Aachen* algorithm [24, 25], while for $p = -1$ we have the *anti-k_t* algorithm [26]. All these variants are infrared and collinear safe to all orders of perturbation theory. Whereas the former two lead to irregularly shaped jet boundaries, the latter results in cone-like boundaries.

Efficient implementations of the above algorithms are available through the *FastJet* package [27], which is also packaged within *SpartyJet* [28].

4.2   QCD Jet Production

Because of its extremely large cross section, the inclusive dijet production ($pp \to 2$ jets + anything) completely dominates over all other expected LHC processes with large momentum transfer. At lowest order in perturbative QCD, it is described as a $2 \to 2$ scattering of partons (quarks and gluons), with only partons in the initial, intermediate and final state. Depending on the exchanged transverse momentum (or generally the energy scale of the scattering process), the final state will consist of more or less energetic jets which arise from the fragmentation of the outgoing partons.

For the measurement of the inclusive jet cross section we simply count the number of jets inside a fixed pseudorapidity region as a function of jet p_T. For a second typical measurement, the dijet cross section, events are selected in which the two highest p_T jets, the leading jets, are both inside a specified pseudorapidity region and counted as a function of the dijet (invariant) mass. Both cases are inclusive processes dominated by the $2 \to 2$ QCD scattering of partons. The distinction between inclusive jets and dijets is only in a different way of measuring the same process. For a common choice of the η region, events selected by the dijet analysis are a subset of the events selected by the inclusive jet analysis, but the number of events in the two analyses coming from QCD is expected to be close at high p_T. The steeply falling cross sections are shown in Fig. 10 (left). For the inclusive jet case, the spectrum roughly follows a power law, however, with increasing power for increasing p_T, i.e., the power increases from about 6 at $p_T = 150$ GeV to about 13 at $p_T = 3$ TeV and keeps on increasing with jet p_T.

Even for very small integrated luminosities the statistical uncertainties will be negligible, up to very high jet momenta. Thus the Tevatron reach in terms of highest momenta and therefore sensitivity to new physics, such as contact interactions or heavy resonances, will be quickly surpassed. For 1 fb^{-1}, the inclusive cross section for central jet production (i.e., jet pseudorapidities below ~ 1) will be known statistically to better than 1% up to a p_T of 1 TeV, and the statistical errors on the dijet cross section will be below 5% up to dijet masses of 3 TeV.

The real challenge for these measurements will be the determination and control of the jet energy scale. As mentioned above, the cross sections are steeply falling as a function of jet p_T. Therefore any relative uncertainty

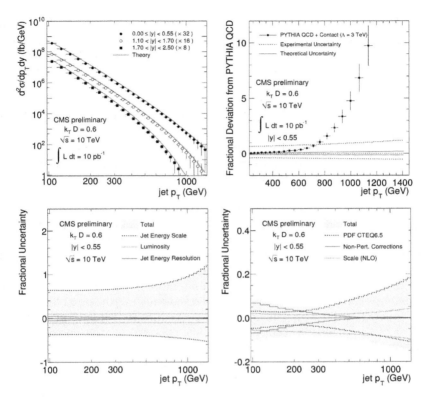

Figure 10. *Inclusive jet cross section measurements (upper) and related systematic uncertainties (lower), as foreseen by CMS [29].*

on the jet p_T will translate into a n-times larger relative uncertainty on the cross section, where n indicates the power of the spectrum in a specified p_T region, i.e., $d\sigma/dp_T \propto p_T^{-n}$. For example, a 5% uncertainty on the energy scale for jets around 100–200 GeV of transverse momentum induces a 30% uncertainty on the inclusive jet cross section. This is shown in Fig. 10 (right), here for the case of a 10% JES uncertainty. As a comparison, the expected theoretical uncertainties on the inclusive jet cross section from the propagation of pdf uncertainties are below the 10% level up to a jet p_T of 1 TeV, thus much smaller than the experimental systematics from the JES. Therefore it is obvious that a measurement of the inclusive jet cross section will not allow constraint of the pdfs, unless the JES is known to 2% or better.

Obviously, the knowledge of the JES also has a strong impact on the achievable precision of the dijet cross-section measurement. However, the problem can be avoided by performing relative instead of absolute cross-section measurements. A well-suited observable is the dijet ratio $N(|\eta| <$

$|\eta_{\mathrm{in}}|)/N(|\eta_{\mathrm{in}}| < |\eta| < |\eta_{\mathrm{out}}|)$, i.e., the ratio of the number of dijet events within an inner region $|\eta| < |\eta_{\mathrm{in}}|$ to the number of dijet events within an outer region $|\eta_{\mathrm{in}}| < |\eta| < |\eta_{\mathrm{out}}|$. Both leading jets of the dijet event must satisfy the $|\eta|$ cuts, with typical values of $\eta_{\mathrm{in}} = 0.7$ and $\eta_{\mathrm{out}} = 1.3$. The dijet ratio has two interesting features. First, it is very sensitive to new physics, such as contact interactions or the production of a heavy resonance, because those lead to jets at more central rapidities than in genuine QCD dijet events. Second, in the ratio we can expect many systematic uncertainties to cancel. For example, the luminosity uncertainty completely disappears in the ratio. More importantly, the JES uncertainty is also strongly reduced, since the dijet ratio is sensitive only to the relative knowledge of the scale as a function of rapidity, but not to the absolute scale any more.

As we have seen above, the JES is the dominant source of uncertainty in jet cross-section measurements. Obviously, it is also important for many other analyses and searches which involve jet final states and possibly invariant mass reconstructions with jets. Therefore major efforts are devoted by the experimental collaborations to prepare the tools for obtaining JES corrections, both from the Monte Carlo simulations and, more importantly, from the data themselves. Currently approaches are followed which are inspired by the Tevatron experience [30, 31]. The correction procedure is split into several steps, such as offset corrections (noise, thresholds, pile-up), relative corrections as a function of η, absolute corrections within a restricted η-region, corrections to the parton level, flavour-specific corrections etc. At the LHC startup we will have to rely on Monte Carlo corrections only, but with the first data coming in it will be possible to switch to data-driven corrections. At a later stage, after a lot of effort will have gone into the careful tuning of the Monte Carlo simulations, it might be feasible to use Monte Carlo corrections again. For example, a rough estimate for the early JES uncertainty evolution in CMS might be 10% at start-up, 7% after 100 pb^{-1} and 5% after 1 fb^{-1}. Certainly it will be difficult and require time to obtain a detailed understanding of the non-Gaussian tails in the jet energy resolution.

Concerning data-driven JES corrections, one of the best channels is $\gamma +$ jet production. At leading order, the photon and the jet are produced back-to-back, thus the precisely measured photon energy can be used to balance the jet energy. Real life is more difficult, mainly because of additional QCD radiation and the large background from jets faking a photon. These can be suppressed very strongly with tight selection and isolation cuts (e.g., no additional third jet with a transverse energy beyond a certain threshold and tight requirements on additional charged and neutral energy in a cone around the photon). The need to understand well the photon-faking jet background and the photon fragmentation is avoided by using the channel $Z(\to \ell\ell) +$ jet, with electrons or muons, however, at the price of a lower cross section.

4.3 Vector Boson Production

The production of vector bosons (W and Z), triggered on with their subsequent leptonic decays, will be among the most important and most precise tests of the SM at the LHC. The leptonic channels, mainly electrons and muons, can be reconstructed very cleanly, at high statistics, with excellent resolution and efficiency and very small backgrounds. At the same time, the theoretical predictions are known to high accuracy, as discussed in more detail below. This precision will be useful for constraining pdfs, e.g., by measuring the rapidity dependence of the Z production cross section, in particular when going to large rapidities and thus probing low x values. As proposed in [32], this process will serve as a standard candle for determining to high precision (at the few per-cent level) the proton-proton luminosity or alternatively the parton-parton luminosity. Finally, it will be attempted to improve on the current precision of the W mass. Besides that, W and Z production will be an important experimenter's tool. As mentioned already earlier, Z and W decays to leptons will be used to understand and calibrate various subdetectors, measure the lepton reconstruction efficiencies and control even the missing transverse energy measurement.

Below I will first discuss the inclusive case, concentrating on resonant production. Then I will highlight some issues for the W and Z production in association with jets. Although being highly interesting processes, di-boson production will not be discussed here, since for integrated luminosities up to 1 fb^{-1} the statistical precision will be the limiting factor for these measurements and only allow for a first proof of existence and rough validations of the model expectations.

4.4 Inclusive W and Z Production

Inclusive W and Z production currently is and probably will remain the theoretically best-known process at the LHC. Predictions are available at next-to-next-to-leading order (NNLO) in perturbative QCD, fully differential in the vector boson and even the lepton momenta [33]. Figure 11 shows the Z rapidity distribution at various orders in perturbation theory. We see that the shape stabilizes when going to higher orders and that the NNLO prediction nicely falls within the uncertainty band of the next-to-leading order (NLO) expansion, giving confidence in the good convergence of the perturbation series. More importantly, the renormalization scale uncertainty is strongly reduced at NNLO, to a level of about 1% for Z rapidities below 3. A renormalization scale uncertainty even below 1% can be obtained for ratio observables such as $\sigma(W^+)/\sigma(W^-)$ and $\sigma(W)/\sigma(Z)$, possibly as a function of rapidity. Again, ratio measurements are interesting also from the experimental point of few, since many systematic uncertainties cancel completely or to a large extent. The prospect of a precise measurement and knowing the hard scattering part of the process so well means that we have a tool for precisely constraining pdfs (or couplings and masses, in a more general sense). Indeed, when taking the full theoretical prediction for the W and Z production cross section, i.e., the convolution of pdfs and hard scattering part, its uncertainty is dominated by the limited knowledge of the pdfs, estimated to be below 5% [34]. This will then also limit the proton-proton luminosity to a precision of this size, unless the pdfs are further constrained, mainly by the rapidity dependence of the cross section, as for example shown in Ref. [35].

In this context one should highlight the importance of having differential cross-section predictions. If we take resonant W and Z production at central vector boson rapidity, we probe x values of around 0.006, a region rather well constrained by the current pdf fits. However, for larger rapidities we probe more and more the small x region, which is less well known, e.g., at leading order and for a Z rapidity of 3 we need (anti-)quark pdfs at $x = 0.12$ and $x = 0.0003$. Experimentally, because of the detector acceptance, we can only access a limited sub-region of the full phase–space. This means that when measuring a total cross section, we have to extrapolate the measurement to the full acceptance (e.g., full rapidity), which introduces a model dependence, especially on the poorly known low-x region. On the other hand, having differential predictions, we can compute exactly the same quantity as we measure, thus eliminating any extrapolation uncertainty. Similarly, for constraining NLO (NNLO) pdfs, exactly the same acceptance cuts (on the leptons) as in the data can now be applied to the available NLO (NNLO) predictions. Of course, with more and more differential higher-order predictions becoming available, this kind of argument applies to any cross-section measurement (and/or deduced determination of physics quantities such as couplings, masses, pdfs), namely that we should compare the measurements and predictions for the experimentally accessible acceptance and avoid unnecessary extrapolations, which will not teach us anything new and only introduce additional uncertainties.

As mentioned above, the experimental reconstruction of W and Z production is rather straightforward. Leptons are required to have a minimum p_T of about 20 GeV, within a pseudorapidity of 2.5. In the Z case the mass peak allows for further event selections and background estimations. However, the neutrino in the W decay leads to missing energy, which obviously is reconstructed less precisely. Instead of an invariant mass peak only the transverse W mass can be reconstructed, with larger backgrounds than for the Z.

4.5 W/Z+jets Production

Vector bosons produced in association with jets lead to final states with high-p_T leptons, jets and possibly missing transverse energy. Such a topology is also expected for many searches, in particular for squark and gluino production and subsequent cascade decays. Obviously it will be important to understand these SM processes as quickly as possible and validate the available Monte Carlo generators, which typically combine LO matrix elements with parton

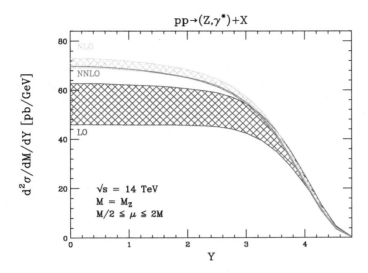

Figure 11. *QCD predictions at various orders of perturbation theory for the Z rapidity distribution at the LHC. The shaded bands indicate the renormalization scale uncertainty. Plot taken from [36].*

showers. A standard observable will be the W/Z cross section as a function of the associated leading jet transverse momentum or the number of additional jets. Obviously, such measurements will suffer from the same JES uncertainties as the QCD measurements discussed above, and thus constitute only limited calibration tools during the early data taking. The problem can be reduced by defining clever ratios of cross sections, involving different vector bosons and/or number of additional jets, or by normalizing the predictions to the data in limited regions of the phase–space (e.g., for small jet multiplicity and extrapolating to larger multiplicities). A completely different approach is to take a more inclusive look at this process, in the sense that the Z transverse momentum is measured from the lepton kinematics, which is possible at high statistical and, more importantly, high experimental accuracy. This distribution can be understood as the convolution of the $Z + 0/1/2/\ldots$jets distributions, therefore any model intended to describe $Z +$ jets production has necessarily to reproduce the Z p_T distribution over its full range.

5 Top Pair Production

The top quark is produced very abundantly at the LHC. With 1 fb^{-1} of integrated luminosity, we should already have a couple of thousand clean signal events on tape in the di-lepton channel, and a factor of 10 more in the single lepton channel (lepton+jets channel). The physics case for the study of top production is very rich and cannot be discussed in detail here. For example, a recent review can be found in Ref. [37]. Combining many different channels, a top mass measurement with a precision of 1 GeV might be achieved, which together with a precise W mass measurement constitutes an important indirect constraint of SM predictions and its extensions. The production cross section (for single and top-pair production) will be an important measurement, again for testing the SM predictions and because top production is a copious background to a large number of new physics searches. In the single muon + X channel, the top-pair production cross section will soon (i.e., with about 1 fb^{-1}) be measured with a statistical precision of 1%. The total uncertainty of 10–15% (excluding the luminosity uncertainty) will be dominated by systematics, most notably due to the knowledge of the b-tagging efficiency. Finally, top production will become an extremely valuable calibration tool. The mass peak can already be reconstructed with much less than 1 fb^{-1}, even without b-tagging requirements. With a clean sample in hand, it can be exploited for controlling the b-tagging efficiency and serve as closure test for the JES corrections determined from other processes. Concerning the JES, the mass of the hadronically decaying W serves as calibration handle.

6   Conclusions

It has been an unprecedented challenge to design and construct the LHC experiments, as well as to put them into operations. Here an attempt was made to sketch the most important criteria, which were the basis of the many design choices, as well as to give a rough comparison of the expected performance of the ATLAS and CMS detectors. The quality of the data, which resulted from first LHC collisions in late 2009 as well as the runs at 7 TeV in 2010 and 2011 gives confidence that exciting physics results will soon emerge from the LHC.

7   Acknowledgements

I would like to thank my colleagues in the LHC experiments who helped me in the preparation of the lectures and these proceedings, in particular D. Treille, D. Froidevaux, S. Cittolin, A. Herve, P. Jenni and R. Tenchini. I would like to thank the organizers of this school for the invitation and their great hospitality during my stay. It was with great sorrow that I received the announcement, at the time of writing this summary, that one of the organizers, Thomas Binoth, tragically died in an avalanche accident.

References

[1] G. Kane and A. Pierce, "Perspectives on LHC physics," *Hackensack, USA: World Scientific (2008) 337 p*

[2] G. Aad *et al.* [The ATLAS Collaboration], arXiv:0901.0512 [hep-ex].

[3] G. L. Bayatian *et al.* [CMS Collaboration], J. Phys. G **34** (2007) 995.

[4] G. Aad *et al.* [ATLAS Collaboration], JINST **3** (2008) S08003.

[5] R. Adolphi *et al.* [CMS Collaboration], JINST **0803** (2008) S08004 [JINST **3** (2008) S08004].

[6] D. Froidevaux and P. Sphicas, Ann. Rev. Nucl. Part. Sci. **56** (2006) 375.

[7] J. Stirling, private communication, 2009.

[8] K. Aamodt *et al.* [ALICE Collaboration], Eur. Phys. J. C **65** (2010) 111 [arXiv:0911.5430 [hep-ex]].

[9] G. P. Salam, Eur. Phys. J. **C67** (2010) 637 [arXiv:0906.1833 [hep-ph]].

[10] S. Cittolin, private communication, 2009.

[11] C. Amsler *et al.* [Particle Data Group], Phys. Lett. B **667** (2008) 1.

[12] http://cms.web.cern.ch/cms/Resources/Website/
Media/Videos/Animations/files/CMS_Slice.gif

[13] A. Hoecker, "Trigger and Data Analysis," lecture given at the HCP Summer School, CERN, June 2009.

[14] P. Charpentier, "Grid computing," these proceedings.

[15] G. L. Bayatian *et al.* [CMS Collaboration], "CMS physics: Technical design report," CMS-TDR-008-1.

[16] F. Gianotti and M. L. Mangano, arXiv:hep-ph/0504221.

[17] P. Sphicas, Nucl. Phys. Proc. Suppl. **117** (2003) 298.

[18] S. Moretti, L. Lonnblad and T. Sjostrand, JHEP **9808** (1998) 001 [arXiv:hep-ph/9804296].

[19] S. D. Ellis, J. Huston, K. Hatakeyama, P. Loch and M. Tonnesmann, Prog. Part. Nucl. Phys. **60** (2008) 484 [arXiv:0712.2447 [hep-ph]].

[20] G. P. Salam and G. Soyez, JHEP **0705** (2007) 086 [arXiv:0704.0292 [hep-ph]].

[21] S. Catani, Y. L. Dokshitzer, M. Olsson, G. Turnock and B. R. Webber, Phys. Lett. B **269** (1991) 432.

[22] S. Catani, Y. L. Dokshitzer, M. H. Seymour and B. R. Webber, Nucl. Phys. B **406** (1993) 187.

[23] S. D. Ellis and D. E. Soper, Phys. Rev. D **48** (1993) 3160 [arXiv:hep-ph/9305266].

[24] Y. L. Dokshitzer, G. D. Leder, S. Moretti and B. R. Webber, JHEP **9708** (1997) 001 [arXiv:hep-ph/9707323].

[25] M. Wobisch and T. Wengler, arXiv:hep-ph/9907280.

[26] M. Cacciari, G. P. Salam and G. Soyez, JHEP **0804** (2008) 063 [arXiv:0802.1189 [hep-ph]].

[27] M. Cacciari and G. P. Salam, Phys. Lett. B **641** (2006) 57 [arXiv:hep-ph/0512210].

[28] P. A. Delsart, K. Geerlins and J. Huston,
http://www.pa.msu.edu/~huston/SpartyJet/SpartyJet.html.

[29] CMS Collaboration, Physics Analysis Summary CMS PAS QCD-08-001.

[30] A. Bhatti *et al.*, Nucl. Instrum. Meth. A **566** (2006) 375 [arXiv:hep-ex/0510047].

[31] B. Abbott *et al.* [D0 Collaboration], Nucl. Instrum. Meth. A **424** (1999) 352 [arXiv:hep-ex/9805009].

[32] M. Dittmar, F. Pauss and D. Zurcher, Phys. Rev. D **56** (1997) 7284 [arXiv:hep-ex/9705004].

[33] K. Melnikov and F. Petriello, Phys. Rev. D **74** (2006) 114017 [arXiv:hep-ph/0609070].

[34] G. Watt, PoS **HCP2009** (2009) 014 [arXiv:1001.3954 [hep-ph]].

[35] M. Dittmar *et al.*, arXiv:hep-ph/0511119.

[36] C. Anastasiou, L. J. Dixon, K. Melnikov and F. Petriello, Phys. Rev. D **69** (2004) 094008 [arXiv:hep-ph/0312266].

[37] J. D'Hondt, arXiv:0707.1247 [hep-ph].

Forward Physics at the LHC

Albert De Roeck

CERN, Geneva, Switzerland, Antwerp University, Antwerp, Belgium and University of California Davis, California, USA

1   Introduction

The Large Hadron Collider (LHC) will ultimately collide protons with a total Centre of Mass (CM) system energy of 14 TeV, and will open up a new high energy frontier. Initially the energy will be lower, however. The commissioning phase of the LHC in 2009 included collisions at a CM energy of 900 GeV, and 2.36 TeV. About half a million events with all the main subdetectors on have been recorded by the experiments at 900 GeV, and a few 10,000 events have been collected at 2.36 TeV, just before the short winter shutdown of the LHC and the experiments. In 2010 and 2011 the LHC ran at a CM energy of 7 TeV. In the next few years and after adding some modifications to the machine, the design energy of 14 TeV or close to this could be reached.

In the first phase of the operation the luminosity was still modest, of order of 10^{28}–10^{32} cm^{-2}s^{-1}, leading to around 45 pb^{-1} and 5 fb^{-1} of integrated luminosity in 2010 and 2011, respectively. The anticipated ultimate high luminosity of the LHC will be a challenge for both machine and experiments: namely 10^{33} cm^{-2}s^{-1} in the next run and 10^{34} cm^{-2}s^{-1} for the high luminosity mode. This will lead to event samples of the order of 10–100 fb^{-1} a year, but also to a considerable number of overlay events in a single bunch crossing, so-called pile-up.

The high energy and luminosity of the collider allow for new physics opportunities in the field of diffraction and low-x QCD, two typical examples of topics associated with forward physics. The parton distributions in the proton can in principle be explored for scaled parton momentum values x down to 10^{-7}. LHC collisions will allow us to probe the structure of the diffractive exchange down to scaled momentum values of the partons, β, of less than 10^{-3}.

In these lectures we discuss opportunities for forward physics at the LHC. We start with the classical forward physics topics, such as total and elastic cross sections and general diffraction. Then we discuss the so-called "new forward physics" topics. The forward physics program at the LHC presently contains the following topics:

Soft and hard diffraction

- Total cross section and elastic scattering, single and double diffractive dissociation

- Gap survival dynamics, multi-gap events; proton light cone studies $(pp \rightarrow 3\,\mathrm{jets}+p)$, Odderon studies

- Diffractive structure: production of jets, $W, J/\psi$, b-quarks, top quarks, hard photons; Generalized Parton Distributions

- Central exclusive production as a gluon factory

Exclusive production of new mass states

- Exclusive Higgs production, exclusive Radion production

- Supersymmetry and other (low mass) exotics, long-lived gluinos, anomalous W, Z production

Low-x dynamics

- Parton saturation, BFKL/CCFM dynamics, proton structure, multi-parton scattering

New phenomena in forward physics

- New phenomena such as DCCs, incoherent pion emission, Centauros

Strong interest from cosmic rays community, and heavy ions

- Forward energy and particle flows/minimum bias event structure

- Two-photon interactions and peripheral collisions

- Forward physics in pA and AA collisions

and QED processes to determine the luminosity to $\mathrm{O}(1\%)$ $(pp \rightarrow ppee, pp\mu\mu)$.

Most of these topics will be briefly discussed in the following. Many of these can be studied best with luminosities of $10^{33}\,\mathrm{cm}^{-2}\mathrm{s}^{-1}$ or lower.

2 Forward Detectors

On a historical note, the first full proposal for an experiment at the LHC that discussed forward physics was FELIX [1]. FELIX was proposed as a dedicated experiment that could measure the full event phase space, but was put on ice in 1997. Its LOI is a good source of information on pioneering studies for

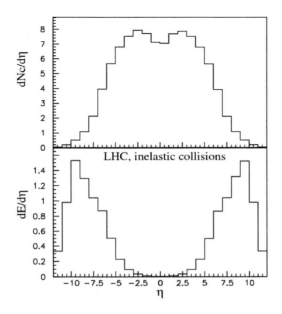

Figure 1. *Pseudorapidity distribution of the charged particles and of the energy flow at the LHC.*

forward physics. Since then several topical workshops on forward physics at the LHC have been organized in [2], and in particular the Blois meetings [3], LISHEP meetings [4], Diffraction meetings [5] and DIS workshops [6] have kept track of the most modern progress.

Seven experiments have been installed at the LHC. Two of these, CMS in IP5 (interaction point 5) and ATLAS in IP1, are general purpose experiments with, for the so-called central detectors, an acceptance in pseudorapidity η, where $\eta = -\ln\tan\theta/2$ with θ the polar angle of the particle, of roughly $|\eta| < 2.5$ for tracking information and $|\eta| < 5$ for calorimeter information. Figure 1 shows the pseudorapidity distribution of the charged particles and of the energy flow at the LHC, demonstrating that with an acceptance limited to $|\eta| < 5$ most of the energy in the collision will not be detected. ALICE (IP2) is also a central detector type of experiment, optimized for heavy ion collisions, while LHCb (IP8) is tailored to measure b-quark hadrons in the forward region. Furthermore two dedicated experiments for forward physics have been approved, TOTEM and LHCf. These use the same interaction points as CMS and ATLAS, respectively. Finally there is the Moedal experiment [8] which will search for heavy ionizing particles such as monopoles.

Figure 2. *Roman pot station layout for the TOTEM experiment.*

Detectors which have been key for forward physics are so-called Roman pot detectors [7] positioned at tens or hundreds of metres away from the interaction region, and have become a "standard" detector to use at modern colliders. UA4 installed such detectors to measure elastic and diffractive cross sections, and these detectors were later used in conjunction with the UA2 detector by the UA8 experiment at the end of the 1980s at the CERN $Sp\bar{p}S$. HERA has Roman pots installed along the proton beam line and also the experiments at the Tevatron have Roman pot detectors in their set-up. The LHC central detectors do not yet have such detectors but these are being planned, as discussed below.

Several of the experiments have so-called Zero Degree Calorimeters (ZDCs). These detectors are located at 140 m from the interaction point, where the proton beams are separated in their own beampipe. The prime goal of the ZDC is to measure the centrality in AA collisions. The so-called ultra-peripheral events can also be tagged. In pp interactions it will allow the study of events with charge exchange and consequently a forward high-energy neutron. Its ability to see low-energy (≈ 50 GeV) photons is important for exclusive diffractive studies. For cosmic ray physics the measurement of the high-energy π^0 component in pp and pA collisions at the LHC will also be very important to tune the air shower models.

The different experiments and their forward detector capabilities are discussed in the next section.

2.1   TOTEM

The main scientific program of TOTEM [9, 10] is to measure the total and elastic pp cross section at the highest energies, with a precision of order 1%, and soft diffractive dissociation. The experiment will use Roman pot (RP) detectors to measure the low-t^1 scattered protons, and detectors for tagging inelastic events in the regions to measure the total cross section (T1,T2), on both sides of IP5. The layout of the Roman pots is shown in Fig. 2. Both the presence of these Roman pots and inelastic event tagging detectors around CMS offers an excellent opportunity for a "combined experiment" which will have excellent coverage of both the central and forward regions. Such experiment will allow for unique measurements.

---

[1] t is the momentum transfer squared between the incoming and scattered proton: $t = (p - p')^2$.

The T1 and T2 telescopes consist of CSC (Cathode Strip Chambers) and GEM (Gas Electron Multipliers) chambers, respectively, and will detect charged particles in the η regions $3.1 < |\eta| < 4.7$ and $5.3 < |\eta| < 6.6$. The latter overlaps in acceptance with the CASTOR calorimeter of CMS.

The TOTEM RP stations are placed at a distance of ± 147 m and ± 220 m from IP5. These stations can measure protons with a momentum loss $\xi = \Delta p/p$ in the range $0.02 < \xi < 0.2$ for the nominal collision optics. For other optics with larger β^*, and hence lower luminosity, much smaller values of ξ can be reached. After the LHC has accelerated the beams and switched to collision optics, the Roman pots will be lowered to typically $15\sigma_{beam}$ away from the beam axis, which could be as little as 1–2 mm! Detectors (e.g., silicon detectors or fibers) will measure the coordinates of the elastically scattered protons. The Roman pots are placed at special locations around the machine for so-called parallel to point focusing, such that the measured coordinates can be directly related to the scattering angle of the proton.

Some particles form the beam halos that were seen in the Roman pots that were operational during the LHC run in 2009.

2.2 CMS

There are two plans to extend the coverage in the forward region of CMS:

- Two calorimeters on either side of the interaction region which will cover higher $|\eta|$ values, called CASTOR and the Zero Degree Calorimeter (ZDC). These calorimeters are of interest for measurements in pp, pA and AA collisions.

- Capitalizing on the opportunity to have common runs with the TOTEM experiment, which uses the same interaction region as CMS (IP5). This common physics programme has been reported in a document, released by the CMS/TOTEM working group [11]. At present the experiments do not yet operate with a common data-taking mode.

CASTOR is an electromagnetic/hadronic calorimeter, azimuthally symmetric around the beam and divided into 16 sectors. It is situated in the collar shielding at the very forward region of CMS, starting at 14.5 m from the interaction point, as shown in Fig. 3. The pseudorapidity range covered is $5.3 < |\eta| < 6.6$. This η-coverage would close almost hermetically the CMS pseudorapidity range over almost 13 units. CMS also has a ZDC on each side, consisting of tungsten absorber/quartz fibers. The EM section is 10 cm long which corresponds to 22 radiation lengths or ~ 1 interaction length. The opening angle is ± 4.4 cm/140 m = 300 μrads.

The Roman pot detectors of TOTEM aim to detect the protons in diffractive interactions of the type $pp \rightarrow p + X$ and $pp \rightarrow p + X + p$. When used in conjunction with the central CMS detector interesting phenomena such as hard diffractive scattering can be studied, where the system X can consist

Figure 3. *Schematics of the CMS forward region.*

of jets, W, Z bosons, high E_T photons, top quark pairs or even the Higgs particle, as discussed, e.g., in [12, 13]. Some groups in CMS plan to propose Roman pots also for operation at higher luminosities, as a result of the FP420 project discussed below.

The combination of T2 and CASTOR will allow the study of phenomena at lower Bjorken-x than otherwise reachable. Drell–Yan measurements will enable the parton distributions to be probed down to $x \approx 10^{-6} - 10^{-7}$. The energy and particle flows in the forward region are also of prime interest for tuning Monte Carlo simulation programs used in cosmic ray studies. CASTOR is designed especially to hunt for "strangelets" in AA collisions, which are characterized by very atypical fluctuations in hadronic showers.

One side of CASTOR and two ZDCs were available for the first data-taking run in 2009.

Furthermore a proposal for early diffractive measurements at low luminosity with scintillators placed at locations between 60 and 140 m from the IP, just surrounding the beampipe, was formulated in [14]. These detectors have however not been installed yet.

2.3 ATLAS

ATLAS plans its own Roman pots primarily for luminosity measurements, the ALFA project. Detectors, consisting of scintillating fibres, will be placed at 240 m in Roman pots. These detectors, in the present design, are not ideal for diffractive physics studies at the nominal high luminosity. Instead the option to put radiation hard detectors in Roman pots or other near beam detector mechanics at 220 m [15] is being studied. ALFA consists of 4 stations (8 Roman pots). For the 2010 run one detector was installed (with 1400 channels for one Roman pot).

ATLAS also installed a Cerenkov detector for relative luminosity measurements (LUCID with an acceptance of $5.6 < \eta < 5.9$) and ZDCs at a position of 140 m. The ZDCs are made of tungsten absorber and quartz fibers. The hadronic sections are already installed while the electromagnetic part will be installed after the LHCf (see below) modules are removed. LUCID can be used to help define a rapidity gap in the events from low luminosity collisions.

2.4 ALICE

The ALICE detector has (on one side) a muon spectrometer that covers the region $2.5 < \eta < 4$ and a ZDC. The muon acceptance is larger than for the ATLAS and CMS experiments, allowing for a more forward range for the detection of heavy flavors.

For some time ALICE has had a program for the study of minimum bias pp collisions (see, e.g., [16]). Recently [17] ALICE also studies specific diffractive channels with the base-line detector, making use of rapidity gaps for the selection.

2.5 LHCb

LHCb is a collider experiment but with the set-up of a fixed target experiment, namely a single-side forward spectrometer covering the range $1.9 < \eta < 4.9$. In particular very forward heavy flavour production can be studied in LHCb. So far LHCb has no specific diffractive program, but an idea to equip the beamline with forward shower counters 20–100 m from the IP, for diffractive measurements, was recently proposed in [18] by non-LHCb members.

2.6 LHCf

LHCf is a recently approved experiment for forward physics, consisting only of two forward electromagnetic calorimeters at zero degrees, positioned at 140 m. The aim is to measure the very forward π^0 and γ energy spectrum for pp collisions with an equivalent E_{lab} of 10^{17} eV. LHCf also plans to take data during heavy ion runs. These results will help calibrate high-energy cosmic ray spectra.

The detectors used are based on a tungsten absorber with scintillating fibres (one side) or silicon microstrips (the other side) as active elements. The detectors should measure energy and position of the γs from the π^0 decays. The detector was operational for the first data period end 2009.

2.7 FP420

The FP420 project, started in 2004, proposes to complement the experiments CMS and ATLAS by installing additional near-beam detectors at 420 m away from the interaction region [19]. The presence of these detectors will allow the measurement of exclusive production of massive particles, such as the Higgs particle, as discussed in Section 5.

Roman pots at 300–420 m encounter two major difficulties: these detectors are in the cold section of the machine and therefore need special care in their integration and they must be compact. Furthermore the event buffers of CMS are about 3 μs long, thus a signal from the Roman pots further than approximately 200 m distance cannot arrive in time for a trigger (91 bunch crossings) at the first trigger level: these events will need to be triggered by information in the central detectors.

The FP420 collaboration has members from ATLAS, CMS, and "independent" physicists, with excellent contacts with the LHC machine group. In the emerging design the principle of FP420 is based on moving "pockets" which contain tracking and timing detectors. The tracking detectors that have been developed are 3D silicon pixel detectors, which are radiation hard and can detect particles close to the edge. Timing detectors include both gas and crystal radiators. Technical details on FP420 are available in [20].

For the next phase, FP420 has now become part of the specific forward detector projects HPS and APS in CMS and ATLAS, respectively. Within the experiments these studies have been extended to include detectors in the region of 220 m, which is in the warm section of the machine.

3   Total and Elastic pp Cross Sections

Most proton–proton interactions are due to collisions at a large distance between the incoming protons, and the protons interact as a whole with small momentum transfer in the interaction. The particles in the final state have large longitudinal momenta but generally small transverse momenta, typically of order of a few hundred MeV. Such events are called soft events and make up the large part of a so-called minimum bias event sample, i.e., a sample which has as little as possible trigger bias (the exact definition of a minimum bias sample depends on the experimental acceptance and trigger).

The total cross section for events at the LHC is humongous, of the order of 100 mb, which is nine orders of magnitude or more larger than e.g., the Higgs cross section at the LHC. The exact shape of the final state of these events at the LHC energy is not well known and the experiments will have to measure their characteristics at the LHC start-up. The total cross section cannot be predicted accurately and needs to be measured. It is expected to be about 100 mb, with a typical uncertainty of about 20% [21]. The total cross section is highly non-perturbative and cannot be predicted easily from QCD. A fit through all data and using a multitude of models has been performed by the COMPETE collaboration [22, 23], and the resulting predictions are shown in Fig. 4. It predicts $84 \leq \sigma_{tot} \leq 112$ mb for 10 TeV and $90 \leq \sigma_{tot} \leq 117$ mb for 14 TeV. A careful Regge-based analysis predicts the cross section to be about 90 mb at 14 TeV [21].

We can easily derive how many events of this kind will be produced during a single bunch crossing at the LHC. Assume a luminosity of 10^{34} cm^{-2}s^{-1}

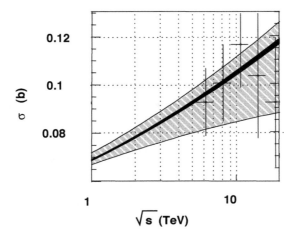

Figure 4. *Predictions of the total cross section from the COMPETE collaboration [23].*

which is $10^7 \, \mathrm{mb}^{-1}\mathrm{Hz}$. The non-elastic component (see below) of the pp cross section is roughly 70 mb. Thus we have an inelastic interaction rate of $7 \cdot 10^8 \, \mathrm{Hz}$. The number of bunch crossings is 40 MHz, but taking into account that only about 80% of the available bunch spaces will be filled (3564/2835), one expects **23** soft inelastic overlap events per bunch crossing, at the highest luminosity, which is a huge number and will lead to additional experimental challenges.

The total cross section consists of several components:

- The elastic cross sections, which is the reaction $pp \to pp$.

- The non-elastic diffractive cross sections (which contains single and double diffractive dissociation events). At high energies these processes are characterized by rapidity gaps, i.e., regions devoid of particles. The processes are shown in Fig. 5 and discussed in Section 4.

- The non-diffractive inelastic cross section. Such events do not have any rapidity gaps other than those compatible with fluctuations in the hadronization.

TOTEM plans to measure the total and elastic cross section during special LHC runs in so-called high β^* optics mode, i.e., $\beta^* = 1000 - 1500 \, \mathrm{m}$ (and also medium β^*) and luminosity $\sim 10^{28} \, \mathrm{cm}^{-2}\mathrm{s}^{-1}$.

Measuring the total cross section is naively speaking very simple: just count the number of interactions for a given luminosity. To this end TOTEM has the inelastic event taggers, which allow recording of the total rate of inelastic events. However, in order to determine the cross section one needs

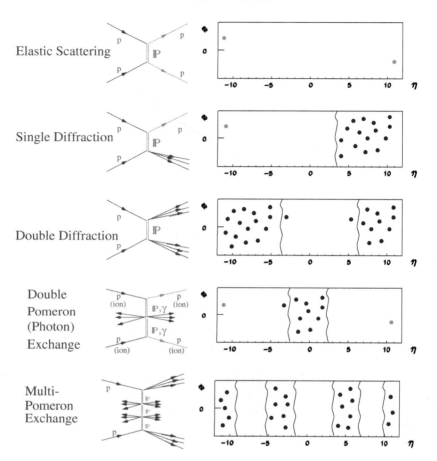

Figure 5. *Diagrams and $\eta - \phi$ pictures for various diffractive processes in pp scattering.*

to determine the luminosity of the data sample. It is as yet unclear on how precise the luminosity can be determined at the LHC but it is likely to be more in the range of 3–5% using methods that involve the proton parton distributions, and around 5% using Van der Meer scans from the machine itself.

One can use however the so-called luminosity independent method to determine the cross section whereby the optical theorem is used which relates the elastic scattering at $t = 0$ to the total cross section:

$$L\sigma_{tot}^2 = \frac{16}{1 + \rho^2} \times \frac{dN(t = 0)}{dt}. \tag{1}$$

Here L is the luminosity and ρ the ratio of the real to imaginary part of the forward scattering amplitude. The latter can be measured experimentally in principle by using elastic data in the Coulomb-Nuclear interference range, i.e., $|t| \sim 10^{-3}\,\mathrm{GeV^2/c^2}$ and below, as has been done at lower energy experiments, but it may be out of reach for the LHC. The expected value of ρ at the LHC energy is in the range of 0.1–0.2 so its impact on the measurement uncertainty is below the percent level. Since the number of elastic events is $L\sigma_{tot}$ the equation can be solved by knowing the number of events and the forward elastic cross section. Conversely σ_{tot} can then be used to determine the luminosity at the LHC.

ATLAS proposes a different approach to measure the total cross section, by explicitly trying to measure the Coulomb-Nuclear interference region [15]. They propose using an optics of 2625 m which would allow the measurement elastic scattering down to $|t| = 0.0003\,\mathrm{GeV^2}$.

TOTEM will cover a $|t|$ range from a few times 10^{-3} to $10\,\mathrm{GeV^2/c^2}$ using the high and medium β^* optics. Note that during the high luminosity operation ATLAS/CMS will run with a β^* of 0.5 m.

The measurement of the total cross section is not only an important number to know and to determine the luminosity. There are some exotic models which, based on anti-shadowing arguments, predict unusually large cross sections, up to 200 mb. Note that if these models were correct the number of minimum-bias overlay events would double at the LHC, compared to what is expected!

Elastic scattering is sensitive to granular structures of the proton, as proposed in some models. Here the key measurement is a precise determination of the $|t|$ distribution up to $10\,\mathrm{GeV^2/c^2}$, which would allow us to distinguish between these models.

It may, however, take some time before the machine will deliver the special optics, of 1540 m to make this precision measurement. Recently TOTEM has explored the possibility to determine σ_{tot} using an optics of 90 m. This may lead to an early estimate with a precision of 5%.

4   Diffraction

Recent data from HERA and Tevatron have re-emphasized the interest in the study of diffraction. Diffractive events can be pictured as shown in Fig. 5: the characteristic is the exchange of a colorless object, often termed the pomeron. After the exchange, the beam particles can either stay intact or dissociate into states with (generally) a low invariant mass M according to the distribution $d\sigma/dM \sim 1/M^2$. This dissociation is termed "diffractive dissociation." At sufficiently high CM system energies a large rapidity gap is formed between the two dissociative states or particles. Phenomenologically the dynamics is described as the exchange of an object called pomeron which has quantum numbers of the vacuum and originates from Regge theory [24]. In more modern language one tries to understand the phenomenon of diffraction and the

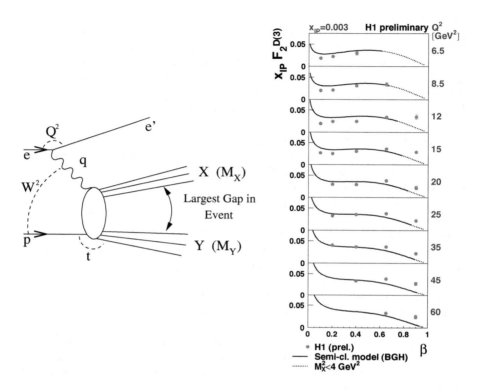

Figure 6. *Left: Schematic diagram for diffractive ep scattering. Right: Example of a structure function $F_2^{D(3)}$ measurement extracted from diffractive ep scattering.*

pomeron in terms of QCD. For an overview on the work on diffraction, see, e.g., [25].

In particular the observation and study of diffractive events in ep scattering, as shown in Fig. 6, has provided a lot of information on the diffractive exchange. It was found that the exchange could be described by partons, dominantly gluons, and within ep scattering the measured structure could be transported from, e.g., inclusive deep inelastic $ep \rightarrow eX$ scattering to the production of charm or di-jets in deep inelastic diffractive scattering. An example of a structure function measurement is shown in Fig. 6. This means that the diffractive process in ep scattering is factorizable: it is the convolution of the partonic cross sections and the diffractive parton distributions, the same as for normal parton distributions discussed in the lectures at this school [26].

It was, however, surprising that when the diffractive PDFs are transported to the Tevatron for $p\bar{p}$ diffractive scattering, they do not give the correct predictions, i.e., factorisation is broken in diffractive scattering. The data are below the predictions which suggests that some mechanism destroys the rapidity gap. The most obvious candidate is rescattering of the proton remnants of the underlying event, and calculations based on this kind of dynamics can indeed describe the data. Diffraction remains a bizarre phenomenon and is far from being understood. The LHC will contribute to its further understanding.

Diffractive events at a pp collider can be experimentally selected by the observation of a rapidity gap in the event or by detecting and measuring the non-dissociated proton. The standard rapidity gap technique is expected not to be applicable at the highest luminosity at the LHC due to the large amount of overlap events per bunch crossing, as discussed previously. However, at the startup luminosity this selection method can be used. For a luminosity of $10^{33}\,\mathrm{cm^{-2}s^{-1}}$ and all bunches "loaded," still in about 22% of the cases the bunch crossing will contain only one interaction. At $2 \cdot 10^{33}\,\mathrm{cm^{-2}s^{-1}}$ this number is reduced to 4% only. With a good control and tagging of the bunch crossings which have single collisions, as already demonstrated to be feasible at the Tevatron, one could select and use these events for diffractive studies. At higher luminosities the usage of Roman pots or similar detectors will be imperative.

A first study on diffraction can be made using the rapidity gap techniques at low bunch intensity, i.e., low luminosity and low pile-up, as expected in the first year. Such studies should establish diffraction in pp collisions at the LHC and allow the measurement of the diffractive components in the multi-TeV energy range. Evidence for hard diffraction should be seen via jet production. Also during the TOTEM runs at low luminosity the diffractive components can be measured. While running in high β^* mode, the acceptance of the TOTEM Roman pot spectrometer is very large: it has basically a full acceptance for diffraction. The luminosity obtainable is, however, low: a few times $10^{29}\,\mathrm{cm^{-2}s^{-1}}$, up to a few $10^{30}\,\mathrm{cm^{-2}s^{-1}}$. Hence with, e.g., two weeks of running this leads to data samples of 100–$500\,\mathrm{nb^{-1}}$. Such samples are useful for studies for soft diffraction, such as a precise measurement of, e.g., the different diffractive components (single diffraction, double diffraction, double pomeron exchange...) of the cross section, which can probably be measured with a precision of a few per cent. Some aspects of hard diffractive scattering can be studied, in those channels where the cross section is sufficiently high.

The combination of the central detectors and LHC Roman pots could open other new opportunities for diffractive studies. The diffractive cross sections and in particular the t dependence of the cross section can be considered as a source of information on the size and shape of the interaction region. It will be of interest to see how these quantities evolve, e.g., in the presence of a short-time perturbation which results in production of jets in $pp \to p + jet + \mathrm{jet} + p$ interactions.

An interesting and still somewhat unexplained phenomenon is the pomeron structure measurement by UA8. This experiment measured di-jet events at

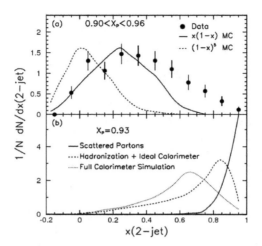

Figure 7. *(a) Observed x(di − jet) distribution for UA8 events. The curves present predictions for a hard and a soft pomeron structure function after detector smearing. (b) Curves assuming the full pomeron participates into the hard scattering, before and after simulation.*

630 GeV CM system energy in $p\bar{p}$ collisions at the $Sp\bar{p}S$ collider, for events with one proton or anti-proton tagged in the Roman pots. The measured variable shown in Fig. 7 is the total longitudinal momentum of the di-jet system in the pomeron-proton CM system normalized to $\sqrt{s\xi}/2$, namely $x(di-jet) = x(pomeron) - x(proton)$ with $x(pomeron)$ approximately the same as β defined before and $x(proton)$ the momentum fraction of the parton in the proton. The figure shows the data compared with predictions for a hard $x(1-x)$ and soft $(1-x)^5$ parton distribution in the pomeron. It appears that the data are even harder than these hard PDFs. A test with POMWIG shows that the old parton distributions of H1 are very close to the hard PDFs used here and would thus also undershoot the data. In fact in order to describe the data the authors had to introduce an additional component, which they term superhard, which is basically a delta-function at $x = 1$, i.e., the parton in the pomeron takes the *full* energy of the pomeron. The origin of this superhard component is still mysterious.

5   Central Exclusive Higgs Production

Central exclusive Higgs (CEP) production $pp \rightarrow p + H + p$ is of special interest. The diagram is shown in Fig. 8. One of the key advantages of CEP is that the $gg \rightarrow b\bar{b}$ process is strongly suppressed in LO, hence the decay

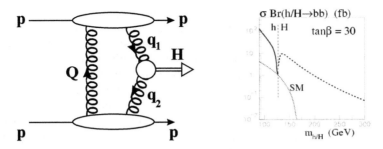

Figure 8. *Left: Diagram for the CEP process. Right: Cross section for SM and MSSM exclusive Higgs production.*

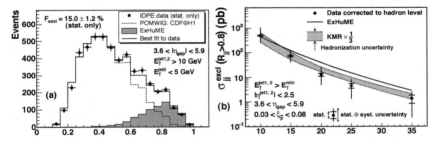

Figure 9. *Left: Di-jet mass fraction in data (points), POMWIG MC and exclusive di-jet (shaded histogram) MC events. Right: Exclusive di-jet cross sections for events with two jets and with $R_{jj} > 0.8$.*

$H \to bb$ in exclusive production has less background and becomes potentially observable. The Higgs to b-quark Yukawa coupling is otherwise very difficult to access at the LHC. The inclusive $H \to bb$ channel is not accessible due to the very large QCD backgrounds. Recently, the ttH channel was analysed with detailed simulation in [27] and found not to be accessible even with $60 \, \text{fb}^{-1}$. Also the WH associated channel was found to be marginally observable in the $b\bar{b}$ decay mode.

The cross section for the production of an SM CEP Higgs and for an MSSM CEP Higgs (for $\tan \beta = 30$) is shown in Fig. 8 (right). A crucial point here for the exclusive models is that they should not be in conflict with the exclusive diffractive jet rate measured by CDF. Tevatron data help discriminate or tune parameters in these models and will reduce the theoretical spread of these models. Recently CDF [28] and D0 [29] established and measured an exclusive di-jet component in the diffractive data, as shown in Fig. 9. Some issues with the theoretical uncertainties in comparing the calculations with data coming from the soft regime were reported in [30].

Generator-level calculations, including detector and trigger cuts, and estimates of selection efficiencies, show that the decay channels $H \to bb$ and $H \to WW$ are accessible [13]. Studies using both detailed [11] and fast [31] simulations show, however, that the measurement of the SM Higgs decay into $b\bar{b}$ will be very challenging, even at the highest luminosities.

The rate is much larger for MSSM Higgs production as shown in Fig. 8 (right), thus leading to a much more favourable signal-to-background ratio than for the SM Higgs. The cross section can be a factor 10 or more larger than the SM model one. This has been explored recently in a systematic way in [32]. A typical result is shown in Fig. 10 (left), for a Higgs decaying into $b\bar{b}$. The lines in the plot show the relative cross section increase w.r.t. the SM cross section. In some regions of the phase space the CEP process could be a discovery channel. Figure 10 (right) shows an example of a signal for $60\,\text{fb}^{-1}$ after acceptance cuts, trigger efficiencies, etc., for an MSSM Higgs with a cross section that is a factor 8 enhanced w.r.t. the SM Higgs. This is the so-called m_h^{max} scenario [33], with $m_A = 120\,\text{GeV}$ and $\tan\beta = 40$. A clear signal over background is observable.

Furthermore, to a very good approximation the central system in CEP is constrained to be a colour singlet, $J_Z = 0$ state, and, due to the strongly constrained three particle final state, the measurement of azimuthal correlations between the two scattered protons will allow the determination of the CP quantum numbers of the produced central system [34]. Hence this is a way to get information on the spin of the Higgs, and gives added value to the LHC measurements.

It was pointed out [35] that in the case of CPV models the h, A, H may mix into states h_1, h_2, h_3 which may be quasi-degenerate in mass, with mass difference of the order of a few GeV or less. Due to the interference these will show up as one broad mass distribution, with a structure that is sensitive to the underlying parameters. Analysing the three-way mixing scenario [35] it was found that the different peaks can be detected with $1\,\text{GeV}$ mass resolution, but would need a few hundred fb^{-1} of accumulated luminosity. Other CP-violating benchmark scenarios may lead to larger differences between the Higgs peaks and may be easier to detect.

With the standard TOTEM/CMS set-up it is not possible to detect both scattered protons of the exclusively produced Higgs, at low mass. This is demonstrated in Fig. 11 (right), which shows the acceptance for detectors at 220 m, at 420 m and at both 420 m and 220 m. The additional detectors at 300 and 400 m will be necessary for this study. Even including all detectors the acceptance for a 120 GeV Higgs will be about 40% only. For the exclusive channel $H(120) \to b\bar{b}$ with a cross section of 3 fb, the S/B after 30 fb^{-1}, taking into account b-tag efficiencies and event selection efficiencies (at the parton level) is about 3. This is a priori not so bad if one compares with several other fully inclusive channels for the Higgs discovery, which have been reviewed in [36].

As mentioned, exclusive Higgs production has the advantage of the spin selection rule $J_Z = 0$, which suppresses the QCD $b\bar{b}$ production processes

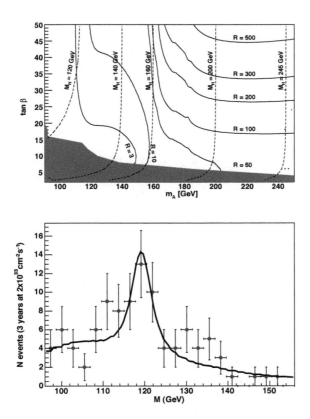

Figure 10. *Upper: Contours for the ratio of signal events in the MSSM to those in the SM in the $H \to bb$ channel in CEP production in the M_A–$\tan\beta$ plane. The ratio is shown in the no-mixing scenario with $\mu = +200\ GeV$. The values of the mass of the heavier CP-even Higgs boson, m_H, are indicated by dashed contour lines. Lower: A typical mass fit for 3 years of data taking at $2 \times 10^{33}\ cm^{-2}\ s^{-1}$ (60 fb^{-1}). The significance of the fit is 3.5σ and uses only events with both protons tagged at 420 m.*

at LO, allowing the $H \to b\bar{b}$ decay mode to be observed. Also it becomes possible to reconstruct the mass of the Higgs particle with the information of the protons only via the missing mass method: $M_H^2 = (p_1 + p_2 - p_3 - p_4)^2$ where p_1, p_2 are the four-momenta of the incident protons and p_3, p_4 those of the scattered protons. Studies show that a resolution of about 1–2% on the mass can be achieved. Inclusive diffractive production does not have these advantages.

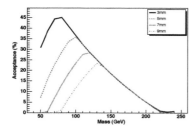 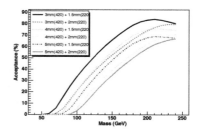

Figure 11. *The acceptance of the forward detectors as a function of central system mass for both protons tagged at 420 m (left) or an asymmetric tag of one proton at 220 m and one at 420 m (right). The dependence is shown of the acceptance on the distance of each detector from the beam.*

6   New Physics

It has been pointed out that the mass of long-lived gluinos, as predicted in split SUSY models, can be determined with CEP events to better than 1%, with 300 fb^{-1} for masses up to 350 GeV [37]. More spectacular are the predictions presented in [38], where a very high cross section of CEP WW and ZZ events is expected, in a color sextet quark model.

Since the pomeron is dominantly a gluonic object, central exclusive production could be also a way to find more exotic or other SUSY particles. An example is the radion, a scalar particle that appears in Randall–Sundrum theories. Its main difference with the Higgs particle is its stronger coupling to gluons. First studies show that large cross sections could be obtained in exclusive pp. Another exciting possibility is the production of exclusive gluino–gluino states. These could become observable if the gluino would be sufficiently light, less than 200–250 GeV, with a few 10 events for the full high LHC luminosity sample, i.e., 300 fb^{-1}. Large enough samples of exclusive gluino-gluino events could allow for a precise determination of the gluino mass and properties, much like at an e^+e^- linear collider, e.g., for very light gluinos (25–30 GeV) as discussed in some scenarios.

Finally we like to note that, since the incoming particles in the interaction are overwhelmingly dominated by gluons (by a ratio of roughly $q/g = 1/3000$), it is basically a gluon factory. These double pomeron exchange events will be the largest clean gluon jet sample available and allow for a plethora of QCD studies. A year at the LHC would, e.g., give 100,000 gluon jets with a $p_T > 50$ GeV with high purity. It is also a playground for searching for and/or studying new resonant states, like glueballs, or 0^{++} quarkonia (e.g., χ_b), in a relative background free environment.

7 Low-x

One of the most important results from HERA is the observed rise of the parton densities at small Bjorken-x, i.e., the fractional momentum of the parton w.r.t. the proton. Presently a debate is still ongoing whether the low-x data, reaching down to $x = 10^{-4} - 10^{-5}$ for scales Q^2 above or around $1\,\mathrm{GeV}^2$, has reached the region of parton saturation, i.e., a region where the parton-parton interaction probability becomes very large. In such a region one would expect to see a reduced growth of the partons due to parton recombination and shadowing mechanisms. Perhaps saturation already occurs in a very small region around valence quarks in the proton, which leads to the formation of hotspots. A saturation region will be a new regime to study QCD where the parton densities are large but α_s is still small enough to perform perturbative calculations. Pioneering studies were done by Gribov–Levin–Ryskin (GLR) in the formulation of non-linear corrections to the Dokshitzer–Gribov–Lipatov–Altarelli–Parisi (DGLAP) equation. Low-x phenomenology is reviewed in [39, 40].

A naive saturation limit based on geometrical scaling arguments leads to the saturation condition $xg(x) = 6Q^2$. Such values could be reached in the region $x = 10^{-6} - 10^{-7}$ for scales of a few GeV^2. In the case of hotspot formation, the effects should already become visible earlier. The LHC will be the first accelerator which can access this region!

As an example we show shadowing corrections which have been estimated with GLR type of corrections to the standard parton evolution equations, using the results from triple pomeron vertex calculations. These results are shown in Fig. 12 (right), for different values of the saturation radius. The effects could be very large: at $x = 10^{-6}$, and $Q^2 = 4\,\mathrm{GeV}^2$, the effect of shadowing is as large as a factor of two. For larger Q^2 values it is reduced to a 20–30% effect. An estimate of saturation effects based on different models leads to similar conclusions: the effect may become strongly visible in the region for $x = 10^{-6}$.

The LHC kinematics are shown in Fig. 12 (left). It shows the x, Q^2 plane together with the region where direct DIS measurements exist. Also indicated are the lines of rapidity in the centre of mass system of the produced heavy object (jet pair, Higgs, ...) with a mass $M = Q$. The scale extends to small $x \sim 10^{-7}$, hence it is opportune to study the small x physics potential at the LHC. Information in pp collisions on parton distributions in the lowest possible x region can come from low mass Drell–Yan production, direct photon production and Jet production.

The Drell–Yan process $q\bar{q} \to \mu^+\mu^-$ or $q\bar{q} \to e^+e^-$ has a simple experimental signature. The $x_{1,2}$ values of the two incoming quarks relate to the invariant mass of the two electron or muon system $M_{\mu\mu}$ as $x_1 \cdot x_2 \cdot s \simeq M_{\mu\mu}^2$, hence when one of the $x_{1,2}$ is large ($x > 0.1$), low-x can be probed with low mass Drell–Yan pairs.

From Fig. 12 (left) we observe that in order to reach small masses (small

LHC parton kinematics

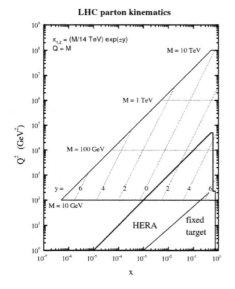

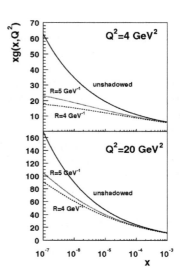

Figure 12. *Left: The kinematic plane* (x, Q^2) *and the reach of the LHC, together with that of the existing data (HERA, fixed target). Lines of constant pseudo-rapidity are shown to indicate the kinematics of the produced objects in the LHC centre of mass frame. Right: Prediction of the gluon distribution at several scales, with and without saturation effects.*

scales) and low-x will require probing large values of η. Hence the resulting electrons will dominantly go in the very forward direction. This is further illustrated with Fig. 13, made with the PYTHIA event generator, using MRST-LO parton distributions. Drell–Yan events were generated with a mass larger than 4 GeV. Figure 14 further shows the $\ln(x)$ distribution for the events with both muons accepted.

Drell–Yan processes depend on the quark densities, in which the onset of saturation effects could be less visible. It has been recently argued, however, that production of virtual photons, which also can produce a muon or electron pair when selected in proper kinematic regions, can be dominated by qg processes and hence be sensitive to the gluon distribution in the proton. Other processes which can be used to probe the gluon content of the proton are prompt photon (+jet) production and di-jet production. To reach the low-x regime, however, processes with low scales need to be accessible, e.g., jet production with E_T^{jet} of around 5 GeV.

Intimately related with low-x phenomena is the question of high-energy QCD. The domain of low-x is also the domain of BFKL theory. The BFKL equation resums multiple gluon radiation of the gluon exchanged in the t

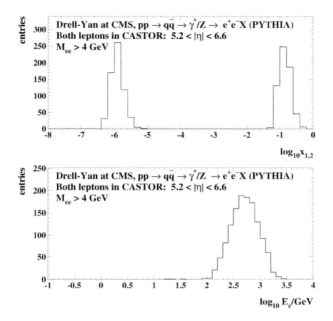

Figure 13. *Distributions of $x_{1,2}$ (top) and energy (bottom) for Drell–Yan electrons with invariant mass $M_{ee} > 4\,GeV$, both within the acceptance of CASTOR.*

channel (resums $\alpha_s \ln 1/x$ terms). It predicts a power increase of the cross section. HERA, Tevatron and even LEP have been searching for BFKL effects in the data. Presently the situation is that in some corners of the phase space the NLO DGLAP calculations are found to undershoot the QCD activity measured (e.g., forward jets and neutral pions at HERA), but BFKL has so far not been unambiguously established.

The large energy and rapidity span of the final state at the LHC may allow for a new (and possibly decisive) attempt to establish BFKL. The "golden measurements" identified so far are azimuthal decorrelations in the production of two jets far apart in rapidity, W production and heavy flavour production (e.g., production of four b-quarks). In particular the di-jet measurements require detectors which cover as large a region in rapidity as possible.

8   Two-Photon Physics

Two-photon physics has been traditionally studied at e^+e^- colliders owing to large fluxes of virtual photons associated with the beams. However, at the LHC the effective luminosity of $\gamma\gamma$ collisions, from photons radiated from

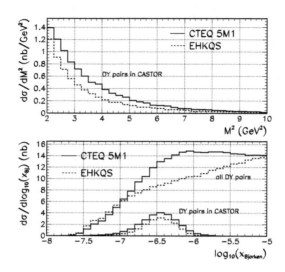

Figure 14. *The differential cross section for Drell–Yan production of e^+e^- pairs is shown for a standard parameterisation of the parton density (CTEQ 5M1) and for a "saturated" parameterisation (EHKQS) as a function of the dilepton invariant mass (top) and of Bjorken-x (bottom).*

the protons, will permit meaningful experiments to be performed. The photon spectrum can be described by the equivalent photon (or Weizsäcker–Williams) approximation (EPA). The spectrum is strongly peaked at low photon energies ω, therefore the photon–photon centre of mass energy $W \simeq 2\sqrt{\omega_1 \omega_2}$ is usually much smaller than the total centre of mass energy. For the elastic production the photon virtuality is usually low, $\langle Q^2 \rangle \approx 0.01 \, \text{GeV}^2$, therefore the proton scattering angle is very small, $\simeq 20 \, \mu\text{rad}$.

As an example, Fig. 15 shows the luminosity spectrum. The integrated spectrum directly gives a fraction of the pp LHC luminosity available for the photon–photon collisions at $W > W_0$.

The same reaction, $pp \to ppX$, occurs also in strong interactions, via double pomeron exchange as discussed earlier and will therefore interfere with the two-photon fusion. However, central diffraction usually results in much larger transverse momenta of the scattered protons, following the distribution $\exp(-bp_T^2)$, with the expected diffractive slope $b \simeq 4 \, \text{GeV}^{-2}$ at the LHC. Soft pomeron–pomeron interactions have several orders of magnitude larger cross sections than the $\gamma\gamma$ interactions, but for the hard processes the cross sections are of similar size. Therefore, the measurement of proton p_T is vital for extracting the $\gamma\gamma$ signal.

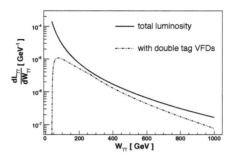

Figure 15. *Relative elastic luminosity spectrum of photon-photon collisions at the LHC in the range $Q^2_{min} < Q^2 < 2$ GeV^2 (solid line) compared to the corresponding luminosity if the energy of each photon is restricted to the forward detector (VFD) tagging range 20 GeV $< E_\gamma <$ 900 GeV (dashed curve) [41].*

Coupling limits [10^{-6} GeV^{-2}]	$\int \mathcal{L} dt = 1$ fb^{-1}	$\int \mathcal{L} dt = 10$ fb^{-1}
$|a_0^Z/\Lambda^2|$	0.49	0.16
$|a_C^Z/\Lambda^2|$	1.84	0.58
$|a_0^W/\Lambda^2|$	0.54	0.27
$|a_C^W/\Lambda^2|$	2.02	0.99

Table 1. *Expected one-parameter limits for anomalous quartic vector boson couplings at 95% CL [41].*

The physics program for two-photon physics at the LHC contains topics such as exclusive Higgs production, WW production and SUSY particle production. A summary of various unique photon–photon and photon–proton interactions accessible to measurement at the LHC, is discussed in detail in [41]. Interesting studies and searches can be performed for initial integrated luminosities of about 1 fb^{-1}, such as exclusive dimuon production in two-photon collisions tagged with forward large rapidity gaps. At higher luminosities, the efficient selection of photon-induced processes is greatly enhanced with dedicated forward proton taggers such as FP420. Photon-induced reactions can provide much higher sensitivity than partonic reactions for various BSM signals such as, e.g., anomalous quartic $\gamma\gamma WW$ gauge couplings, as shown in Table 1.

The associated photoproduction of a top quark or a W boson is also very large, offering a unique opportunity to measure the fundamental Standard

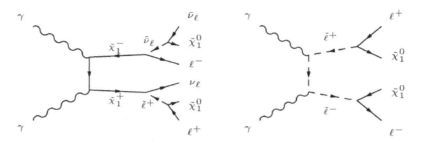

Figure 16. *Relevant Feynman diagrams for SUSY pair production with leptons in the final state: chargino disintegration in a charged/neutral scalar and a neutral/charged fermion (left); slepton disintegration (right) [41].*

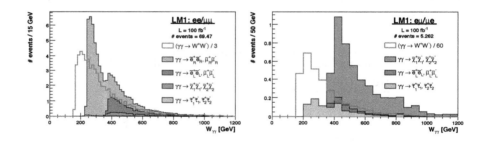

Figure 17. *Photon-photon invariant mass for benchmark point LM1 with $\int \mathcal{L}dt = 100\,fb^{-1}$. Cumulative distributions for signal with two detected leptons ($p_T > 3\,GeV$, $|\eta| < 2.5$), two detected protons, with same (left) or different flavour (right). The WW background has been down-scaled by the quoted factor [41].*

Model parameters, such as the top quark charge or the V_{tb} element of the quark mixing matrix. Anomalous γqt couplings might also be uniquely revealed in single top photoproduction. Larger integrated luminosity, of about one hundred inverse femtobarns, will open complementary ways to search for production of supersymmetric particles in photon-photon interactions. The diagrams for chargino and slepton production are shown in Fig. 16. The photon-photon invariant mass for a benchmark point [27] is shown in Fig. 17.

Even larger luminosities might help access important information on the Higgs boson coupling to b quarks and W bosons. FP420 detectors are mandatory for the determination of the masses of the centrally produced particles, and to increase the sensitivity to new anomalous couplings contributions in two-photon interactions.

Studying the photon-induced processes in the early LHC runs can provide valuable checks of the various components of the general formalism used to predict the cross sections of central exclusive reactions [42]. Thus, the photon-exchange-dominated W-boson production with rapidity gaps on either side provides information on the gap survival factor S^2. As discussed in [42], such studies can be performed even without tagging of the forward proton. Another example is exclusive Υ photoproduction induced by the process $\gamma p \to \Upsilon p$ [43], which has now been observed by CDF [44]. The study of such processes will not only reduce the theoretical uncertainties associated with the generalised, unintegrated gluon distributions f_g, e.g., by testing models based on diffusion in transverse momentum as incarnated in the Balitsky–Fadin–Kuraev–Lipatov (BFKL) equation [45], but will also be of help to calibrate and align the forward proton detectors at 420 m.

Two-photon events could also allow for a more precise determination of the luminosity at the LHC. The methods proposed so far are based on processes that involve parton distributions in the proton, which are not known to better than a few per cent, hence one can hope for a precision of 4–5% at best from, e.g., measuring the luminosity using the production of W's. Two-photon processes are, however, QED processes and can be calculated at the per cent level or better (since the proton form factors have been measured sufficiently precisely). The processes $pp \to ppe^+e^-$ or $pp \to pp\mu^+\mu^-$ have been shown to be good candidates for luminosity measurements, if sufficiently clean data samples can be extracted.

In addition, two-photon physics, γp interactions also become available, and can be studied at higher energies than available at HERA.

9 Forward Physics and Cosmic Rays

The forward LHC data are also very relevant for cosmic rays analyses. The energy of the collisions at the LHC correspond to an incident energy of a proton on a fixed target proton of 10^{17} eV, as shown in Fig. 18. This is (on a logarithmic scale) exactly in the middle between the so-called knee and the ankle of the measured cosmic ray particle energy spectrum.

The development of a cosmic ray shower is depicted in Fig. 19. A primary particle penetrates the Earth's atmosphere and starts a shower of particles which can in turn interact further in the atmosphere. The starting point and shape of the shower depends on the incoming particle, as shown in Fig. 19. One important method to reconstruct the incident particle type and energy is to use surface measurements of the electromagnetic and muon content of the showers, and then reconstruct the shower "bottom-up." Such a reconstruction relies on a model for the interaction (number of particles produced, energies of the particles etc.). In particular the forward part of the shower is of extreme importance. Also the poorly understood cross section decomposition into diffractive and non-diffractive events is a key ingredient in these models. Several different models are used in the cosmic ray community, tuned

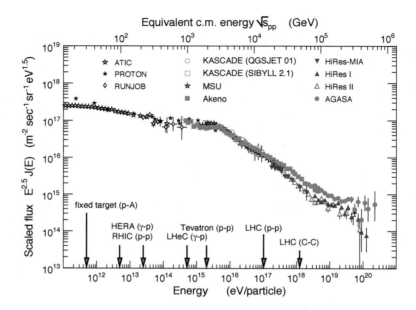

Figure 18. *Cosmic ray energy spectrum with present and future accelerators indicated [46].*

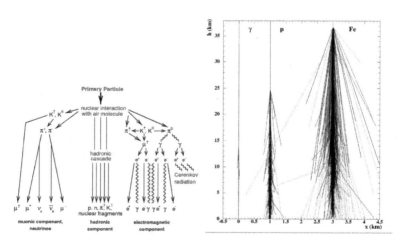

Figure 19. *Left: Pictoral development of a cosmic ray air shower with its different components. Right: Shower shapes and indicative starting points or photon-, proton- and iron-induced showers.*

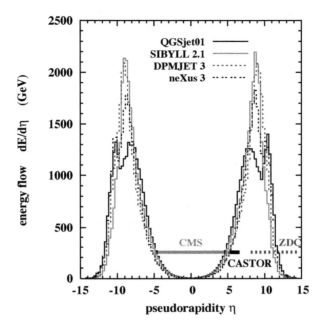

Figure 20. *Pseudorapidity spectra of particles as predicted by air shower models, with the CMS acceptance overlaid [47].*

to as much relevant accelerator data as is available. When comparing the predictions of the models in Figs. 20 and 21 at the energy of the LHC, e.g., the momentum fraction (Feynman-x) taken by the leading particle, one finds differences larger than a factor of 2. Note that the Auger experiment will measure and reconstruct cosmic rays with energies of 10^{20} eV! Hence accurate data on particle production in the forward region at the highest energies reachable at accelerators will be very useful to constrain these models. This explains the strong interest of the cosmic ray community in this part of the program at the LHC.

A wish list produced by the cosmic ray community for measurements includes

- Measurement of the total diffractive and elastic cross section

- Leading hadron distributions

- General event features such as event multiplicities, correlations and low p_T jets in both the forward and central regions

Cosmic rays also form a motivation for forward physics. In very high-energy cosmic rays some exotic events have been observed. Most notable are

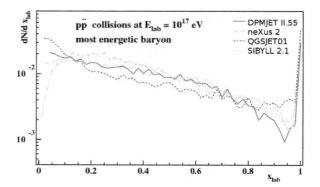

Figure 21. *Comparison of different x_F spectra predictions for air shower models for pp collisions at the LHC centre of mass energy.*

the so-called centauro events, which seem to be hadronic showers with a very small or no electro-magnetic content. Such events have never been observed at an accelerator. Perhaps the energy thus far has not been high enough? While the signal for these kinds of events is surely not unambiguous, several possible mechanisms to explain these events have been proposed: exotic extraterrestrial globs of matter, diffractive fireballs, disoriented chiral condensates, strange quark matter.... Hence, forward physics at the LHC could well lead to a few "unexpected" surprises.

10   Running LHC at Lower Energies

At the time of the lectures the LHC was supposed to start up at 7 TeV and then move to 10 TeV for most of the time. The time plan has changed as a result of the increased safety imposed on the machine, and the immediate consequences were that LHC started end of 2009 and took about $10\,\mu b^{-1}$ of 900 GeV CM energy data, and about $0.4\,\mu b^{-1}$ of 2.36 TeV data. Since March 30th 2010, the machine ran at a CM energy at 7 TeV and continued to do so until end of 2011. Then the machine will go for at least a year in a shutdown to repair the connectors between the magnets, and come back at a CM energy of 14 TeV or close to that value. Possibly also pp data at 2.8 TeV CM energy could be envisaged, which would be very useful for direct calibration of the AA collisions. These data below 14 TeV allow for interesting measurements to study the energy dependences of processes or comparisons of data with the Tevatron, which produces $p\bar{p}$ collisions instead of pp collisions at the LHC.

An interesting topic is remeasuring the total cross section at a CM system energy of 2 TeV. The measurements at Tevatron from CDF and E710 show a discrepancy of several sigmas. Hence TOTEM could go back and remeasure the total cross section with a precision of 1–2 mb and thus referee these results. At the same time it would be a good consistency check of the TOTEM method at the LHC machine. The total cross section at intermediate energies can also be measured.

BFKL studies can be made in a similar way as was done at the Tevatron by comparing di-jet production with a jet separation in rapidity chosen such that the parton distribution effects cancel in the ratio of the rates at two different CM system energy measurements. Taking into account the acceptance of the detectors, it appears that 14 TeV and 7 TeV are the excellent energies for such measurements.

Other topics include the energy dependence of gap survival probabilities, inclusive jets and many others.

11   Conclusions

The LHC has come on line, with the first 7 TeV collisions being produced since the end of March 2010. The first signatures of diffraction at the highest energies will certainly be shown to the world soon. This will be the start of an exciting physics program; especially at the start-up luminosities the conditions are favourable for many diffractive studies.

Forward physics at the LHC came a long way during the last years. Two forward physics experiments have been approved (TOTEM, LHCf). ATLAS and CMS have extended the detector coverage in the forward direction, with ZDCs, CASTOR (CMS) and LUCID (ATLAS). ATLAS also plans to add RPs at 240 m and studies additional near-beam detectors at 220 m. CMS and TOTEM have in principle a common physics program on diffraction. The R&D for FP420 is completed and discussions with the ATLAS and CMS management have started.

In summary, forward physics is now in the blood of the LHC experiments.

References

[1] FELIX Collab., http:/felix.web.cern.ch/FELIX/.

[2] Proceedings of the workshop on Forward Physics and Luminosity Determination at the LHC, Editors K. Huiti, V. Khoze, R. Orava, S. Tapprogge, (Singapore, World Scientific) (2001).

[3] 13th International Conference on Elastic and Diffractive Scattering (Blois Workshop): Moving Forward into the LHC Era; M. Deile, A. De Roeck and D. D'Enterria Eds. Feb. 2010, arXiv:1002.3527.

[4] http://www.lishep.uerj.br/lishep2009/.

[5] http://www.cs.infn.it/diff2008/.

[6] Proc. of XVII Int. Workshop on Deep-Inelastic Scattering and Related Topics, Madrid, Spain, April 2009.

[7] U. Amaldi et al., Phys. Lett. **B36** (1971) 504.

[8] http://web.me.com/jamespinfold/MoEDAL_site/Welcome.html.

[9] TOTEM Collaboration, TOTEM Technical Design Report, CERN/LHCC 2004-002 (2004).

[10] TOTEM Collaboration, Addendum to the TOTEM-TDR, CERN/LHCC 2004-020 (2004).

[11] M. G. Albrow et al., *Prospects for diffractive and forward physics at the LHC.* CERN-LHCC-2006-039, CERN-LHCC-G-124, CERN-CMS-NOTE-2007-002, Dec 2006. 156 pp.

[12] A. De Roeck, V. A. Khoze, A. D. Martin, R. Orava and M. G. Ryskin, Eur. Phys. J. **C25** (2002) 391.

[13] B. E. Cox et al., Eur. Phys. J. **C45** (2006) 401.

[14] M. Albrow et al., JINST **4** (2009) P10001.

[15] M. Heller, 13th International Conference on Elastic and Diffractive Scattering (Blois Workshop), M. Deile, A. De Roeck and D. D'Enterria Eds. Feb. 2010, arXiv:1002.3527.

[16] J. P. Revol, talk at the TeV4LHC workshop at CERN 28-30 April 2005.

[17] R. Schicker, 13th International Conference on Elastic and Diffractive Scattering (Blois Workshop), M. Deile, A. De Roeck and D. D'Enterria Eds. Feb. 2010, arXiv:1002.3527.

[18] J. Lamsa and R. Orava, JINST **4** (2009) P11019.

[19] M. G. Albrow et al. *FP420: An R& D proposal to investigate the feasibility of installing proton tagging detectors in the 420-m region at LHC*, CERN-LHCC-2005-025, Jun 2005.

[20] M. G. Albrow et al., JINST **4** (2009) T10001.

[21] M. G. Ryskin et al., J. Phys. G. **36** (2009) 093001.

[22] C. Amsler et al. (Particle Data Group), Phys. Lett. **B667** (2008) 1 (2008).

[23] J. R. Cudell, arXiv:0911.3508.

[24] *An Introduction to Regge Theory and High Energy Physics*, P.D.B. Collins, (Cambridge) (1977).

[25] *Soft multihadron dynamics*, W. Kittel, E. A. De Wolf (Hackensack, NJ, USA: World Scientific) (2005).

[26] K. Ellis, these proceedings

[27] CMS Collaboration *CMS physics, Technical Design Report v.2, Physics performance* J. Phys. G: Nucl. Part. Phys. **34** (2007) 995.

[28] CDF Collaboration, T. Aaltonen et al., Phys. Rev. **D77** (2008) 052004.

[29] D0 Collaboration, http://www-d0.fnal.gov/Run2Physics/WWW/results/prelim/QCD/Q17/.

[30] J. Cudell et al., Eur. Phys. J. **C6** (2009) 390.

[31] B. E. Cox, F. K. Loebinger and A. D. Pilkington, JHEP **0710** (2007) 090.

[32] S. Heinemeyer et al., Eur. Phys. J. **C53** (2008) 231.

[33] M. S. Carena, S. Heinemeyer, C. E. M. Wagner and G. Weiglein, Eur. Phys. J. **C26** (2003) 601.

[34] V. A. Khoze, A. D. Martin and M. G. Ryskin, Eur. Phys. J. **C34** (2004) 327.

[35] J. Ellis, J. S. Lee and A. Pilaftsis, Phys. Rev. **D71** (2005) 075007.

[36] A. De Roeck et al., Eur. Phys. J. **C25** (2002) 391.

[37] P. J. Bussey, T. D. Coughlin, J. R. Forshaw and A. D. Pilkington, JHEP **0611** (2006) 027.

[38] A. R. White, Phys. Rev. **D72** (2005) 036007.

[39] A. M. Cooper-Sarkar, R. C. E. Devenish, A. De Roeck, Int. J. Mod. Phys. **A13** (1998) 3385.

[40] D. G. d'Enterria, arXiv:0708.0551.

[41] J. Favereau de Jeneret et al., arXiv:0908.2020.

[42] V. A. Khoze, A. D. Martin and M. G. Ryskin, Eur. Phys. J. **C55** (2008) 363.

[43] J. Hollar, S. Ovyn and X. Rouby, CMS PAS DIF-07-001 (CMS AN-2007/032).

[44] A. Abulencia et al. [CDF Collaboration], in preparation.

[45] Y. Y. Balitski and L. N. Lipatov, Sov. J. Nucl. Phys., **28** (1978) 822.

[46] R. Engel, Nucl. Phys. Proc. Suppl. 151 (2006) 437.

[47] R. Engel and H. Rebel, Acta Phys. Polon. **B35** (2004) 321.

Heavy-Ion Physics

Raimond Snellings

Nikhef, Amsterdam, The Netherlands

1 Introduction

One of the fundamental questions in the field of subatomic physics is what happens to matter at extreme densities and temperatures as may have existed in the first microseconds after the big bang and exist, perhaps, in the core of dense neutron stars. The aim of heavy-ion physics is to collide nuclei at very high energies and thereby create such a state of matter in the laboratory. The experimental program started in the 1990s with collisions made available at the AGS and SPS with energies up to 20 GeV per nucleon in the centre of mass, and continued at the Relativistic Heavy Ion Collider in Brookhaven, USA with energies of up to 200 GeV per nucleon. Collisions of heavy ions at the unprecedented energy of 5.5 TeV will soon be made available at the LHC collider at CERN, Geneva, Switzerland. In these lectures I will give a brief introduction to the physics of ultra-relativistic heavy-ion collisions and review some selected highlights of the current and future experimental program. The material covered in these lectures is published in, amongst others, [1–8], to which I refer the reader for more details.

2 The QCD Phase Transition and Equation of State

In the last 30 years particle physics has led to a profound understanding of the world around us, summarized in the so-called "Standard Model." It provides a coherent and precise description of the building blocks of matter and the three fundamental interactions: the weak, the strong and the electromagnetic. However, at the same time we realize that this model is far from complete. In fact we have learned that we do *not* know what most of the universe is made of. To answer the questions, what the universe is made of and how it works, is the ultimate challenge of particle physics.

In our current understanding, the universe went through a series of phase transitions after the Big Bang. These phase transitions mark the most important epochs of the expanding universe. At 10^{-11} s after the Big Bang and at a temperature $T \sim 100$ GeV ($\sim 10^{15}$ K) the electroweak phase transition took place. At this time most of the known elementary particles acquired their Higgs masses. At 10^{-5} s and at a temperature three orders of magnitude lower (170 MeV $\sim 10^{12}$ K), the strong phase transition took place. During the strong phase transition the quarks and gluons became confined in hadrons. At the same time the approximate chiral symmetry was spontaneously broken. This symmetry is crucial in the standard model and gives rise to the presence of the light pions.

The underlying theory of the strong force, QCD, is well established even though its fundamental degrees of freedom, the quarks and gluons, cannot be observed as free particles due to confinement. The known QCD Lagrangian provides in principle the complete picture but the QCD field equations are notoriously hard to solve. In fact, the two most important and interesting properties, confinement and chiral symmetry breaking, are still poorly understood from first principles.

One of the key features of QCD is that the strength of the coupling between quarks and gluons depends on their relative momenta. At higher momenta and thus smaller distances the coupling becomes weaker, leading to so-called asymptotic freedom. Therefore, in a QCD system at very high temperatures the quarks and gluons are expected to become quasi-free so that the bulk properties can be described by an ideal gas Equation of State (EoS). This deconfined dense state of matter is called a Quark Gluon Plasma (QGP). Properties like energy density and pressure provide direct information about the EoS and thus about the basic degrees of freedom.

Dimensional arguments allow us to estimate the critical energy density $\epsilon_c \sim 1$ GeV/fm^3 and temperature $T_c \sim 170$ MeV. However, these values imply that the transition occurs in a regime where the coupling constant is of order unity, casting doubts on results of perturbative calculations. Better understanding of the non-perturbative domain comes from lattice QCD calculations, where the field equations are solved numerically on a discrete spacetime grid. Lattice QCD provides quantitative information on the QCD phase transition between confined and deconfined matter and the EoS. Figure 1a shows the calculated energy density as a function of temperature. It is seen that the energy density changes rapidly at the critical temperature $T_c \sim 170$ MeV, which is due to the rapid increase in the effective degrees of freedom. From these lattice calculations it follows that at T_c not only deconfinement sets in but that also chiral symmetry is restored. The pressure, shown in Fig. 1b, changes slowly at T_c compared to the rapid increase of the energy density. Therefore the pressure gradient in the system, $dP/d\epsilon$, is significantly reduced during the phase transition.

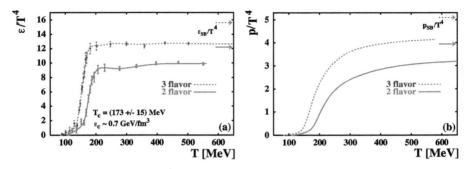

Figure 1. *(a) Energy density ϵ as a function of temperature from lattice calculations. For an ideal gas the energy density is proportional to the number of thermal degrees of freedom (g). This causes the sharp increase at T_c where the system changes from approximately a pion gas, g = 3, into a quark gluon plasma where g = 37 in the case of two quark flavours. The arrows in the figure indicate the ideal Stefan–Boltzmann values. (b) The pressure from lattice calculations versus the temperature. The pressure also reflects the number of degrees of freedom but changes slowly at the phase boundary.*

In the limit of an ideal Stefan–Boltzmann gas the EoS of a QGP is given by:

$$P_{\mathrm{SB}} = \frac{1}{3}\epsilon_{\mathrm{SB}}, \quad \epsilon_{\mathrm{SB}} = g\frac{\pi^2}{30}T^4, \tag{1}$$

$$g = n_f \times 2_s \times 2_q \times 3_c \times \frac{7}{8} + 2_s \times 8_c, \tag{2}$$

where P_{SB} is the pressure, ϵ_{SB} the energy density and T the temperature. Each bosonic degree of freedom contributes $\frac{\pi^2}{30}T^4$ to the energy density; each fermionic degree of freedom contributes $\frac{7}{8}$ of this value. The value of g is obtained from the sum of the appropriate number of flavours $\times$ spin $\times$ quark/antiquark $\times$ colour factors for the quarks and spin $\times$ colour for the gluons. The energy density for a two (three) flavour QGP, where $g = 37$ ($g = 47.5$) is an order of magnitude larger than for a hadron gas where $g \sim 3$ (π^+, π^- and π^0). The corresponding Stefan–Boltzmann values of the energy density and pressure are plotted in Figs. 1(a) and 1(b) and show that the lattice results reach a significant fraction (0.8) of these values. The deviation from the Stefan–Boltzmann limit shows that the QCD system around T_c does *not* behave like a weakly interacting parton gas.

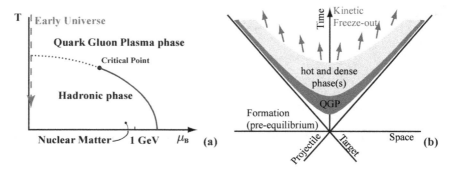

Figure 2. *(a) Theoretical phase diagram of nuclear matter for two massless quarks as function of temperature T and baryon chemical potential μ_B. (b) Illustration of the characteristic periods in time for a heavy-ion collision.*

3   Heavy-Ion Collisions

Figure 2(a) shows a theoretical phase diagram of nuclear matter for two mass-less quarks as function of temperature and baryon chemical potential. Relativistic heavy-ion collisions are a unique tool to test this phase diagram by studying deconfinement and the EoS of hot QCD matter under controlled conditions. Like the early universe, the hot and dense system created in a heavy-ion collision will expand and cool down. In this time evolution the system probes a range of energy densities and temperatures, and possibly different phases. The evolution of the created system can be divided in two characteristic periods; see Fig. 2(b). During the formation of the system ($\leq 3 \times 10^{-24}$ sec) collisions with large momentum transfer occur. During this period the largest energy density is created. The system will thermalize and form the QGP provided that the quarks and gluons undergo multiple interactions. Due to the thermal pressure, the system undergoes a collective expansion and eventually becomes so dilute that it hadronizes. In the hadronic phase it further cools down via inelastic and elastic interactions until it becomes non-interacting (the freeze-out stage).

To study QCD at extreme densities ultra relativistic heavy-ion experiments have been performed at the Brookhaven Alternating Gradient Synchrotron (AGS), the CERN Super Proton Synchrotron (SPS) and the Brookhaven Relativistic Heavy Ion Collider (RHIC) with maximum center of mass energies of $\sqrt{s_{\mathrm{NN}}} = 4.75$, 17.2 and 200 GeV respectively. The future Large Hadron Collider (LHC) will make Pb–Pb collisions available at an unprecedented energy of $\sqrt{s_{\mathrm{NN}}} = 5.5$ TeV. Some of the main probes available in heavy-ion collisions will be described in the next section, together with a few selected results which are considered to be the highlights of the experimental program so far.

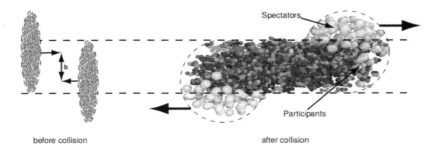

Figure 3. *Left: The two heavy-ions before collision with impact parameter* **b**. *Right: The spectators continue unaffected, while in the participant zone particle production takes place.*

3.1 Probes and Observables

3.1.1 Event Characterization

Heavy ions are extended objects and the system created in a head-on collision is different from that in a peripheral collision. Therefore, collisions are categorized by their centrality. Theoretically the centrality is characterized by the impact parameter **b** (see Fig. 3) which is, however, not a direct observable. Experimentally, the collision centrality can be inferred from the measured particle multiplicities if one assumes that this multiplicity is a monotonic function of **b**. Another way to determine the event centrality is to measure the energy carried by the spectator nucleons (which do not participate in the reaction) with Zero Degree Calorimetry (ZDC). A large (small) signal in the ZDC thus indicates a peripheral (central) collision.

Instead of by impact parameter, the centrality is often characterized by the so-called number of wounded nucleons or by the number of equivalent binary collisions. These measures can be related to the impact parameter **b** using a realistic description of the nuclear geometry in a Glauber calculation; see Fig. 4. Phenomenologically it is found that soft particle production scales with the number of participating nucleons whereas hard processes scale with the number of binary collisions.

3.1.2 Global Observables

Examples of global observables which provide important information about the created system are the particle multiplicity and the transverse energy. Figure 5 shows the transverse energy versus the collision centrality as measured at $\sqrt{s_{NN}} = 130$ GeV by the PHENIX collaboration. This measurement allows for an estimate of the energy density as proposed by Bjorken for head-on collisions

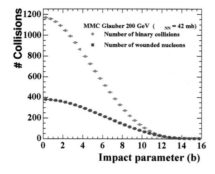 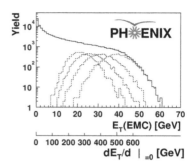

Figure 4. *Number of wounded nucleons and binary collisions versus impact parameter.*

Figure 5. *Transverse energy as a function of centrality as measured by PHENIX.*

$$\epsilon = \frac{1}{\pi R^2} \frac{1}{c\tau_0} \frac{\mathrm{d}E_\mathrm{t}}{\mathrm{d}y},$$

were R is the nuclear radius and τ_0 is the effective thermalization time (0.2–1.0 fm/c). From the measured $\langle \mathrm{d}E_\mathrm{t}/\mathrm{d}\eta \rangle = 503 \pm 2$ GeV it follows that ϵ is about 5 GeV/fm³ at RHIC. This is much larger than the critical energy density of 1 GeV/fm³ obtained from Lattice QCD (see Fig. 1).

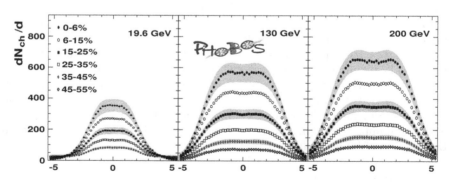

Figure 6. *Multiplicity versus pseudo-rapidity for 19.6, 130 and 200 GeV measured by PHOBOS.*

Figure 6 shows the charged particle multiplicity distributions versus the pseudorapidity η measured by PHOBOS at three different energies. Notice that in total about 5000 charged particles are produced in the most central Au + Au collisions at the top RHIC energy.

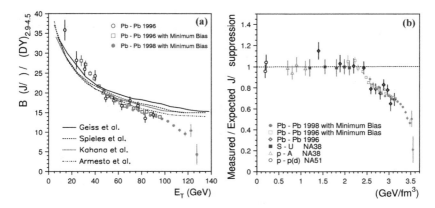

Figure 7. *(a) Comparisons between the NA50 Pb–Pb measured J/Ψ data and several calculations of the suppression. (b) The J/Ψ anomalous suppression as a function of the energy density reached in the collisions. Suppression is obtained from the measured cross sections divided by the values expected from nuclear absorption. For higher-energy densities an increase in the amount of suppression is observed, as can be explained by charmonium melting due to deconfinement.*

3.1.3 J/Ψ Suppression

One of the most promising QGP signatures at SPS energies has been the J/Ψ suppression predicted by T. Matsui and H. Satz. This prediction is based on the idea that in the plasma phase the interaction potential is expected to be screened beyond the Debye length. This will prevent $c\bar{c}$ states with a radius greater than the Debye length, such as the J/Psi resonance, from forming. Observations of the suppression by the NA50 experiment are shown in Fig. 7. Figure 7(a) shows the J/Ψ production normalized to the Drell–Yan yield and compared to calculations of the suppression expected in nuclear matter. Figure 7(b) shows the ratio of measured to expected J/Ψ suppression. For higher energy densities an anomalously large suppression is observed as is expected from charmonium melting in a QGP.

The PHENIX collaboration at RHIC also observed a significant J/Ψ suppression in central Au-Au collisions. They have shown that the J/Ψ suppression at these higher energies is indeed larger than that expected from a extrapolating of cold nuclear matter effects as measured in d-Au collisions. Contrary to expectations the suppression at RHIC is found to be similar to that observed at the SPS. The models that described the J/Ψ suppression at the SPS all predicted a significantly larger suppression at RHIC. At these RHIC energies, however, central collisions produce multiple pairs of heavy quarks. These multiple c and $\bar{c}$ quarks, originally produced in separate incoherent interactions, might coalesce and form a J/Ψ. This additional

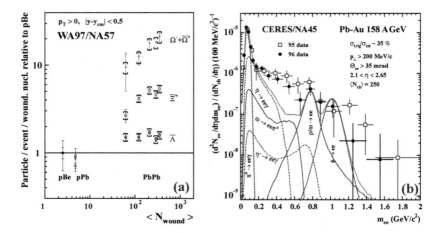

Figure 8. *(a) Strangeness enhancements measured by the NA57 experiment. The enhancements are defined as the particle yields normalized by the number of participating nucleons in the collision, and divided by the observed yield in proton-beryllium collisions. The yields expected from a simple superposition of nucleon-nucleon collisions would then lie on a straight line positioned at unity. (b) Normalized invariant-mass spectra of dilepton pairs. The measured yield is compared to the known hadronic decay sources, showing the individual contributions. At low invariant mass a clear excess of the dilepton yield is observed.*

production mechanism complicates the interpretation, and can at RHIC be responsible for some regeneration of the J/Ψ yield. At LHC energies this production mechanism could actually become so important that there is no longer a J/Ψ suppression but instead a J/Ψ enhancement. This would be an unambiguous signature of quark coalesence.

3.1.4 Strangeness Production

Strange particles produced in heavy-ion collisions give important information on the collision mechanism. In particular, if a phase transition to a QGP state takes place, one would expect an enhancement in the yields of strange and multi-strange particles in nucleus-nucleus reactions compared to those from proton-nucleus interactions. In fact, the formation of such a state will lead to equilibration of strange quarks on a time scale of a few fm/c, and to the formation of multi-strange baryons and antibaryons close to thermal and chemical hadronic equilibrium. Their abundances will be frozen at the critical temperature T_c since hadronic reactions are too slow to compete with the rapid collective expansion of the fireball at temperatures below T_c. It is expected that the enhancement should be more pronounced for multi-strange than for singly strange particles.

The experimental results on Λ, Ξ and Ω production from the WA97/NA57 collaboration are plotted in Fig. 8(a) and indeed show this predicted enhancement. However, the NA49 collaboration has found that this enhancement is already present in small colliding systems, e.g., Si-Si collisions. Since QGP formation in such small systems is perhaps less likely, an alternative enhancement mechanism could be the reduction of canonical suppression in an extended system.

3.1.5 Dileptons

Correlated electron–positron pairs (dileptons) provide a probe of the expanding system at an early stage. The absence of any final state interaction conserves the primary information within the limits imposed by the space-time folding over the emission period. In the low-mass region, the thermal radiation is dominated by the decays of the light vector mesons ρ, ω and ϕ. The ρ is of particular interest, due to its short lifetime of 1.3 fm/c, therefore its in-medium behavior around the critical temperature provides a direct link to chiral symmetry restoration. The shape of the measured dilepton yield by NA45/CERES, as shown in Fig. 8(b), can be explained by a strong medium modification of the intermediate ρ. This modification can theoretically be described by a reduction in mass (as a precursor of chiral symmetry restoration), known as Brown–Rho scaling, or by a spreading of the width in a hadronic medium. More recent experimental results from the NA60 collaboration, with much-improved accuracy, show that the space-time averaged ρ spectral function is strongly broadened, but not shifted in mass.

3.1.6 Particle Yields

The measurement of the integrated yields of the various hadron species produced in the collisions (chemical composition) provides information on the chemical freeze-out temperature and baryon chemical potential. These two parameters are obtained from a fitting a thermal model to the yields. In Fig. 9(a) the yields measured at RHIC are compared to those obtained from a the fit to the thermal model, where all the particles yields are characterized by a single chemical freeze-out temperature of 164 MeV and a single baryon chemical potential of 30 MeV. This freeze-out temperature is very close to the critical temperature predicted by lattice QCD calculations.

The statistical description of the hadron yields is not only successful for heavy-ion collisions at RHIC energies but also at lower energies. This allows us to plot the chemical freeze-out temperature versus μ_B obtained at different energies in a phase diagram (see Fig. 9(b)). It is seen that at chemical potentials of less than about 400 MeV, the temperatures trace the phase boundary predicted by lattice QCD (dashed curve in Fig. 9(b)). This observation of a limiting temperature is considered to be a strong indication that the relevant degrees of freedom have changed from that of hadronic matter.

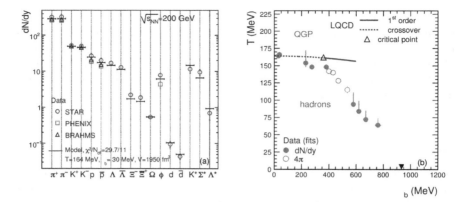

Figure 9. *(a) The measured hadron yields at RHIC compared to the best fit from a thermal model. (b) The phase diagram of QCD matter in the $T - \mu_B$ plane. The full and open symbols show the obtained chemical freeze-out temperature and μ_B obtained from fitting the thermal model to the measured hadron yields. A lattice QCD estimate of the location of the critical point is indicated by the open triangle. The dashed and full lines represent the lattice QCD prediction for the crossover and first-order phase transition from a QGP to a hadron gas, respectively.*

3.1.7   Spectra

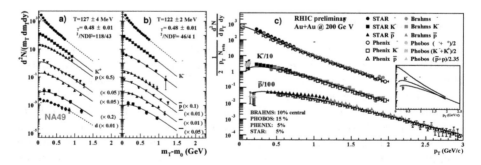

Figure 10. *NA49 (SPS) and RHIC low-p_t spectra. (a) and (b) show the transverse momentum measured by NA49 at $\sqrt{s_{NN}} = 17$ GeV. The lines are a fit of a hydrodynamic model to the transverse momentum spectra. (c) shows the transverse momenta measured at RHIC.*

The particle spectra provide much more information than the integrated particle yields alone. The particle yields as a function of transverse momentum reveal the dynamics of the collision, characterized by the temperature and

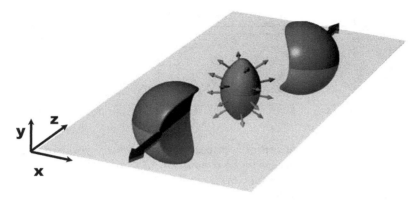

Figure 11. *Almond-shaped interaction volume after a non-central collision of two nuclei. The spatial anisotropy with respect to the x–z plane (reaction plane) translates into a momentum anisotropy of the particles produced in the collision (anisotropic flow).*

transverse flow velocity of the system at freeze-out. The kinetic freeze-out corresponds to the final stage of the collision when the system becomes so dilute that all interactions between the particles cease to exist so that the momentum distributions do not change anymore. Figures 10(a) and (b) show the transverse momentum distributions at $\sqrt{s_{NN}} = 17$ GeV from NA49. The lines are a fit to the particle spectra with a hydrodynamically inspired model (blast wave). The fit describes all the particle spectra rather well which shows that these spectra can be characterized by the two parameters of the model: a single kinetic freeze-out temperature and a common transverse flow velocity. Fig. 10(c) shows the combined pion, Kaon and proton p_t-spectra from the four RHIC experiments. Also at these energies, a common fit to all the spectra shows that the system seems to freeze out with a temperature and with a transverse flow velocity similar to those observed at SPS energies.

3.1.8 Anisotropic Flow

Flow is an ever-present phenomenon in nucleus–nucleus collisions, from low-energy fixed-target reactions up to $\sqrt{s_{NN}} = 200$ GeV collisions at the Relativistic Heavy Ion Collider (RHIC), and is expected to be observed at the Large Hadron Collider (LHC). Flow signals the presence of multiple interactions between the constituents and is an unavoidable consequence of thermalization.

The usual theoretical tools used to describe flow are hydrodynamic or microscopic transport (cascade) calculations. Flow depends in the transport models on the opacity, be it partonic or hadronic. Hydrodynamics becomes valid when the mean free path of particles is much smaller than the system size and allows for a description of the system in terms of macroscopic quantities.

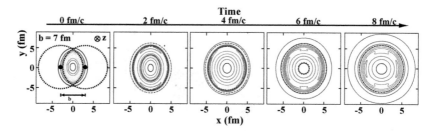

Figure 12. *The created initial transverse energy density profile and its time dependence in coordinate space for a non-central heavy-ion collision. The z-axis is along the colliding beams, the x-axis is defined by the impact parameter* **b** *(the vector connecting the centers of the colliding heavy-ions, perpendicular to the beam axis).*

This can be used to determine the equation of state of the flowing matter and, in particular, on the value of the sound velocity, which is a measure of the magnitude of the collective flow. In both types of models it may be possible to deduce from a flow measurement whether the flow originates from partonic or hadronic matter or from the hadronization process.

Experimentally the most direct evidence of flow comes from the observation of anisotropic flow, which is the anisotropy in the particle momentum distributions correlated with the reaction plane. A sketch of a heavy ion collision is shown in Fig. 11. The reaction plane is defined by the impact parameter, x, and the beam direction z. The resulting spatial anisotropy with respect to the reaction plane translates into a momentum anisotropy of the particles produced in the collision. The energy density profile and its evolution with time in a non-central collision is shown in Fig. 12.

A convenient way of characterizing the various patterns of anisotropic flow is to use a Fourier expansion of the triple differential invariant distributions:

$$E\frac{\mathrm{d}^3N}{\mathrm{d}^3\mathbf{p}} = \frac{1}{2\pi}\frac{\mathrm{d}^2N}{p_t\mathrm{d}p_t,\mathrm{d}y}\left\{1 + 2\sum_{n=1}^{+\infty} v_n \cos[n(\varphi - \Psi_R)]\right\},$$

where φ and Ψ_R are the azimuthal angles of a particle produced in the collision and the reaction-plane, respectively. The sine terms in such an expansion vanish due to reflection symmetry with respect to the reaction plane. The Fourier coefficients are given by

$$v_n(p_t, y) = \langle\cos[n(\varphi - \Psi_R)]\rangle,$$

where the angular brackets denote an average over the particles, summed over all events, in the (p_t, y) bin under study. In this parameterization, the first two coefficients, v_1 and v_2, are known as directed and elliptic flow, respectively.

Figure 13 shows the measured dependence of v_2 as a function of the centre-of-mass energy. From this figure it is seen that a positive v_2 measured at low

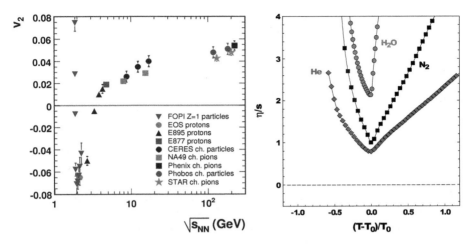

Figure 13. *Elliptic flow at midrapidity as function of beam energy.*

Figure 14. *The ratio η/s versus temperature for several liquids.*

energies becomes negative and reaches its lowest value at about 2 GeV. Above this energy v_2 rises again to positive values. The positive v_2 at lower energies reflects the bounce-off of the two colliding heavy ions that subsequently fragment in the reaction plane. At these energies, beam rapidity and mid-rapidity have significant overlap. As the energy increases, particle emission from the strongly compressed matter in the centre of the collision is shadowed by the passing spectator nucleons. This causes the particles produced to emerge perpendicular to the reaction plane leading to a negative value of v_2 (squeeze-out). At top AGS and SPS energies, due to Lorentz contraction, the timescale for the spectator nucleons to pass the created hot and dense system becomes much shorter than the characteristic time for the buildup of transverse flow. At these energies the elliptic flow becomes in-plane again (positive v_2) with a magnitude that is proportional to the initial spatial anisotropy of the created system and the sound velocity in the medium.

The large elliptic flow measured at RHIC is surprisingly well described by ideal hydrodynamics, which applies to fluids that are in thermal equilibrium and have no viscosity, a perfect fluid. The reason for this perfect fluid behavior of the matter created at RHIC energies is not fully understood. It may indicate that it is a strongly coupled QGP, or a state of matter with as yet unknown properties.

Viscous hydrodynamical models were recently developed to investigate to what extent the matter created at RHIC is indeed a perfect fluid. Here the quantity of interest is the ratio of the shear viscosity η to the entropy density s (this ratio is inversely proportional to the Reynolds number). Figure 14 shows the value of η/s versus temperature for different liquids. Water close to the triple point reaches a value of $\eta/s = 2$, while for liquid helium the ratio is as

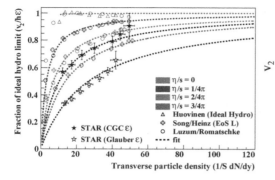

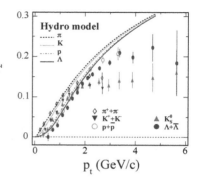

Figure 15. *Elliptic flow v_2 normalized to the spatial eccentricity ε and the limit of this ratio derived from ideal hydro h versus the transverse particle density. The full and open stars are derived from elliptic flow measurements by STAR.*

Figure 16. *Elliptic flow for various particle types as function of transverse momentum.*

low as $\eta/s = 0.7$. This raises the more fundamental question: is there a lower bound that can be derived from first principles on how perfect a liquid can be? Such a bound has been obtained by Kovtun, Son and Starinets, who showed that conformal field theories with gravity duals (anti-de Sitter/Conformal Field Theory) yield a ratio of $\eta/s = 1/4\pi$ (in natural units). They conjectured that this value is a lower bound for any relativistic thermal field theory.

Figure 15 shows elliptic flow, divided by the spatial eccentricity of the created system (ε), normalized to the ideal hydro limit (h) as a function of the transverse particle density. The open symbols are theoretical calculations with different values of η/s. The figure clearly shows that even small shear viscosities, of the order of the bound, already significantly reduce the elliptic flow. Also plotted in Fig. 15 are data on elliptic flow from STAR that show a reduction which is compatible with an η/s of a few times the bound. However, the comparison between data and theory depends intimately on the equation of state and the value of the spatial eccentricity, both of which are poorly known for the data obtained at RHIC. For instance, the data would indicate a value of $\eta/s < 3/4\pi$, assuming a soft equation of state and colour Glass Condensate initial conditions (full stars in Fig. 15). On the other hand, a larger value of η/s is obtained when a hard equation of state and Glauber wounded nucleon initial conditions are assumed (open stars). These estimates of η/s include possible contributions from the pre-equilibrium phase and from the hadronic phase. Therefore they can be considered to be upper limits on η/s in the QGP phase.

The measurement of elliptic flow as a function of transverse momentum provides more detailed constraints on the properties of the system at kinetic freeze-out, in addition to the information obtained from hadron spectra. Figure 16 compares the measured elliptic flow of various particle species as a function of transverse momentum with that from ideal hydrodynamical calculations. At low transverse momenta the strong characteristic mass dependence of elliptic flow is rather well described by ideal hydrodynamics. This description can be improved by introducing a non-zero viscosity via a hybrid description of ideal hydrodynamics and a hadronic cascade model (not shown). The good agreement between the measured integrated and differential elliptic flow and these theoretical descriptions lends strong support to the underlying assumption that the system is partonic and is approaching thermalization at the early stage of the collision.

The particle dependence of v_2 at higher transverse momenta falls, within uncertainties, into two distinct classes: that of baryons and that of mesons. In the proposed scenario of constituent quark coalescence the mesons (baryons) carry twice (three times) the constituent quark v_2 which naturally leads to the observed baryon/meson scaling. This observation may indicate that the flow of partonic degrees of freedom is observed.

3.1.9 Jet Quenching

The products of initial state hard scatterings, e.g., mesons containing charm or bottom, high-p_t photons and high-p_t hadrons provide a detailed probe of the created system. While initial state production of these probes is relatively unaffected by the presence of the created system it was predicted that partons would lose energy while traversing the system, mainly by induced gluon radiation. The amount of energy loss in this picture is directly related to the parton density. Thus by studying these high-p_t probes the density of the created system can be determined. In heavy-ion collisions at RHIC, jets with transverse energies above 40 GeV are produced in abundance; however, the abundant soft particle production in heavy-ion collisions tends to obscure the characteristic jet structures. At sufficient high-p_t the contribution from the tails of the soft particle production becomes negligible and jets can be identified by their leading particles. When the parton loses energy, the hadrons into which it fragments will, on average, have lower transverse momenta. Therefore the parton energy loss leads to a suppression of the yield of high-p_t hadrons compared to nucleon-nucleon collisions where there is no dense medium. One of the observables suggested for measuring energy loss is the so-called nuclear modification factor defined by

$$R_{\mathrm{AA}}(p_t) = \frac{\mathrm{d}^2\sigma_{\mathrm{AA}}/\mathrm{d}y\mathrm{d}p_t}{\langle N_{\mathrm{binary}}\rangle \mathrm{d}^2\sigma_{\mathrm{pp}}/\mathrm{d}y\mathrm{d}p_t},$$

where $\mathrm{d}^2\sigma_{\mathrm{pp}}/\mathrm{d}y\mathrm{d}p_t$ is the inclusive cross section measured in p + p collisions and $\langle N_{\mathrm{binary}}\rangle$ accounts for the geometrical scaling from p + p to nuclear collisions. In the case that a Au + Au collision is an incoherent superposition of p + p collisions this ratio R_{AA} would be unity. Energy loss and shadowing would reduce this ratio below unity while anti-shadowing and the Cronin effect

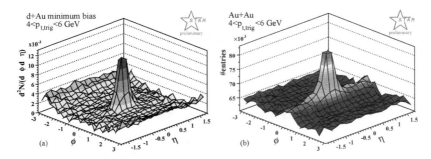

Figure 19. *Preliminary di-hadron correlations in $\Delta\eta$ and $\Delta\phi$ for (a) d+Au and (b) Au+Au. The trigger particle is between $4 < p_t < 6$ GeV/c and the associated particles are between 2 GeV/c $< p_t < p_t^{trig}$.*

The parton energy loss shows up even more dramatically in the azimuthal correlations as shown in Fig. 18. Azimuthal correlations between two high-p_t hadrons in a nucleon-nucleon collision exhibit strong peaks at 0 and π because the two partons from a hard collision traverse the system back to back in azimuth. In the dense medium created in heavy-ion collisions the energy loss reduces the yield of high-p_t particles; therefore, the high-p_t particles which do escape the system are on average produced close to the surface. The recoiling associated parton is maximally affected by the energy loss causing a suppression of the away-side hadron jet. This jet-quenching effect is clearly observed in the gold-gold measurements shown in Fig. 18(b).

The magnitude of the observed suppression at the top RHIC energy indicates, in the jet quenching picture, densities which are at least a factor 30 higher than in nuclear matter. To become more quantitative the energy loss mechanism and the response of the medium have to be understood in more detail.

One recent striking result, which might provide additional information on the response of the medium, is the di-hadron correlation at intermediate p_t. The di-hadron correlation, in azimuth and in pseudo-rapidity η, shows that the near-side correlation (small $\Delta\phi$ which was already shown in Fig. 18(b)) extends over large $\Delta\eta$. Figure 19 shows this correlation for d + Au and Au + Au collisions. It is clear that this long-range $\Delta\eta$ correlation is unique for heavy-ion collisions. The long-range correlation is approximately independent of $\Delta\eta$ and is therefore referred to as *the ridge*.

4 Heavy-Ions at the LHC

The LHC at CERN will provide colliding Pb ions with an energy of $\sqrt{s_{NN}} = 5.5$ TeV. This exceeds the maximum energy of 200 GeV available at RHIC by a factor 30 and will open up a new physics domain. Qualitative new features

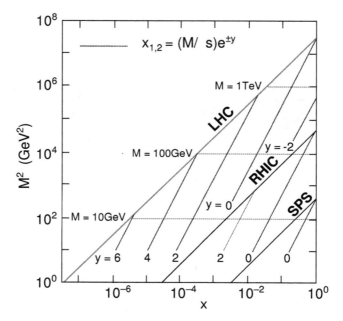

Figure 20. *The range of Bjorken-$x_{1,2}$ and M^2, relevant for particle produc-tion in nucleus-nucleus collisions at the top SPS ($\sqrt{s_{NN}} = 17.2$ GeV), RHIC ($\sqrt{s_{NN}} = 200$ GeV) and LHC ($\sqrt{s_{NN}} = 5.5$ TeV) energies.*

of the heavy-ion collisions at the LHC include:

- Particle production is determined by high-density saturated parton dis-tributions.

  The LHC heavy-ion program accesses a novel range of low Bjorken-x values (see Fig. 20), where strong nuclear shadowing is expected. The initial density of gluons is expected to be close to saturation of the avail-able phase space. These very high initial densities allow us to describe important aspects of the subsequent time evolution in terms of classical chromodynamics. The ALICE detector will probe a continuous range of x as low as $\sim 10^{-5}$.

- Hard processes become abundant.

  The abundance of hard processes at the LHC will allow for precision test of perturbative QCD. In addition, the large jet rates at the LHC permit detailed measurements of jet quenching to study the early stages of the collision.

- Access to weakly interacting hard probes.

  Direct photons as well as Z^0 and $W^{\pm}$ bosons produced in hard processes will provide information about nuclear parton distributions at high Q^2.

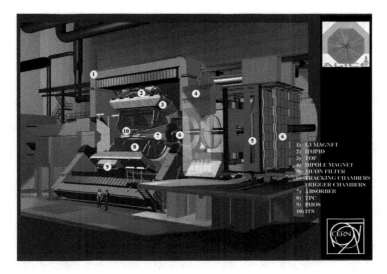

Figure 21. *The ALICE detector layout.*

Jet tagging with such probes yields a calibrated energy scale for jet quenching studies.

- Fireball expansion is dominated by parton dynamics.

  Due to the expected longer lifetime of the QGP, the parton dynamics will dominate over the hadronic contribution to the fireball expansion and therefore collective features such as elliptic flow provide much stronger constraints on the EoS of the partonic phase.

Of course, the large increase in centre of mass energy provided by the LHC will offer unique opportunities for new and unexpected discoveries.

It is expected that the LHC can deliver luminosities of 10^{27} cm^2 s^{-1} for Pb–Pb collisions, which results in a minimum-bias interaction rate of 8 kHz. Lighter ions can be delivered with higher luminosities of up to 10^{29} cm^2 s^{-1}, corresponding to an interaction rate of several 100 kHz.

5 The ALICE Detector

At the LHC three collaborations (ALICE, ATLAS and CMS) have a heavy-ion program. The ALICE collaboration has designed a detector optimized for heavy-ion collisions. The apparatus will detect and identify hadrons, leptons and photons over a wide range of momenta. The requirement to study the various probes of interest in a very high multiplicity environment, which may be as large as 8000 charged particles per unit of rapidity in central Pb–Pb collisions, imposes severe demands on the tracking of charged particles. A

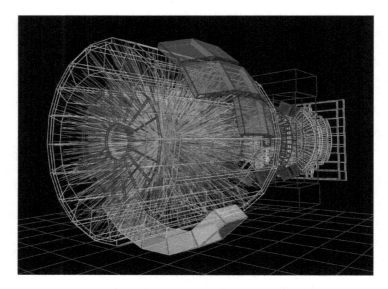

Figure 22. *Particle tracks from a single simulated heavy-ion collision in ALICE.*

schematic view of the ALICE detector is shown in Fig. 21. Figure 22 shows the particle tracks originating from a single heavy-ion collision in the ALICE detector.

In ALICE, the excellent PID capabilities, momentum resolution and complete azimuthal coverage of the central detectors allow comprehensive measurements of particle ratios, momentum spectra, particle correlations, anisotropic flow and event-by-event fluctuations. These observables do not require large amounts of data and will either quickly confirm our current understanding of high-density QCD or provide fundamental new insights.

The LHC will be the first machine where heavy quarks are produced abundantly in heavy-ion collisions. Due to the excellent impact parameter resolution and particle identification capabilities, ALICE is well suited to study charm and beauty. From detailed simulation studies of the benchmark channel $D^0 \to K^- \pi^+$ it is found that in one LHC year at nominal luminosity, we cover the transverse momentum range $1 < p_t < 18$ GeV/c in the central barrel acceptance of $|\eta| < 0.9$. Beauty production can be measured from semi-leptonic decays in the range of $2 < p_t < 30$ GeV/c. Single muons and opposite-sign dimuon pairs detected in the muon spectrometer allow for measurements of open-beauty production with high statistics in the pseudorapidity region $-4 < \eta < -2.5$. The measurement of heavy flavour production down to very low transverse momenta is sensitive to the collective motion of heavy quarks in the medium and will provide strong constraints on the thermalization of light quarks. At higher momenta, on the other hand, the measurement of heavy flavour production will provide detailed information on the energy loss mechanism.

In addition to open charm and open beauty, the complete spectrum of heavy quarkonia states (J/Ψ, Ψ', Υ, Υ' and Υ") is accessible at the LHC. In ALICE quarkonia are detected at mid-rapidity ($-0.9 < \eta < 0.9$) in the dielectron channel, and at ($-4.0 < \eta < -2.5$) in the dimuon channel, which will allow for detailed studies of suppression effects due to deconfinement.

The jet rates in central Pb–Pb collisions at the LHC in the ALICE acceptance are sufficient to map out the energy dependence of jet fragmentation over a wide kinematic range up to $E_t \simeq 200$ GeV. Jet reconstruction in nuclear collisions has to cope with the large background from the underlying event, therefore jet reconstruction has to be limited to a small cone of fixed radius in azimuth and pseudorapidity ranging between 0.3–0.5. In addition, a transverse momentum cut in the range 1–2 GeV/c has to be applied to reduce the background. As a consequence, even for perfect calorimetry, the transverse energy resolution is limited to $\simeq 20\%$. Another very promising approach to study jet fragmentation is using prompt photons to tag charged jets emitted in the opposite direction. Prompt photons allow the study of the hard interaction without any final state modifications and with this tag the in-medium modification of the fragmentation function will be measured with an accuracy of the order of a few percent. The combined tracking capabilities of the ALICE detector combined with electromagnetic calorimetry represent an ideal tool for jet structure modifications at the LHC.

6 Summary

The relativistic heavy-ion collision program provides a unique tool to test the phase diagram of strongly interacting matter under controlled conditions. The program so far has tremendously improved our understanding of the probes and we start to quantitatively characterize the created system. The heavy-ion program at the LHC will allow for a more precise characterization of the QGP and will allow us to explore new aspects of the structure of strongly interacting nuclear matter. With the LHC machine currently delivering the first proton–proton collisions and its detectors fully operational we are looking forward to the first heavy-ion collisions.

Acknowledgements

I would like to thank the organizers of the school for their hospitality and creating this wonderful stimulating environment. I would also like to thank M. Botje for his help with preparing this document.

References

[1] R. Stock, "Relativistic Nucleus–Nucleus Collisions and the QCD Matter Phase Diagram," arXiv:0807.1610 [nucl-ex].

[2] H. Satz, "The States of Matter in QCD," arXiv:0903.2778 [hep-ph].

[3] P. Braun-Munzinger and J. Stachel, "Charmonium from Statistical Hadronization of Heavy Quarks: A Probe for Deconfinement in the Quark-Gluon Plasma," arXiv:0901.2500 [nucl-th].

[4] H. Oeschler, H. G. Ritter and N. Xu, arXiv:0908.1771 [nucl-ex].

[5] U. W. Heinz, "Early collective expansion: Relativistic hydrodynamics and the transport properties of QCD matter," arXiv:0901.4355 [nucl-th].

[6] S. A. Voloshin, A. M. Poskanzer and R. Snellings, "Collective phenomena in non-central nuclear collisions," arXiv:0809.2949 [nucl-ex].

[7] R. Rapp and H. van Hees, "Heavy Quarks in the Quark-Gluon Plasma," arXiv:0903.1096 [hep-ph].

[8] U. A. Wiedemann, "Jet Quenching in Heavy Ion Collisions," arXiv:0908.2306 [hep-ph].

New Physics Searches

Gustaaf Brooijmans

Columbia University, New York, USA

1 Introduction

At the time of this writing, all experimental results are consistent with a model of particle physics which contains six quarks (up, down, charm, strange, top and bottom) and six leptons (electron neutrino, electron, muon neutrino, muon, tau neutrino and tau). The properties of these particles, with the exception of the neutrino masses, have been measured accurately, and the standard model describes their weak, electromagnetic and strong interactions as mediated by the W and Z bosons, photons and gluons, respectively, with remarkable accuracy [1]. Lacking in this description are a consistent quantum theory of the gravitational interaction, and any understanding of the pattern suggested by the properties of the fermions.

Many questions thus remain to be answered, e.g., what is spin or colour or electric charge? Are these static "properties" of particles, or do they result from a hidden dynamic? Are there only three generations? Why are there, for example, no neutral, coloured fermions? What is the link between particle and nucleon masses?

There is also one known problem in the standard model: at a center-of-mass energy of 1.7 TeV, the longitudinal W boson scattering cross-section violates the unitarity bound [2].[1] One solution to this problem is the introduction of the so-called Higgs boson [3–7], which in the standard model can also generate the fermion masses. The latter allows the decoupling of the mechanism responsible for fermion masses from the standard model interactions. However, quadratically divergent radiative corrections suggest the standard model Higgs boson mass is close to the limit of validity of the theory, and experimental constraints [8] imply its mass is less than approximately 200 GeV. This suggests the scale of new physics is at or below 1 TeV, although

---

[1]A second problem is with the $f\bar{f} \to W^+W^-$ cross section, but this is less severe and is also addressed by the standard model Higgs boson.

in principle, if one accepts a high level of fine tuning, with the addition of a $m_H \approx 150$ GeV Higgs boson the list of existing particles could be "complete," in analogy with Mendeleev's table in chemistry. No new physics would then appear below the Planck scale of 10^{19} GeV. This would of course be a very unsatisfactory outcome, as it would not help us answer any questions as to the nature of the fermions. The high degree of fine tuning needed in this scenario is a strong motivation for the presence of new particles and/or interactions at or just above the electroweak scale.

2   Unravelling the Mystery: The Tools

Assuming that there is indeed new physics to help us understand the observed patterns, then the path to its discovery goes through both the collection of additional information (experiments), and searching for the underlying pattern (theory). Experimental searches for new physics can broadly be categorized into (a) precision measurements of particle properties and interactions in searches for deviations from standard model predictions, and (b) searches for new particles or interactions in new areas in "phase space," typically energy domains that have not yet been probed. On the theoretical front, hypotheses lead to models whose internal consistency as well as compatibility with existing data needs to be verified, and if confirmed leads to new suggestions on where to look. Given the time available, these lectures will focus on the experimental search for new physics at the energy frontier.

The new physics must couple to the standard model in some way, but the lack of any deviations so far implies that that coupling must be either weak or hidden through some mechanism, for example near-cancellation of competing amplitudes. It could be "standard model-like" in the sense that it consists of new short-lived, massive particles decaying to known fermions or bosons, or there might be some new longer-lived particles with unusual properties, or no new particles may be in reach, but new interactions might manifest themselves. The situation may even be more extreme and the concept of "particles" or "interactions" may need to be replaced by a fundamentally different paradigm. In any case, the number of possibilities is certainly too vast to explore here, and in the following it is the search for anomalous production of standard model particles, in either resonant or non-resonant mode by short-lived new particles that will be discussed. While in many cases this can be done in a model-independent way, it is important to keep the implications of known constraints in mind, and to understand that some scenarios require the development of new experimental techniques.

The tools used in the searches that will be described are the Fermilab Tevatron and CERN LHC, proton-(anti)proton colliders at 1.96 and 7–14 TeV, respectively. Since these are hadron colliders, the incoming longitudinal[2]

---

[2]Longitudinal and transverse denote directions w.r.t. the direction of the incoming colliding beams.

momentum of the quarks and/or gluons engaged in the "hard" (large $\sqrt{\hat{s}}$) scatter is not known, and the large momentum of these particles leads to a Lorentz contraction in the lab frame of all distances in the longitudinal direction. For these two reasons, most selections are based on quantities measured in the transverse direction, and the larger the $\sqrt{\hat{s}}$, the more "central" the events will be in the detector. The detectors are designed to make the best possible measurements of all particles produced in the collisions. A detailed description of the detectors installed at the LHC is given elsewhere in this volume. The Tevatron detectors, CDF and D0, are very similar.

3   The Higgs Hunt

The search for the Higgs boson at the Tevatron will be used to give a detailed description of the experimental techniques used in the search for new physics. This is an interesting example since it is a search for a small signal in a large background dominated by vector boson plus jets events, and thus requires the use of the most advanced tools on both the experimental and theoretical fronts.

The Higgs boson production cross section at the Tevatron ranges from approx. 1 ($m_H = 120$ GeV) to 0.3 pb ($m_H = 180$ GeV) in the gluon fusion channel, and 0.2 ($m_H = 120$ GeV) to 0.03 pb ($m_H = 180$ GeV) for production in association with a vector boson. At the lower end of this mass region the Higgs boson dominantly decays to a $b\bar{b}$ pair, so that in the face of the overwhelming QCD production of $b\bar{b}$ events, only Higgs bosons produced in association with vector bosons (which themselves decay leptonically) are detectable. At higher mass, where decays to a W^+W^- pair become sizable, the higher rate gluon fusion process becomes viable provided at least one of the W bosons decays leptonically. Since each experiment has collected about 7 fb^{-1} of data, the total number of detectable Higgs bosons that have been produced is at best a few thousand. Decays with small branching ratios are thus not accessible at the Tevatron.

3.1   Dilepton Plus Missing Transverse Energy Channel

$H \to W^+W^- \to \ell^+\nu\ell^-\nu$ decays, where ℓ represents an electron or a muon, form the "golden channel" at the Tevatron: the main background, $Z \to \ell^+\ell^-$ is also a great reference signal to determine efficiencies and the sample composition, and the missing transverse energy ($\not{E}_T$) present in the signal provides an excellent means to suppress the Z boson contribution. Both the dilepton invariant mass and $\not{E}_T$ are shown at preselection level in Fig. 1 for the most recent D0 search [9]. After the preselection, which mainly consists of requiring the presence of two good quality charged leptons (electrons or muons) with transverse momentum (p_T) greater than 10 (15) GeV for muons (electrons) and invariant mass larger than 15 GeV, cuts are applied to en-

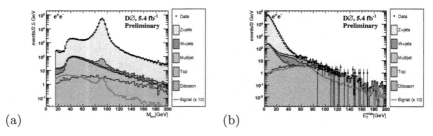

Figure 1. *(a) Dielectron invariant mass and (b) missing transverse energy at the preselection level in the D0 search for the Higgs boson in the dilepton plus missing transverse energy channel.*

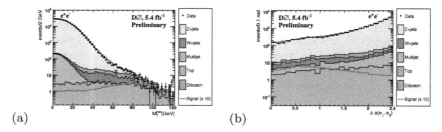

Figure 2. *(a) Minimum transverse mass and (b) azimuthal angle between the leptons at preselection level in the D0 search for the Higgs boson in the dilepton plus missing transverse energy channel.*

hance the signal/background ratio. These selection cuts are applied on the following variables: $\not{E}_T$, scaled $\not{E}_T$ (which is a measure of the probability that the $\not{E}_T$ originates from poor measurement of other objects in the event), $M_T^{min}(\ell, \not{E}_T)$, the minimum of the transverse masses calculated from each of the charged leptons and the $\not{E}_T$, and $\Delta\phi(\ell, \ell)$, the azimuthal angle between the charged leptons. Two of these variables, $M_T^{min}(\ell, \not{E}_T)$ and $\Delta\phi(\ell, \ell)$, are illustrated at preselection level in Fig. 2.

After the enhancement cuts, the signal/background ratio is still relatively small (1/30, 1/50 and 1/1000 in the $e\mu$, ee and $\mu\mu$ channels, respectively). While this could be improved further, cuts are left loose since optimization is done in a next step: multivariate tools are used to exploit the different correlations between variables for signal and background. In this particular analysis, 14 variables with good agreement between data and expectation, based on lepton transverse momenta, $\not{E}_T$ and the angles between these, are injected into a neural network for each of the three channels. The neural network output distributions are then combined into one (Fig. 3), and in the absence of signal that distribution will be used to set a limit on the Higgs boson production cross section.

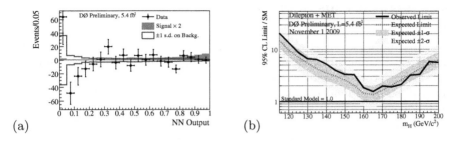

(a)

(b)

Figure 3. *(a) Neural network output distribution, and (b) limit on the Higgs boson production cross section in units of the standard model prediction (b) for the D0 search for the Higgs boson in the dilepton plus missing transverse energy channel.*

The systematic uncertainties are propagated through the full analysis chain to the neural network output distribution: the analysis is repeated with, for example, the jet energy scale shifted up, then down by one standard deviation. Some of these uncertainties, like the jet reconstruction efficiency, affect the shape of the distribution, while others, such as the uncertainty on the integrated luminosity, affect the normalization only. The uncertainties are treated as nuisance parameters which are generally correlated, but not at the 100% level, between different background contributions. These uncertainties are often only known with moderate accuracy, and the data can be used to further constrain them. To do this, pseudo-experiments are produced by generating events for each of the neural network output distribution's bins according to a Poisson distribution, with mean equal to the number of expected events. For each pseudo-experiment, the nuisance parameters are then varied within the expected range, leading to variations in both the signal and signal + background distributions. The results can then be compared to the data using a log-likelihood ratio, which can be maximized as a function of the nuisance parameters, effectively constraining their values. Basically, the full shape of the neural network output distribution is used to "profile" the systematics, in other words determine which background uncertainties are over- or underestimated. Bins with large signal/background ratio can be removed to avoid any signal-induced bias.

Finally, the full shape of the neural network output distribution is compared to both background and signal+background templates, including all uncertainties, to determine the limit on the Higgs boson production cross section. This is usually shown as a ratio to the standard model prediction, so that the region where the observed limit curve is below one is excluded at 95% C.L. Both the neural network output distribution and limit are shown in Fig. 3.

3.2   Lepton Plus Jets Channel

The final state consisting of a W boson (decaying to $\ell\nu$) plus two jets is critical at both low ($WH \to \ell\nu b\bar{b}$) and high mass ($H \to WW \to \ell\nu jj$.) In the former case we have $m_{b\bar{b}} = m_H$, and in the latter $m_{jj} = m_W$ and $m_{WW} = m_H$. However, the di-jet mass resolution is intrinsically much worse than the dilepton mass resolution, and the backgrounds are substantially larger than in the dilepton channel.

In this case, the sample composition after preselection (typically one good lepton with $p_T > 20$ GeV, $\not{E}_T > 20$ GeV and two jets with $p_T > 20$ GeV) is more difficult to determine.

- The contributions from diboson and top quark production are taken from Monte Carlo (MC) simulation since they are both relatively small, and 10%-level uncertainties on these backgrounds have a small impact.

- The contamination from the instrumental QCD multijet background in which one jet is misidentified as a lepton is evaluated directly from the data.

- The Z boson plus jets component is taken from a simulated sample which is corrected based on direct measurements: indeed, as we will see, the difference between the simulation and observation is significant.

- For the major background, W boson plus jets production, simulation is also used, but here the correction to be applied is much harder to determine, in part due to the potential signal contamination. In principle, corrections determined from the Z boson plus jets sample can be mapped to W boson plus jets after correcting for vector boson mass effects, but in practice the correlation between variables makes this highly non-trivial.

Four types of MC generators are used:

- "Calculators" implement (up to next-to-next-to-leading-order) calculations of specific quantities, for example boson p_T distributions. Two commonly used generators of this type are RESBOS [10] and MCFM [11]. These calculators do not generate complete events.

- Traditional $2 \to 2$ generators like PYTHIA [12] or HERWIG [13] implement leading-order calculations of many $2 \to 2$ processes (e.g., $q\bar{q} \to e\nu$, and include a parton shower (PS) model and hadronization, thus allowing the generation of complete events that can be fed into a detector simulation. Any jets beyond those from the $2 \to 2$ matrix element are produced at the parton shower stage.

- "Matrix Element" generators allow the generation of $2 \to n$ ($n < 9$) events (e.g., $q\bar{q} \to e\nu jjjj$). The calculations are also done at leading

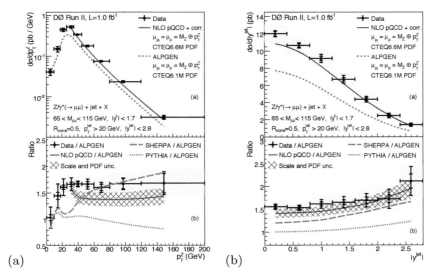

(a)   (b)

Figure 4. *Comparison of Z boson plus jets data with the prediction from various generators after unfolding of detector effects, i.e., at particle level. Left: Z boson p_T and Right: Leading jet rapidity distributions. The curve denoted "NLO pQCD" was produced using* MCFM.

order, but these generators are necessary to simulate events with multiple hard jets. Care needs to be taken not to "double count" events in which additional jets are produced during the parton shower stage. Two approaches exist to handle this: the CKKW method implemented in SHERPA [14] and the MLM method used in ALPGEN [15].

- "NLOwPS" $2 \to 2$ generators include next-to-leading order corrections, so in a sense are $2 \to 3$ generators with virtual corrections. The two generators of this type that are in use are MC@NLO [16] and POWHEG [17].

In the case of the Higgs boson search in the W boson plus jets channel, the matrix element generators are used to simulate the dominant background, but this simulation needs to be "corrected" to address modeling deficiencies arising from non-perturbative effects, their leading order nature, and other aspects for which our understanding is limited, for example the underlying event contribution. The magnitude of these necessary corrections can be estimated from a comparison of Z boson plus jets data to simulated samples. Two example distributions from a D0 analysis [18] are given in Fig. 4.

After the W boson plus jets events produced with ALPGEN have been reweighted using distributions obtained from RESBOS (for the boson p_T) and SHERPA (for the jet angular distributions), good agreement between data and simulation is obtained. The search is then optimized using multivariate tools, and the so-called matrix element technique. This approach, which was suc-

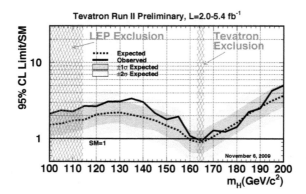

Figure 5. *Combination of the CDF and D0 results in the search for a standard model Higgs boson.*

cessfully used in measuring the top quark mass and observation of single top quark production, is basically an unbinned maximum likelihood fit that gives extra weight to more signal-like events: for each event, signal and background probabilities are calculated based on the compatibility of the physics object (leptons and jets) four-vectors with the process under study. The output of the matrix element calculation is used as an additional input to a neural network, and boosts the sensitivity by 5%, equivalent to an increase in the data set size by 10%.

3.3   Current Result

In the end, 90 mutually exclusive final states are analyzed by CDF and D0 and combined to produce a Tevatron limit on standard model Higgs boson production [19]. The result at the time of this writing is shown in Fig. 5, and excludes the existence of a standard model Higgs boson with mass $163 < m_H < 166$ GeV.

4   Searches for Supersymmetry

The existence of new particles with masses close to the Higgs boson mass could be an effective way to cancel out the quadratically divergent loop corrections to the latter. In supersymmetry, for every standard model fermion (boson) there is a partner boson (fermion), so that the quadratically divergent diagrams are naturally cancelled by their supersymmetric counterparts. "Little Higgs" models take a similar approach, except that partners only exist for the worst offenders, the top quark, W, Z and Higgs bosons, and other new physics is expected to exist at the 10 TeV scale.

Even in the minimal supersymmetric standard model (MSSM), which has the minimal set of supersymmetric particles, there are 105 new free parameters, and the simplest searches performed in this context make a number of basic assumptions: R-parity is conserved so that superpartners are pair-produced (or in some cases explicitly violated and single particle production is considered); the pair-produced superpartners typically each decay to its standard model partner and the lightest supersymmetric particle (LSP), leading to signatures with a pair of jets or leptons and missing (transverse) energy. Results are then presented as limits in the superpartner-LSP mass plane. In this scenario, searches by the LEPII experiments set limits on superpartner masses that are in the range from 90 to 100 GeV, very close to the kinematic limit, and cover just about all of the allowed ($m_{superpartner}, m_{LSP}$) plane below that. Tevatron experiments have a larger kinematic reach (typically $\sim$200 GeV in superpartner mass), but are typically not sensitive in the band $m_{superpartner} - m_{LSP} \leq 15$ GeV.

An exception to the two-body decay case is the pair production of top squarks at the Tevatron. If the top squark is lighter than the top quark then the three-body decay to a bottom quark, a lepton and an LSP sneutrino becomes dominant. The final state differs from the dilepton signature of $t\bar{t}$ decays only in the kinematics, with potentially very small lepton p_T values if the mass difference between the top squark and the sneutrino is small. This can be seen in the left panel of Fig. 6 which shows the electron p_T spectrum in the D0 search for the top squark in the $e\mu\not{E}_T$ channel [20]. The cut-based analysis has low background so that no explicit cut on the number of jets is required, leading to the excluded region given in the right panel of Fig. 6, which also illustrates the contrast between LEP and Tevatron searches in probing the small $m_{superpartner} - m_{LSP}$ region.

Supersymmetry is manifestly broken, since the superpartner masses differ from the standard model particle masses, and various breaking models have been developed. These lead to predictions for superpartner mass hierarchies and the nature of the LSP, thus predicting specific experimental signatures. An added bonus is that in most of these breaking models, the μ^2 term in the Higgs potential is driven negative for a large fraction of parameter space when run down from the GUT to the electroweak scale, thus explaining electroweak symmetry breaking. This is, however, driven by the large value of the (in principle arbitrary) top quark mass.

In the vast majority of these models, strongly interacting superpartners (squarks and gluinos) are substantially heavier than the sleptons and electroweak gauginos, but at hadron colliders this is usually more than compensated for by the larger interaction strength. If they are within the kinematical reach, the produced squarks and gluinos will decay to quarks, gluons and LSPs, leading to a final state consisting of jets and $\not{E}_T$. This is a difficult experimental signature, since most collisions lead to jets, most detector mal-

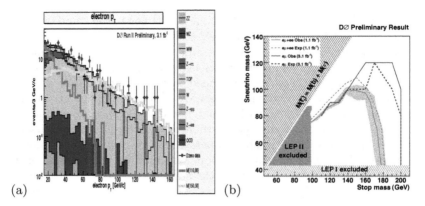

(a)   (b)

Figure 6. *Search for top squarks in $e\mu\not{E}_T$ events by D0: (a) electron p_T distribution and (b) excluded region in the sneutrino-stop mass plane.*

functions will lead to $\not{E}_T$, and the jet energy resolution is intrinsically limited given the large fraction of invisible energy jets deposit. Searches in this final state therefore require close attention to data quality, both on- and offline, but since some noise sources will manifest themselves very infrequently, continuous feedback from the analysis to online operations is a necessity. From past experience, some effects only lead to a few anomalous events per year.

The search sensitivity is as always maximized by treating different final states separately, with specific optimizations of the background suppression cuts. Whereas squarks typically decay to a quark and an LSP, gluinos decay to a squark and a quark yielding a $q\bar{q}$LSP final state. Pair production of squarks then leads to a dijet plus $\not{E}_T$ signature, and pair production of gluinos to four jets plus $\not{E}_T$. Associated production of a squark with a gluino is also possible giving the three jets plus $\not{E}_T$ final state. The recent D0 search [21] treats each channel separately starting at the trigger level. The instrumental QCD multijet background is suppressed requiring that the $\not{E}_T$ not be aligned with any of the jets in the azimuthal plane, and the dominant physics backgrounds are $Z(\to \nu\nu) + jets$, $W(\to \ell\nu) + jets$ and $t\bar{t}$ production. Figure 7 shows the $\not{E}_T$ distribution in the two-jet channel, and the obtained result interpreted in the minimal supergravity (mSUGRA) model. Note that the exclusion reaches further in m_0 than $m_{1/2}$, showing better sensitivity to squarks than gluinos.

If the coloured superpartners are out of kinematical reach, the search for electroweak gaugino pair production is the most promising avenue for discovery at hadron colliders. This can lead to a spectacular trilepton signature, which has an enhanced rate if the sleptons are light enough to mediate the chargino or neutralino decays: charginos (neutralinos) then decay to a slepton and neutrino (charged lepton), with the slepton subsequently decaying to a charged lepton and LSP. The cross section is of course small, but so are the backgrounds from pair production of weak vector bosons. In fact, the

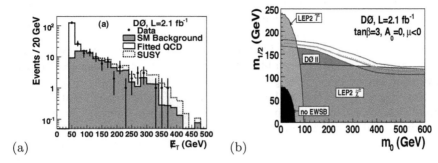

Figure 7. *Search for squarks and gluinos in the jets plus $\not{E}_T$ channel by D0: (a) $\not{E}_T$ distribution in the two-jet channel and (b) excluded region in the mSUGRA interpretation.*

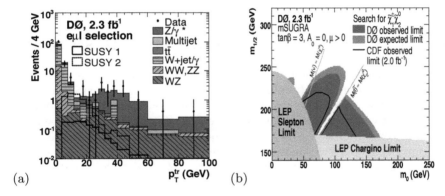

Figure 8. *Search for associated production of charginos and neutralinos in trilepton events by D0: (a) Track p_T distribution in the $e\mu l$ channel and (b) excluded region in the mSUGRA interpretation.*

backgrounds are so small that only two well-identified charged leptons are required, and an isolated track is taken as a candidate for the third. The main difficulty comes from the fact that the mass splittings between the involved supersymmetric particles are presumably small, so that the charged leptons produced in the decays are relatively soft. Figure 8 shows the track transverse momentum distribution in the $e\mu l$ channel (where l can be any charged lepton, identified as an isolated track) and the excluded mSUGRA region in a recent D0 analysis [22] combining multiple final states. Note that in contrast to the squark and gluino search, this analysis is more sensitive in $m_{1/2}$ than m_0, since it searches for SUSY fermions.

The LHC reach for supersymmetric particles will go up to about 4 TeV in m_0 and 1400 GeV in $m_{1/2}$. If supersymmetry is to provide us with an efficient cancellation of the Higgs mass corrections' quadratic divergences and a good dark matter candidate, it will therefore need to be particularly finetuned to escape detection for another decade. If SUSY is found, however, the prospects for measuring its mass spectrum and learning about the way supersymmetry is broken are good, even with LSPs escaping undetected in every event. Kinematic endpoints and new variables like m_{T2} [23] or the contransverse mass [24] have been shown to be very powerful for such measurements.

Supersymmetric theories have a number of very attractive features: they explain the low Higgs boson mass (and sometimes electroweak symmetry breaking) lead to gauge coupling unification at a high scale, feature a candidate dark matter particle, and all of this without the need to introduce new interactions. However, this also comes at a large cost: many new particles are needed with a correspondingly large number of new free parameters, and no answers are given for any questions regarding the nature of particles. Parallels can certainly be drawn between expecting the presence of low scale supersymmetry and dinosaurs on Venus [25].

5 New Gauge Bosons and Parity Restoration

Our assignments of fermions to generations are based on the subjective criteria of mass ordering and keeping the Cabibbo–Kobayashi–Maskawa matrix as diagonal as possible. If we accept this classification, however, it becomes apparent that within a generation, the more a fermion interacts, the heavier it is (with the exception of the up and down quarks for which mass is an ill-defined concept). This pattern suggests that fermion masses may have their origin in a more complex mechanism with an indirect relation to the standard model interactions, as for superpartner masses in gauge-mediated supersymmetry-breaking scenarios for example. The Higgs boson may then only be relevant to regulate the longitudinal vector boson cross section, relaxing existing mass constraints and limits.

The difficulty with fermion masses in the standard model of course originates in the purely left-handed nature of the weak interaction, since massive fermions can change helicity. A deeper understanding of spin would be a major step forward, but there is as yet no indication as to the scale at which any hints in that direction might reveal themselves. A related phenomenon would be a step towards the restoration of parity symmetry.

The primary signals of parity restoration are of course the existence of a heavy right-handed W' boson and corresponding Z' boson, although the couplings need not be purely right-handed. Dilepton decays of these offer clean signals with well-understood backgrounds, and although there is some concern about determining the energy scale for very high p_T leptons at the LHC, the different dependence of electron and muon energy/momentum resolution as a function of p_T should offer a good handle. As opposed to the standard model W and Z bosons, $t\bar{t}$ decays should also be present and observable, and decays to a right-handed neutrino ν_R may be important if it is light enough. Beyond

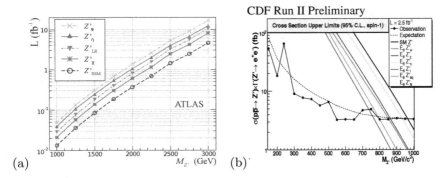

Figure 9. *(a) Sensitivity of the ATLAS experiment to a new Z' boson as a function of luminosity for various models at $\sqrt{s} = 14$ TeV. (b) The current limits on Z' bosons from CDF.*

parity restoration, many models in fact predict the existence of Z' bosons [26].

Z' boson production at the Tevatron and LHC mainly comes from up and down quarks, with model-dependent couplings determining the cross section and width [27]. The decays are somewhat similar to the standard model Z boson, but the branching fraction to light neutrinos is presumably suppressed while the $t\bar{t}$ channel opens up and $\nu_R\bar{\nu}_R$ may exist. The most promising channel for discovery is $Z' \to ee$ since the electron energy resolution at high p_T is dominated by the constant term, leading to a typical value of 10 GeV at 1.5 TeV. This is sufficiently good to be able to measure the Z' boson width in many models. The LHC surpassed the Tevatron reach ($\sim$1 TeV) already in 2010, since the Z boson peak offers an excellent analysis calibration reference and the backgrounds are typically very low. A recent study of the ATLAS sensitivity [28] at $\sqrt{s} = 14$ TeV is shown in the left panel of Fig. 9. Note that at 10 TeV, the Z' boson production cross section is typically about a factor of two smaller. Many searches for resonances are done by counting events in a shifting mass window, leading to the so-called "look elsewhere" effect: an excess will always be found if a sufficient number of distributions is studied. A better approach is to perform a global fit to the Drell–Yan spectrum, with an added Gaussian of free amplitude and width, thus allowing the fit to find the mass. This type of "shape" analysis is naturally more sensitive, but not immune to the look elsewhere effect: in all cases, pseudo-experiments should be run to determine the sensitivity of the experiment or quantify the magnitude of an excess. The current limits on Z' boson masses from CDF [29] are shown in Fig. 9(b). If a new high mass dielectron resonance is found, the study of the angle between the lepton and the beam direction will be very valuable in determining its spin, since spin 1 particles tend to emit leptons closer to the beam. However, experimental acceptance effects largely negate this, because lepton identification in the forward region is significantly more difficult and suffers from larger backgrounds.

It is generally possible to look for a new resonance in the dijet channel as well: while the backgrounds are manifestly much larger, good sensitivity can be reached [30, 31]. For Z' bosons, the sensitivity is generally inferior to the leptonic channels, however [32]. If the ν_R is lighter than $m_{Z'}/2$ then the $\nu_R \bar{\nu}_R$ decay channel opens up. This is presumably followed by $\nu_R \to l W_R^*$ and $W_R^* \to q\bar{q}$ leading to a signature with two charged leptons and four jets, or more leptons if, for example, $m_{\nu'_R} < m_{\nu_R}$ and more complex cascades open up. The ultimate LHC sensitivity to such scenarios is $m_{Z'} \approx 4$ TeV and $m_{\nu_R} \approx 1$ TeV [33]. Note that if $m_{\nu_R} \ll m_{Z'}$, the lepton and jets from the ν_R decay could be collimated. New approaches to such situations will be discussed later.

The W' boson production rate is not very sensitive to its couplings [34], but the interference with the standard model W boson, neglected in most experimental studies, is key in identifying the W' boson coupling helicity. Of course, the absence of $W' \to l\nu$ decays, as expected in the purely right-handed case, would also provide an important indication! The standard transverse mass distribution can be used to search for $\ell\nu$ decays, with a sensitivity reaching 5 TeV at the LHC [35]. An interesting alternative is to search for the decay $W' \to WZ$: at low-to-moderate mass where there is some background, the trilepton channel is the most sensitive, whereas for higher masses the semileptonic channels, where one of the W or Z bosons is allowed to decay hadronically, dominate. The Tevatron reach is expected to be somewhat below $m_{W'} = 1$ TeV for $BR(W' \to WZ) = 1\%$, and the LHC should be able to probe masses close to 3 TeV. It should be noted that in the semileptonic channel, for $m_{W'} \gtrsim 600$ GeV, the quarks from the hadronically decaying vector boson are sufficiently collimated to be reconstructed as a single jet. Techniques to handle this will be discussed below. Of course, the $W' \to tb$ channel is also possible, and would be very valuable in determining branching ratios. The current Tevatron limits impose $m_{W'} \gtrsim 800$ GeV [36, 37].

In general, if extra gauge bosons exist, new exotic fermions are needed to cancel anomalies [26]. Such quarks or leptons could be pair-produced, followed by decays to weak vector bosons and standard model quarks or leptons. The LHC mass reach for such quarks should be in excess of 1 TeV [38].

6   Gravity and Hierarchy

A promising approach to quantum gravity consists in adding space dimensions: this is string theory. The additional space dimensions are then hidden, presumably because they are "compactified" in some way. The radius of compactification was usually assumed to be at the scale of gravity, $\sim 10^{18}$ GeV, until it was realized it could be as low as 1 TeV [39].

6.1 ADD

In the ADD large extra dimension scenario developed in 1998 [40], the standard model fields are confined to a 3 + 1 dimensional subspace ("brane") while gravity is allowed to propagate in all dimensions. Gravity then only appears weak on the standard model brane because it is only felt when gravitons "go through" the brane. Now, since the edges of the extra dimensions are identified by compactification, boundary conditions are generated which lead to quantification of momentum along the extra dimension. Since momentum in those directions looks like mass to observers confined to the brane, and in this scenario many graviton excitations with small mass splittings are present, a Kaluza–Klein tower of gravitons emerges. The coupling of standard model particles to individual gravitons remains very small, but there are so many graviton states that the phase space becomes very large and observable cross sections ensue. Furthermore, since the graviton couples to the energy-momentum tensor, this impacts all processes.

Two classes of processes can be studied. In the case of direct production of gravitons, the bulk space is involved, and since translational invariance in the directions perpendicular to the standard model brane is broken, momentum in that direction is not conserved. Physically, the graviton escapes into the bulk right after production, leading to a $\not{E}_T$ signature. The most sensitive channels are the search for monophoton or monojet events. The other option is to look for high-mass cross-section deviations in standard model processes. Indeed, at high mass the graviton contribution can exceed the standard model one, and furthermore its spin 2 nature can affect the angular distribution of the outgoing particles. Figure 10 shows the $\not{E}_T$ distribution in a CDF search in the monophoton channel [41], and the dielectron and diphoton invariant mass spectrum in a D0 search for a high mass deviation in dielectron and diphoton events [42]. The limits range from ∼1.4 to ∼1 TeV for two to six extra dimensions in the CDF analysis, and ∼2.1 to ∼1.3 TeV for two to seven extra dimensions in the D0 analysis. Note, however, that these limits are not directly comparable since slightly different formalisms are used.

6.2 RS

A different model has a single extra dimension with a warped metric [43]. The standard model fields are still confined to a brane with gravity propagating in the extra dimension, but gravity "originates" on a second brane. The extra dimension is compactified with radius r_c and the two branes are located at $y = r_c\phi = 0, \pi r_c$ along the extra dimension. With the metric warped by a factor $e^{-2kr_c\phi}$, a TeV scale can be generated from a fundamental scale at M_{Pl} if $2kr_c\phi \sim 30$. In this scenario, there are only a few massive graviton excitations, with mass separations corresponding to the zeros of the Bessel function, providing a smoking gun signature for the model if more than one of the excitations can be observed. The lightest excitation is expected to have

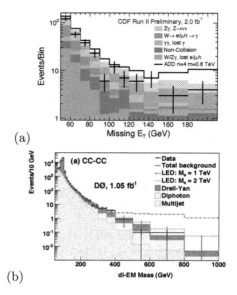

(a)

(b)

Figure 10. *Search for ADD large extra dimensions: (a) $\not{E}_T$ distribution in a CDF search in the monophoton channel and (b) dielectron and diphoton mass spectrum in a D0 interference search.*

mass below a few TeV, and the widths of the excitations depend on the warp factor k [44]. Also note that since this is a graviton, in contrast to a Z' decays to a pair of photons are allowed. Both CDF [45] and D0 [46] have search results setting 95% C.L. limits ranging from $m_G \gtrsim 300$ GeV for $k/M_{Pl} = 0.01$ to $m_G \gtrsim 900$ GeV for $k/M_{Pl} = 0.1$.

The warped extra dimension model can be augmented by localizing particles along the extra dimension: since scales depend on the position of the particle wave function, masses are generated by geometry, and mixing angles originate from particle wavefunction overlaps. The heavier fermions and gauge boson excitations are located close to the "infrared" brane with scale 1 TeV where the Higgs boson is also located. Bounds from precision measurements require that the gauge boson excitations have masses in the multi-TeV range, and because these are located close to the IR brane, they mainly couple to the top quark, and W and Z bosons. These gauge boson excitations represent the most promising channels for discovery, but since their couplings to light fermions are small, the production cross sections are small. In contrast with this, they are expected to be very broad tt, WW or ZZ resonances [47].

This also leads to a new experimental phenomenology: having very heavy particles decaying to top quarks, W and Z bosons means the latter are produced with momenta much larger than their mass, leading to collimated decay products. For leptonic W and Z boson decays this is manageable, since de-

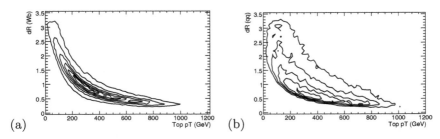

(a)   (b)

Figure 11. *Angular distance between (a) the b quark and W boson and (b) light quarks from hadronic W boson decays as a function of top quark transverse momentum.*

tectors measure charged lepton directions extremely well, however, hadronic decays lead to jets, which are intrinsically relatively wide. This is illustrated in Fig. 11, where the angular ($dR = \sqrt{\Delta\phi^2 + \Delta\eta^2}$) distance between top quark decay products is shown as a function of top quark transverse momentum in simulated decays. Since the typical physical jet radius R_{jet} is ~ 0.5, for top transverse momenta larger than ~ 300 GeV the distance between quarks from W decays starts dropping below $2R_{jet}$, and similarly for the distance between the b quark and the W boson. Hadronically decaying W bosons can then be reconstructed as a single jet, and for leptonic W boson decays, lepton isolation loses its effectiveness as a signal selection variable. But the LHC experiments' calorimeters have very fine granularity, and resolving the substructure of merged jets can be attempted.

A recent ATLAS study [48] has explored this using fully simulated $m = 2, 3$ TeV Z' bosons decaying to top quark pairs. This covers top quark transverse momenta from 500 to 1500 GeV, with only few events in the "transition region" between 200 and 600 GeV. QCD multijet events with $280 < p_T < 2240$ GeV are used as the main background sample. For the fully hadronic top quarks decays, the fundamental idea is that even though the decay hadrons are reconstructed as a single jet, this jet originates from a massive particle decaying to three hard partons, not one. If it were possible to measure each of the partons in the jet perfectly it would be possible to reconstruct the originator's invariant mass and its direct daughters. Of course, since quarks radiate and hadronize there is cross-talk, and the detectors cannot resolve the individual partons. However, the invariant mass of all the jet constituents (typically calorimeter cells) can be calculated, and is expected to be $\gtrsim m_{top}$. This can be seen in Fig. 12(a): the slow increase in jet mass versus top quark p_T is due to increased radiation. The jet mass is not sensitive to jet substructure, however, and for single jets from hadronic top quark decays three "concentrations" of energy are expected. There are multiple ways to exploit this, and the ATLAS study uses $k_\perp$ splitting scales [49]. The $k_\perp$ algorithm, a nearest neighbor jet clustering algorithm based on p_T-weighted angular distance, is much better suited to understanding jet substructure than

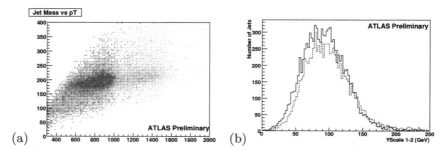

Figure 12. *ATLAS study of high-p_T top quark decays: (a) jet mass as a function of top quark p_T, both in GeV, for $m = 2$ (open) and 3 (solid) TeV Z' bosons, and (b) one-to-two jet splitting scale for top quark "monojets" from $m = 2$ (solid) and 3 (dashed) TeV Z' bosons.*

cone-type algorithms which seek to maximize energy in an $\eta \times \phi$ cone. The splitting, or y-scale, gives the energy scale at which one switches from one to two, two to three, etc. jets. The splitting scale from one to two jets for the signal is given in Fig. 12(b). For the QCD multijet background, both the jet mass and splitting scales take the shape of negative exponential functions. In the study, the jet mass and splitting scales are combined into a likelihood variable y_L, which is shown for both signal and background as a function of jet p_T in Fig. 13. Signal and background efficiencies depend on the chosen value of the likelihood cut: for 90 (65)% signal efficiency at $p_T = 1$ TeV, the QCD multijet pass rate is 15 (7)%.

In the same analysis, ATLAS also studied semileptonic top quark decays. Since the b-jet is close to the lepton, the usual lepton isolation requirement is ineffective. This is replaced by two variables, $x_\mu = 1 - m_b^2/m_{visible}^2$ (with m_b the mass of the jet near the lepton), which represents the fraction of visible top mass carried away by the muon [50], and the relative p_T of the lepton with respect to the jet. The key conclusion from this analysis is that the QCD multijet background can be reduced to almost an order of magnitude less than the irreducible continuum $t\bar{t}$ background. The main challenge in the search for a resonance is then the ability to identify a peak over the background. For narrow resonances, the ATLAS study suggests that a mass resolution of approximately 5% of the resonance mass is achievable.

7 Additional Topics Not Addressed

Due to the limited time available, many topics were not, or barely addressed. Some of these are:

- Long-lived particles, which can decay outside [51], or halfway out [52–54] the detector, or come to rest and decay later [55]

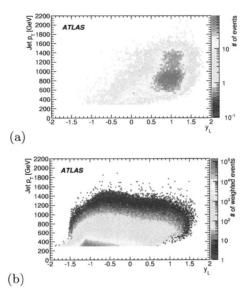

(a)

(b)

Figure 13. *ATLAS study of high-p_T top quark decays: likelihood value as a function of jet p_T for (a) signal and (b) background.*

- "Quirks" [56]

- Lepton jets [57]

- *R*-parity violating supersymmetry [58]

- Model-independent searches [59]

8 Conclusions

So far, searches for new particles or interactions have only succeeded in setting stringent constraints on their existence. However, we do expect to see something new in the next few years, as the LHC breaks the TeV barrier. If there is a Higgs boson, does it generate fermion masses? Do new particles stabilize its mass? Can we learn something about the nature of standard model particles from these? If there is no Higgs boson, how is its role fulfilled? Are there new interactions or space dimensions?

We can hope for a very rich phenomenology which will help us understand more than the question of particle masses, so that we not only have the particle physics equivalent of Mendeleev's table, but also understand how its structure comes about. This may require a new paradigm shift, as during the emergence of the quark model.

Acknowledgements

The author would like to thank the organizers of the school for a wonderful time at St. Andrews. From my colleagues' excellent lectures to haggis and stimulating discussions, it was an experience I will remember fondly.

References

[1] ALEPH Collaboration, The LEP and SLD Collaborations, Phys. Rept. **427** (2006) 257 [hep-ex/0509008].

[2] B. W. Lee, C. Quigg and H. B. Thacker, Phys. Rev. **D16** (1977) 1519.

[3] P. W. Higgs, Phys. Rev. Lett. **13** (1964) 508.

[4] P. W. Higgs, Phys. Lett. **12** (1964) 132.

[5] P. W. Higgs, Phys. Rev. **145** (1966) 1156.

[6] F. Englert and R. Brout, Phys. Rev. Lett. **13** (1964) 321.

[7] G. S. Guralnik, C. R. Hagen and T. W. B. Kibble, Phys. Rev. Lett. **13** (1964) 585.

[8] LEP Collaboration, J. Alcaraz et al., arXiv:0712.0929 [hep-ex].

[9] D0 Collaboration, Tech. Rep. D0 Conference Note 6006-CONF, Nov. 2009.

[10] C. Balazs and C. P. Yuan, Phys. Rev. **D56** (1997) 5558 [hep-ph/9704258].

[11] J. M. Campbell and R. K. Ellis, Phys. Rev. **D65** (2002) 113007 [hep-ph/0202176].

[12] T. Sjostrand, S. Mrenna and P. Skands, JHEP **05** (2006) 026 [hep-ph/0603175].

[13] G. Corcella et al., JHEP **01** (2001) 010 [hep-ph/0011363].

[14] T. Gleisberg et. al., JHEP **02** (2009) 007 [arXiv:0811.4622].

[15] M. L. Mangano, M. Moretti, F. Piccinini, R. Pittau and A. D. Polosa, JHEP **07** (2003) 001 [hep-ph/0206293].

[16] S. Frixione and B. R. Webber,JHEP **06** (2002) 029 [hep-ph/0204244].

[17] S. Frixione, P. Nason and G. Ridolfi, arXiv:0707.3081 [hep-ph].

[18] D0 Collaboration, V. M. Abazov et al., Phys. Lett. **B669** (2008) 278 [arXiv:0808.1296].

[19] CDF and D0 Collaborations, arXiv:0911.3930 [hep-ex].

[20] D0 Collaboration, Tech. Rep. D0 Conference Note 5937-CONF, June 2009.

[21] D0 Collaboration, V. M. Abazov et al., Phys. Lett. **B660** (2008) 449 [arXiv:0712.3805].

[22] D0 Collaboration, V. M. Abazov et al., Phys. Lett. **B680** (2009) 34–43 [arXiv:0901.0646].

[23] C. G. Lester and D. J. Summers, Phys. Lett. **B463** (1999) 99 [hep-ph/9906349].

[24] G. Polesello and D. R. Tovey, JHEP **1003** (2010) 030 [arXiv:0910.0174 [hep-ph]].

[25] C. Sagan, *Cosmos*. Random House, New York, 1st ed., 1980.

[26] P. Langacker, Rev. Mod. Phys. **81** (2009) 1199 [arXiv:0801.1345 [hep-ph]].

[27] T. G. Rizzo, hep-ph/0610104.

[28] ATLAS Collaboration, G. Aad et al., arXiv:0901.0512 [hep-ex].

[29] CDF Collaboration, T. Aaltonen et al., Phys. Rev. Lett. **102** (2009) 031801 [arXiv:0810.2059].

[30] D0 Collaboration, V. M. Abazov et al., Phys. Rev. **D69** (2004) 111101 [hep-ex/0308033].

[31] CDF Collaboration, T. Aaltonen et al., Phys. Rev. **D79** (2009) 112002 [arXiv:0812.4036].

[32] A. Henriques and L. Poggioli, Tech. Rep. ATL-PHYS-92-010. ATL-GE-PN-10, CERN, Geneva, Oct. 1992.

[33] A. Ferrari and J. Collot, Tech. Rep. ATL-PHYS-2000-034, CERN, Geneva, Dec. 2000.

[34] T. G. Rizzo, JHEP **05** (2007) 037 [arXiv:0704.0235].

[35] A. de Roeck, A. Ball, M. Della Negra, L. Fo and A. Petrilli, CMS physics: Technical Design Report, CERN, Geneva, Sep. 2006.

[36] CDF Collaboration, T. Aaltonen et al., Phys. Rev. Lett. **103** (2009) 041801 [arXiv:0902.3276].

[37] D0 Collaboration, V. M. Abazov et al., Phys. Rev. Lett. **100** (2008) 211803 [arXiv:0803.3256].

[38] R. Mehdiyev, A. Siodmok, S. Sultansoy and G. Unel, Eur. Phys. J. **C54** (2008) 507 [arXiv:0711.1116].

[39] I. Antoniadis, Phys. Lett. **B246** (1990) 377.

[40] N. Arkani-Hamed, S. Dimopoulos and G. R. Dvali, Phys. Lett. **B429** (1998) 263 [hep-ph/9803315].

[41] CDF Collaboration, T. Aaltonen et al., Phys. Rev. Lett. **101** (2008) 181602 [arXiv:0807.3132].

[42] D0 Collaboration, V. M. Abazov et al., Phys. Rev. Lett. **102** (2009) 051601 [arXiv:0809.2813].

[43] L. Randall and R. Sundrum, Phys. Rev. Lett. **83** (1999) 3370 [hep-ph/9905221].

[44] H. Davoudiasl, J. L. Hewett and T. G. Rizzo, Phys. Rev. **D63** (2001) 075004 [hep-ph/0006041].

[45] CDF Collaboration, T. Aaltonen et al., Phys. Rev. Lett. **99** (2007) 171801 [arXiv:0707.2294].

[46] D0 Collaboration, V. M. Abazov et al., Phys. Rev. Lett. **100** (2008) 091802 [arXiv:0710.3338].

[47] B. Lillie, L. Randall and L.-T. Wang, JHEP **09** (2007) 074 [hep-ph/0701166].

[48] The ATLAS Collaboration, Tech. Rep. ATL-PHYS-PUB-2009-081. ATL-COM-PHYS-2009-255, CERN, Geneva, May 2009.

[49] J. M. Butterworth, B. E. Cox and J. R. Forshaw, Phys. Rev. **D65** (2002) 096014 [hep-ph/0201098].

[50] J. Thaler and L.-T. Wang, JHEP **07** (2008) 092 [arXiv:0806.0023].

[51] D0 Collaboration, V. M. Abazov et al., Phys. Rev. Lett. **102** (2009) 161802 [arXiv:0809.4472].

[52] D0 Collaboration, V. M. Abazov et al., Phys. Rev. Lett. **97** (2006) 161802 [hep-ex/0607028].

[53] D0 Collaboration, V. M. Abazov et al., Phys. Rev. Lett. **101** (2008) 111802 [arXiv:0806.2223 [hep-ex]].

[54] D0 Collaboration, V. M. Abazov et al., Phys. Rev. Lett. **103** (2009) 071801 [arXiv:0906.1787 [hep-ex]].

[55] D0 Collaboration, V. M. Abazov et al., Phys. Rev. Lett. **99** (2007) 131801 [arXiv:0705.0306 [hep-ex]].

[56] J. Kang and M. A. Luty, JHEP **0911** (2009) 065 [arXiv:0805.4642 [hep-ph]].

[57] C. Cheung, J. T. Ruderman, L.-T. Wang and I. Yavin, JHEP **1004** (2010) 116 [arXiv:0909.0290 [hep-ph]].

[58] J. L. Feng, J.-F. Grivaz and J. Nachtman, Rev. Mod. Phys. **82** (2010) 699 [arXiv:0903.0046 [hep-ex]].

[59] CDF Collaboration, T. Aaltonen et al., Phys. Rev. **D78** (2008) 012002 [arXiv:0712.1311 [hep-ex]].

Section III: Tools

Monte Carlo Tools

Torbjörn Sjöstrand

Department of Theoretical Physics, Lund University, Sweden

1 Introduction

Given the current landscape in experimental high-energy physics, these lectures are focused on applications of event generators for hadron colliders such as the Tevatron and LHC. Much of this would also be relevant for e^+e^- machines like LEP, ILC and CLIC, or $e^{\pm}p$ machines like HERA, but with some differences not discussed here. Heavy-ion physics is not addressed, since it involves rather different aspects, specifically the potential formation of a quark–gluon plasma. Further, within the field of high-energy $pp/p\overline{p}$ collisions, the emphasis will be on the common aspects of QCD physics that occurs in all collisions, rather on those aspects that are specific to a particular physics topic, such as B production or supersymmetry. Many of these topics are covered by other lectures at this school.

Section 2 contains a first overview of the physics picture and the generator landscape. Section 3 describes the usage of matrix elements, Section 4 covers the important topics of initial- and final-state showers, and Section 5 discusses how showers can be matched to different hard processes. The issue of multiparton interactions and their role in mimimum-bias and underlying-event physics is introduced in Section 6, followed by some comments on hadronisation in Section 7. The article concludes with a brief description of ongoing generator-development work in Section 8.

Event generators are also described in Refs. [1–4] which are complementary to this writeup in style.

2 Overview

In real life, machines produce events that are stored by the data acquisition system of a detector. In the virtual reality of simulation, event generators like HERWIG [5] and PYTHIA [6] play the role of machines like the Tevatron and

LHC, and detector simulation programs such as GEANT 4 the role of detectors like ATLAS or CMS. The real and virtual worlds can share the same event reconstruction framework and subsequent physics analysis. An understanding of how the original physical event is distorted by the detection processes and analysis methods can be gained in the better-controlled virtual world that an understanding can be gained of what may be going on in the real world. The initial development of physics analyses or determination of the basic design parameters of a new detector can be done by using the generators themselves.

A number of physics analyses would not be feasible without generators. Specifically, a proper understanding of the (potential) signal and background processes is important to separate the two. The key aspect of generators here is that they provide a detailed description of the final state so that, ideally, any experimental observable or combination of observables can be predicted and compared with data. Generators can be used at various stages of an experiment: when optimizing the detector and its trigger design to the intended physics program, when estimating the feasibility of a specific physics study, when devising analysis strategies, when evaluating acceptance corrections, and so on.

However, it should always be kept in mind that generators are not perfect. They suffer from having to describe a broad range of physics, some of which is known from first principles, while other parts are modelled in different frameworks. (In the latter case, a generator actually acts as a vehicle of ideology, where ideas are disseminated in prepackaged form from theorists to experimentalists.) Given the limited resources, different authors may also have invested more or less time in specific physics topics, and therefore these topics may be more or less well modelled. It always pays to compare several approaches before drawing firm conclusions. Blind usage of a generator is not to be encouraged: then you are the slave rather than the master.

Why then *Monte Carlo* event generators? Basically because Einstein was wrong: God does throw dice! In quantum mechanics, calculations provide the *probability* for different outcomes of a measurement. Event by event, it is impossible to know beforehand what will happen: anything that is at all allowed could be next. It is only when averaging over large event samples that the expected probability distributions emerge — provided we did the correct calculation to high enough accuracy. In generators, (pseudo)random numbers are used to make choices intended to reproduce the quantum mechanical probabilities for different outcomes at various stages of the process.

The buildup of the structure in an event occurs in several steps, and can be summarized as follows:

- Initially two hadrons are coming in on a collision course. Each hadron can be viewed as a bag of partons — quarks and gluons.

- A collision between two partons, one from each side, gives the hard process of interest, be it for physics within or beyond the Standard Model: $ug \to ug$, $u\bar{d} \to W^+$, $gg \to h^0$, etc. (Actually, the bulk of the cross section results in rather mundane events, with at most rather soft jets. Such events usually are filtered away at an early stage.)

- When short-lived "resonances" are produced in the hard process, such as the top, $W^{\pm}$ or Z^0, their decay has to be viewed as part of this process itself, since, e.g., spin correlations are transferred from the production to the decay stages.

- A collision implies accelerated colour (and often electromagnetic) charges, and thereby bremsstrahlung can occur. Emissions that can be associated with the two incoming colliding partons are called Initial-State Radiation (ISR). As we shall see, such emissions can be modelled by so-called space-like parton showers.

- Emissions that can be associated with outgoing partons are instead called Final-State Radiation (FSR), and can be approximated by time-like parton showers. Often the distinction between a hard process and ISR and FSR is ambiguous, as we shall see.

- So far we have only extracted one parton from each incoming hadron to undergo a hard collision. But the hadron is made up of a multitude of further partons, and so further parton pairs may collide within one single hadron–hadron collision. These are multiparton interactions (MPI) and are not to be confused with pileup events, when several hadron pairs collide during a bunch–bunch crossing, but with obvious analogies.

- Each of these further collisions may also be associated with its ISR and FSR.

- The colliding partons take a fraction of the energy of the incoming hadrons, but much of the energy remains in the beam remnants, which continue to travel essentially in the original directions. These remnants also carry colours that compensate the colour taken away by the colliding partons.

- As the partons created in the previous steps recede from one another, confinement forces become significant. The structure and evolution of these force fields cannot currently be described from first principles, so models have to be introduced. One common approach is to assume that a separate confinement field is stretched between each colour and its matching anticolour, with each gluon considered as a simple sum of a colour and an anticolour, and all colours distinguishable from one another (the $N_C \to \infty$ limit).

- Such fields can break by the production of new quark–antiquark pairs that screen the endpoint colours, and where a quark from one break (or from an endpoint) can combine with an antiquark from an adjacent break to produce a primary hadron. This process is called hadronisation.

- Many of those primary hadrons are unstable and decay further with various timescales. Some are sufficiently long-lived that their decays are visible in a detector or are (almost) stable. At this stage, we have reached scales where the event-generator description has to be matched to a detector-simulation framework.

- It is only at this final stage that experimental information can be obtained and used to reconstruct back what may have happened at the core of the process.

The Monte Carlo method allows these steps to be considered sequentially, and within each step to define a set of rules that can be used iteratively to construct a more and more complex state, ending with hundreds of particles moving out in different directions. Since each particle contains of the order of ten degrees of freedom (flavour, mass, momentum, production vertex, lifetime, ...), thousands of choices are involved for a typical event. The aim is to have a sufficiently realistic description of these choices that both the average behaviour and the fluctuations around this average are well decribed.

Schematically, the cross section for a range of final states is provided by

$$\sigma_{\text{final state}} = \sigma_{\text{hard process}} \, \mathcal{P}_{\text{tot,hard process} \rightarrow \text{final state}},$$

properly integrated over the relevant phase-space regions and summed over possible "paths" (of showering, hadronisation, etc.) that lead from a hard process to the final state. That is, the dimensional quantities are associated with the hard process; subsequent steps are handled in a probabilistic approach.

The spectrum of event generators is very broad, from general-purpose ones to more specialized ones. HERWIG and PYTHIA are the two most commonly used among the former ones, with SHERPA [7] becoming more common. Among more specialized programs, many deal with the matrix elements for some specific set of processes, a few with topics such as parton showers or particle decays, but there are, e.g., no freestanding programs that handle hadronisation. In the end, many of the specialized programs are therefore used as "plugins" to the general-purpose ones.

3 Matrix Elements and Their Usage

From the Lagrangian of a theory the Feynman rules can be derived, and from them matrix elements can be calculated. Combined with phase space it allows the calculation of cross sections. As a simple example consider the scattering of quarks in QCD, say $u(1) \, d(2) \rightarrow u(3) \, d(4)$, a process similar to Rutherford scattering but with gluon exchange instead of photon exchange. The Mandelstam variables are defined as $\hat{s} = (p_1 + p_2)^2$, $\hat{t} = (p_1 - p_3)^2$ and $\hat{u} = (p_1 - p_4)^2$. In the cm frame of the collision $\hat{s}$ is the squared total energy and $\hat{t}, \hat{u} = -\hat{s}(1 \mp \cos\hat{\theta})/2$ where $\hat{\theta}$ is the scattering angle. The differential cross section is then

$$\frac{\mathrm{d}\hat{\sigma}}{\mathrm{d}\hat{t}} = \frac{\pi}{\hat{s}^2} \frac{4}{9} \alpha_{\mathrm{s}}^2 \frac{\hat{s}^2 + \hat{u}^2}{\hat{t}^2},$$

which diverges roughly like $\mathrm{d}p_\perp^2/p_\perp^4$ for transverse momentum $p_\perp \to 0$. We will come back to this issue when discussing multiparton interactions; for now suffice it to say that some lower cutoff $p_{\perp \mathrm{min}}$ needs to be introduced. Similar cross sections, differing mainly by colour factors, are obtained for $q\,g \to q\,g$ and $g\,g \to g\,g$. A few further QCD graphs, like $g\,g \to q\,\bar{q}$, are less singular and give smaller contributions. These cross sections then have to be convoluted with the flux of the incoming partons i and j in the two incoming hadrons A and B:

$$\sigma = \sum_{i,j} \iiint \mathrm{d}x_1\,\mathrm{d}x_2\,\mathrm{d}\hat{t}\, f_i^{(A)}(x_1, Q^2)\, f_j^{(B)}(x_2, Q^2)\, \frac{\mathrm{d}\hat{\sigma}_{ij}}{\mathrm{d}\hat{t}}. \tag{1}$$

The parton density functions (PDFs) of gluons and sea quarks are strongly peaked at small momentum fractions $x_1 \approx E_i/E_A$, $x_2 \approx E_j/E_B$. This further enhances the peaking of the cross section at small $p_\perp$ values.

In order to address the physics of interest a large number of processes, both within the Standard Model and in various extensions of it, have to be available in generators. Indeed many can be found in the general-purpose ones, but not enough by far. Often processes are only available to lowest order, while experimental interest may be in higher orders with more jets in the final state, either as a signal or as a potential background. So a wide spectrum of matrix-element-centered programs are available, some quite specialized and others more generic.

The way these programs can be combined with a general-purpose generator is illustrated in Fig. 1. In the study of Supersymmetry (SUSY) it is customary to define a model in terms of a handful of parameters, e.g., specified at some large Grand Unification scale. It is then the task of a spectrum calculator to turn this into a set of masses, mixings and couplings for the physical states to be searched for. Separately, the matrix elements can be calculated with these properties as unknown parameters. Only when the two are combined is it possible to speak of physically relevant matrix-element expressions. These matrix elements now need to be combined with PDFs and sampled in phase space, preferable with some preweighting procedure so that regions with high cross sections are sampled more frequently. The primarily produced SUSY particles typically are unstable and undergo sequential decays down to a lightest supersymmetric particle (LSP), again with branching ratios and angular distributions that should be properly modelled. The LSP, in many models, would be neutral and escape undetected, while other decay products would be normal quarks and leptons.

It is at this stage that general-purpose programs take over. They describe the showering associated with the above process, the presence of additional interactions in the same hadron–hadron collision, the structure of beam rem-

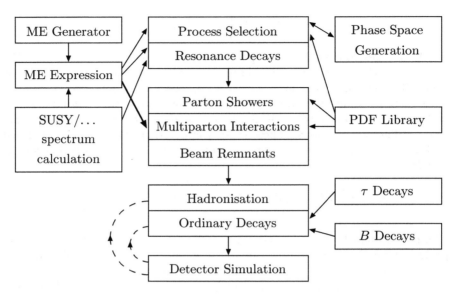

Figure 1. *An example how different programs can be combined in the event-generation chain.*

nants, and the hadronisation and decays. They would still rely on the externally supplied PDFs, and potentially make use of programs dedicated to τ and B decays, where spin information and form factors require special encoding. Even after the event has been handed on to the detector-simulation program some parts of the generator may be used in the simulation of secondary interactions and decays.

Several standards have been developed to further this interoperability. The Les Houches Accord (LHA) for user processes [8] specifies how parton-level information about the hard process and sequential decays can be encoded and passed on to a general-purpose generator. Originally it was defined in terms of two Fortran commonblocks, but more recently a standard Les Houches Event File format [9] offfers a language-independent alternative approach. The Les Houches Accord Parton Density Functions (LHAPDF) library [10] makes different PDF sets available in a uniform framework. The SUSY Les Houches Accord (SLHA) [11, 12] allows a standardized transfer of masses, mixings, couplings and branching ratios from spectrum calculators to other programs. The HepMC C++ event record [13] succeeds the HEPEVT Fortran one as a standard way to transfer information from a generator on to the detector-simulation stage. One of the key building blocks for several of these standards is the PDG codes for all the most common particles [14], also in some scenarios for physics beyond the Standard Model.

The $2 \rightarrow 2$ processes we started out with above are about the simplest one can imagine at a hadron collider. In reality one needs to go on to higher orders. In $\mathcal{O}(\alpha_s^3)$ two new kind of graphs enter. One kind is where one additional parton is present in the final state, i.e., $2 \to 3$ processes. The cross section for such processes is almost always divergent when one of the parton energies vanishes (soft singularities) or two partons become collinear (collinear singularities). The other kind is loop graphs, with an additional intermediate parton not present in the final state, i.e., a correction to the $2 \to 2$ processes. This gives negative divergences that exactly cancel the positive ones above, with only finite terms surviving. For inclusive event properties, such next-to-leading order (NLO) calculations lead to an improved accuracy of predictions, but for more exclusive studies the mathematical cancellation of singularities has to be supplemented by more physical techniques, which is far from trivial.

The tricky part of the calculations is the virtual corrections. NLO is now state of the art, with NNLO still in its infancy. If one is content with Born-level diagrams only, i.e., without any loops, it is possible to go to quite high orders, with eight or more partons in the final state. These partons have to be kept well separated to avoid the phase-space regions where the divergences become troublesome. In order to cover also regions where partons become soft/colliner we therefore next turn our attention to parton showers.

4   Parton Showers

As already noted, the emission rate for a branching such as $q \to qg$ diverges when the gluon either becomes collinear with the quark or when the gluon energy vanishes. The QCD pattern is similar to that for $e \to e\gamma$ in QED, except with a larger coupling, and a coupling that increases for smaller relative $p_\perp$ in a branching, thereby further enhancing the divergence. Furthermore the non-Abelian character of QCD leads to $g \to gg$ branchings with similar divergences, without any correspondence in QED. The third main branching, $g \to q\bar{q}$ with its $\gamma \to e^+e^-$ QED equivalence, does not have the soft divergence and is less important.

Now, if the rate for one emission of a gluon is high, then also the rate for two or more will be high, and thus consideration of high orders is required. With showers we introduce two new concepts that make life easier: (1) An iterative structure that allows simple expressions for $q \to qg$, $g \to gg$ and $g \to q\bar{q}$ branchings to be combined to build up complex multiparton final states, and (2) A Sudakov factor that offers a physical way to handle the cancellation between real and virtual divergences.

Neither of the simplifications is exact, but together they allow us to provide sensible approximate answers for the structure of emissions in soft and collinear regions of phase space.

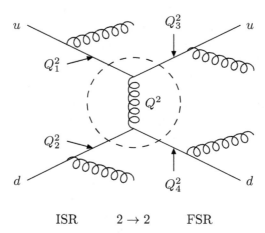

$$\text{ISR} \qquad 2 \to 2 \qquad \text{FSR}$$

Figure 2. *The "factorization" of a 2 → n process.*

4.1   The Shower Approach

The starting point is to "factorize" a complex $2 \to n$ process, where n represents a large number of partons in the final state, into a simple core process, such as $2 \to 2$ convoluted with showers; see Fig. 2. To begin with, in a simple $ud \to ud$ process the incoming and outgoing quarks must be on the mass shell, i.e., satisfy $p^2 = E^2 - \mathbf{p}^2 = m_q^2 \sim 0$, at long timescales. By the uncertainty principle, however, the closer one comes to the hard interaction, i.e., the shorter the timescales considered, the more off-shell the partons may be.

Thus the incoming quarks may radiate a succession of harder and harder gluons, while the outgoing ones radiate softer and softer gluons. One definition of hardness is how off-shell the quarks are, $Q^2 \sim |p^2| = |E^2 - \mathbf{p}^2|$, but we will encounter other variants later. In the initial-state radiation (ISR) part of the cascade of these virtualities is spacelike, $p^2 < 0$, hence the alternative name spacelike showers. Correspondingly the final-state radiation (FSR) is characterized by timelike virtualities, $p^2 > 0$, and hence also called timelike showers. The difference is a consequence of the kinematics in branchings.

The cross section for the whole $2 \to n$ graph is associated with the cross section of the hard subprocess, with the approximation that the other Q_i^2 virtualities can be neglected in the matrix-element expression. In the limit that all the $Q_i^2 \ll Q^2$ this should be a good approximation. In other words, first the hard process can be picked without any reference to showers, and only thereafter are showers added with unit total probability. But, of course, the showers do modify the event shape, so at the end of the day the cross section is affected. For instance, the total transverse energy $E_{\perp \text{tot}}$ of an event is increased by ISR, so the cross sections of events with a given $E_{\perp \text{tot}}$ is increased by the influx of events that started out with a lower $E_{\perp \text{tot}}$ in the

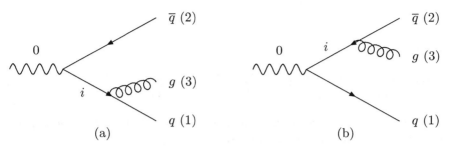

Figure 3. *The two Feynman graphs, (a) and (b), that contribute to* $\gamma^*/Z^0(0) \to q(1)\,\bar{q}(2)\,g(3)$.

hard process.

It is important that the hard-process scale Q^2 is picked to be the largest one, i.e., $Q^2 > Q_i^2$ in Fig. 2. If, e.g., $Q_1^2 > Q^2$ then instead the $ug \to ug$ subgraph ought to be chosen as hard process, and the gluon of virtuality Q^2 ought to be part of the ISR radiated from the incoming d. Without such a criterion one might double count a given graph.

4.2 Final-State Radiation

Let us next turn to a more detailed presentation of the showering approach, and begin with the simpler final-state stage. This is most cleanly studied in the process $e^+e^- \to \gamma^*/Z^0 \to q\bar{q}$. The first-order correction here corresponds to the emission of one additional gluon by either of the two Feynman graphs in Fig. 3. Neglecting quark masses and introducing energy fractions $x_j = 2E_j/E_{\mathrm{cm}}$ in the rest frame of the process, the cross section is of the form

$$\frac{\mathrm{d}\sigma_{\mathrm{ME}}}{\sigma_0} = \frac{\alpha_s}{2\pi}\frac{4}{3}\frac{x_1^2 + x_2^2}{(1-x_1)(1-x_2)}\,\mathrm{d}x_1\,\mathrm{d}x_2,$$

where σ_0 is the $q\bar{q}$ cross section, i.e., without the gluon emission.

Now study the kinematics in the limit $x_2 \to 1$. Since $1 - x_2 = m_{13}^2/E_{\mathrm{cm}}^2$ we see that this corresponds to the "collinear region," where the separation between the q and g vanishes. Equivalently, the virtuality $Q^2 = Q_i^2 = m_{13}^2$ of the intermediate quark propagator i in Fig. 3(a) vanishes. Although the full answer contains contributions from both graphs it is obvious that, in this region, the amplitude of the graph in Fig. 3(a) dominates over the graph in Fig. 3(b). We can therefore view the process as $\gamma^*/Z^0 \to q\bar{q}$ followed by $q \to qg$. We define the energy sharing in the latter branching by $E_q = zE_i$ and $E_g = (1-z)E_i$. The kinematic relationships are

$$1 - x_2 = \frac{m_{13}^2}{E_{\mathrm{cm}}^2} = \frac{Q^2}{E_{\mathrm{cm}}^2} \implies \mathrm{d}x_2 = \frac{\mathrm{d}Q^2}{E_{\mathrm{cm}}^2}$$

$$x_1 \approx z \implies dx_1 \approx dz$$
$$x_3 \approx 1 - z$$

so that

$$d\mathcal{P} = \frac{d\sigma_{\mathrm{ME}}}{\sigma_0} = \frac{\alpha_s}{2\pi} \frac{dx_2}{(1-x_2)} \frac{4}{3} \frac{x_2^2 + x_1^2}{(1-x_1)} dx_1 \approx \frac{\alpha_s}{2\pi} \frac{dQ^2}{Q^2} \frac{4}{3} \frac{1+z^2}{1-z} dz. \quad (2)$$

Here dQ^2/Q^2 corresponds to the "collinear" or "mass" singularity and $dz/(1-z) = dE_g/E_g$ to the soft-gluon singularity.

The interesting aspect of Eq. (2) is that it is universal: whenever there is a massless quark in the final state, this equation provides the probability for the same final state except for the quark being replaced by an almost collinear qg pair (plus some other slight kinematics adjustments to conserve overall energy and momentum). That is reasonable: in a general process any number of distinct Feynman graphs may contribute and interfere in a nontrivial manner, but once we go to a collinear region only one specific graph will contribute, and that graph always has the same structure, in this case with an intermediate quark propagator. Corresponding rules can be derived for what happens when a gluon is replaced by a collinear gg or $q\bar{q}$ pair. These rules are summarized by the DGLAP equations:

$$d\mathcal{P}_{a \to bc} = \frac{\alpha_s}{2\pi} \frac{dQ^2}{Q^2} P_{a \to bc}(z) \, dz \quad (3)$$

where
$$P_{q \to qg} = \frac{4}{3} \frac{1+z^2}{1-z},$$
$$P_{g \to gg} = 3 \frac{(1 - z(1-z))^2}{z(1-z)},$$
$$P_{g \to q\bar{q}} = \frac{n_f}{2} (z^2 + (1-z)^2) \quad (n_f = \text{no. of quark flavours}).$$

Furthermore, the rules can be combined to allow for successive emission in several steps, e.g., where a $q \to qg$ branching is followed by further branchings of the daughters. Thus a whole shower can be developed.

Such a picture should be reliable in cases where the emissions are strongly ordered, i.e., $Q_1^2 \gg Q_2^2 \gg Q_3^2 \ldots$, but it would not be useful if it could only be applied to strongly ordered parton configurations. A further study of the $\gamma^*/Z^0 \to q\bar{q}g$ example shows that the simple sum of the $q \to qg$ and $\bar{q} \to \bar{q}g$ branchings reproduces the full matrix elements, with interference included, to better than a factor of 2 over the full phase space. This is one of the simpler cases, and of course one should expect the accuracy to be worse for more complicated final states. Nevertheless, it is meaningful to use the shower over the whole strictly ordered, but not necessarily strongly ordered, region $Q_1^2 > Q_2^2 > Q_3^2 \ldots$ to obtain an approximate answer for multiparton topologies.

We have not yet resolved the fact that probabilities blow up in the soft and collinear regions. Perturbation theory will certainly cease to be meaningful at such small Q^2 scales that $\alpha_s(Q^2)$ diverges; in these regions confinement effects and hadronisation phenomena take over. Typically, therefore, some lower cutoff at around 1 GeV is used to regulate both soft and collinear divergences: below such a scale no further branchings are simulated. Whatever perturbative effects may remain are effectively pushed into the parameters of the nonperturbative framework. That way we avoid the singularities, but we can still have branching "probabilities" well above unity, which does not seem to make sense.

This brings us to the second big concept of this section, the *Sudakov (form) factor*. In the context of particle physics it has a specific meaning related to the properties of virtual corrections, but more generally we can just see it as a consequence of the conservation of total probability

$$\mathcal{P}(\text{nothing happens}) = 1 - \mathcal{P}(\text{something happens}),$$

where the former is multiplicative in a time-evolution sense:

$$\mathcal{P}_{\text{nothing}}(0 < t \leq T) = \mathcal{P}_{\text{nothing}}(0 < t \leq T_1)\,\mathcal{P}_{\text{nothing}}(T_1 < t \leq T).$$

When these two are combined the end result is

$$\mathrm{d}\mathcal{P}_{\text{first}}(T) = \mathrm{d}\mathcal{P}_{\text{something}}(T)\,\exp\left(-\int_0^T \frac{\mathrm{d}\mathcal{P}_{\text{something}}(t)}{\mathrm{d}t}\mathrm{d}t\right).$$

That is, the probability for something to happen for the *first* time at time T is the naive probability for this to happen, *times* the probability that this did not yet happen.

A common example is that of radioactive decay. If the number of undecayed radioactive nuclei at time t is $\mathcal{N}(t)$, with initial number $\mathcal{N}_0$ at time $t = 0$, then a naive ansatz would be $\mathrm{d}\mathcal{N}/\mathrm{d}t = -c\mathcal{N}_0$, where c parameterises the decay likelihood per unit of time. This equation has the solution $\mathcal{N}(t) = \mathcal{N}_0(1 - ct)$, which becomes negative for $t > 1/c$, because by then the probability for having had a decay exceeds unity. So what we did wrong was not to take into account that only an undecayed nucleus can decay, i.e., that the equation ought to have been $\mathrm{d}\mathcal{N}/\mathrm{d}t = -c\mathcal{N}(t)$ with the solution $\mathcal{N}(t) = \mathcal{N}_0\exp(-ct)$. This is a well-behaved expression, where the total probability for decays goes to unity only for $t \to \infty$. If c had not been a constant but varied in time, $c = c(t)$, it is simple to show that the solution instead would have become

$$\mathcal{N}(t) = \mathcal{N}_0\exp\left(-\int_0^t c(t')\,\mathrm{d}t'\right) \implies \frac{\mathrm{d}\mathcal{N}}{\mathrm{d}t} = -c(t)\mathcal{N}_0\exp\left(-\int_0^t c(t')\,\mathrm{d}t'\right).$$

For a shower the relevant "time" scale is something like $1/Q$, by the Heisenberg uncertainty principle. That is, instead of evolving to later and later times we evolve to smaller and smaller Q^2. Thereby the DGLAP Eq. (3) becomes

$$d\mathcal{P}_{a \to bc} =$$

$$\frac{\alpha_s}{2\pi} \frac{dQ^2}{Q^2} P_{a \to bc}(z)\, dz \, \exp\left(-\sum_{b,c} \int_{Q^2}^{Q^2_{max}} \frac{dQ'^2}{Q'^2} \int \frac{\alpha_s}{2\pi} P_{a \to bc}(z')\, dz' \right),$$

where the exponent (or simple variants thereof) is the Sudakov factor. As for the radioactive-decay example above, the inclusion of a Sudakov ensures that the total probability for a parton to branch never exceeds unity. Then you may have sequential radioactive decay chains, and you may have sequential parton branchings, but that is another story.

It is a bit deeper than that, however. Just as the standard branching expressions can be viewed as approximations to the complete matrix elements for real emission, the Sudakov is an approximation to the complete virtual corrections from loop graphs. The divergences in real and virtual emissions, so strange-looking in the matrix-element language, here naturally combine to provide a physical answer everywhere. What is not described in the shower, of course, is the non-universal finite parts of the real and virtual matrix elements.

The implementation of a cascade evolution now makes sense. Starting from a simple $q\bar{q}$ system, the q and $\bar{q}$ are individually evolved downwards from some initial Q^2_{max} until they branch. At a branching the mother parton disappears and is replaced by two daughter partons, which in their turn are evolved downwards in Q^2 and may also branch. Thus the number of partons increases until the lower cutoff scale is reached.

This does not mean that everything is uniquely specified. In particular, the choice of evolving in $Q^2 = |p^2|$ is by no means obvious. Any alternative variable $P^2 = f(z)\, Q^2$ would work equally well, since $dP^2/P^2 = dQ^2/Q^2$. Other evolution variables include the transverse momentum, $p_\perp^2 \approx z(1-z)m^2$, and the energy-weighted emission angle $E^2\theta^2 \approx m^2/(z(1-z))$.

Both these two alternative choices are favourable when the issue of *coherence* is introduced. Coherence means that emissions do not occur independently. For instance, consider $g_1 \to g_2\, g_3$, followed by an emission of a gluon either from 2 or 3. When this gluon is soft it cannot resolve the individual colour charges of g_2 and g_3, but only the net charge of the two, which of course is the charge of g_1. Thereby the multiplication of partons in a shower is reduced relative to naive expectations. As it turns out, evolution in $p_\perp$ or angle automatically includes this reduction, while one in virtuality does not.

In the study of FSR, e.g., at LEP, three algorithms have been commonly used. The HERWIG angular-ordered and PYTHIA mass-ordered ones are conventional parton showers as described above, while the ARIADNE [15, 16] $p_\perp$-ordered one is based on a picture of dipole emissions. That is, instead of considering $a \to bc$ one studies $a\, b \to c\, d\, e$. One aspect of this is that, in addition to the branching parton, ARIADNE also explicitly includes a "recoil parton" needed for overall energy–momentum conservation. Additional emissions off a and b are combined in a well-defined manner.

All three approaches have advantages and disadvantages. As already mentioned, PYTHIA does not inherently include coherence, but has to add that approximately by brute force. Both PYTHIA and HERWIG break Lorentz invariance slightly. The HERWIG algorithm cannot cover the full phase space but has to fill in some "dead zones" using higher-order matrix elements. The ARIADNE dipole picture does not include $g \to q\bar{q}$ branchings in a natural way.

When all is said and done, it turns out that all three algorithms do quite a decent job of describing LEP data, but typically ARIADNE does best. In recent years the $p_\perp$-ordered approach has also gained ground, having been introduced in PYTHIA [17] and being developed for SHERPA [18, 19]. The angular-ordered showers are also being further developed [20].

4.3 Initial-State Radiation

The structure of initial-state radiation (ISR) is more complicated than that of FSR, since the nontrivial structure of the incoming hadrons enter the game. A proton is made up of three quarks, uud, plus the gluons that bind them together. This picture is not static, however: gluons are continuously emitted and absorbed by the quarks, and each gluon may in its turn temporarily split into two gluons or into a $q\bar{q}$ pair. Thus a proton is teeming with activity, and much of it in a nonperturbative region where we cannot calculate. We are therefore forced to introduce the concept of a parton density $f_b(x, Q^2)$ as an empirical distribution, describing the probability of finding a parton of species b in a hadron, with a fraction x of the hadron energy–momentum when the hadron is probed at a resolution scale Q^2.

While $f_b(x, Q^2)$ itself cannot be predicted, the change of f_b with resolution scale can, once Q^2 is large enough that perturbation theory can be applied:

$$\frac{\mathrm{d}f_b(x, Q^2)}{\mathrm{d}(\ln Q^2)} = \sum_a \int_x^1 \frac{\mathrm{d}z}{z} \, f_a(x', Q^2) \frac{\alpha_s}{2\pi} \, P_{a \to bc}\left(z = \frac{x}{x'}\right). \qquad (4)$$

This is actually nothing but our familiar DGLAP equations. Before they were written in an exclusive manner: given a parton a, what is the probability that it will branch to bc during a change $\mathrm{d}Q^2$? Here the formulation is instead inclusive: given that the probability distributions $f_a(x, Q^2)$ of all partons a are known at a scale Q^2, how is the distribution of partons b changed by the set of possible branchings $a \to b$ ($+c$, here implicit). The splitting kernels $P_{a \to bc}(z)$ are the same at leading order, but differ between ISR and FSR for higher orders. Additionally, for higher orders the concept of $f_b(x, Q^2)$ as a positive definite probability is lost. We will not discuss these additional complications any further here.

Even though Eqs. (3) and (4) are equivalent, the physics context is different. In FSR the outgoing partons have been kicked to large timelike virtualities by the hard process and then cascade downwards towards the mass shell. In ISR we start out with a simple proton at early times and then allow the

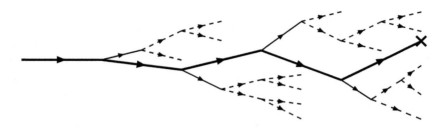

Figure 4. *A cascade of successive branchings. The thick line represents the main chain of spacelike partons leading to the hard interaction (marked by a cross). The thin lines are partons that cannot be recombined, while dashed lines are further fluctuations that may (if spacelike) or may not (if timelike) recombine. In this graph lines can represent both quarks and gluons.*

incoming partons to have increasing spacelike virtualities as we get closer to the hard interaction.

So, when the hard scattering occurs, in some sense the initial-state cascade is already there, as a virtual fluctuation. Had no collision occured the fluctuation would have collapsed back, but now one of the partons of the fluctuation is kicked out in a quite different direction and can no longer recombine with its sister parton from its last branching, nor with other partons in the cascade that lead up to this particular parton. Post facto we therefore see that a chain of branchings with increasing Q^2 values built up an ISR shower; see Fig. 4.

The obvious way to simulate this situation would be to pick partons in the two incoming hadrons from parton densities at some low Q^2 scale, and then use the exclusive formulation of Eq. (3) to construct a complete picture of partons available at higher Q^2 scales, event by event. The two sets of incoming partons could then be weighted by the cross section for the process under study. A problem is that this may not be very efficient. We have to evolve for all possible fluctuations, but at best one particular parton will collide and most of the other fluctuations will collapse back. The cost may become prohibitive when the process of interest has a constrained phase space, like a light-mass Higgs which has to have the colliding partons matched up in a very narrow mass bin.

There are ways to speed up this "forwards evolution" approach. However, the most common solution is instead to adopt a "backwards evolution" point of view. Here one starts at the hard interaction and then tries to reconstruct what happened "before." To be more precise, the cross-section formula in Eq. (1) already includes the summation over all possible incoming shower histories by the usage of Q^2-dependent parton densities. Therefore what remains is to pick one exclusive shower history from the inclusive set that went into the Q^2-evolution. To do this, recast Eq. (4) as

$$\mathrm{d}\mathcal{P}_b = \frac{\mathrm{d}f_b}{f_b} = |\mathrm{d}(\ln Q^2)| \sum_a \int \mathrm{d}z \, \frac{x' f_a(x', t)}{x f_b(x, t)} \, \frac{\alpha_\mathrm{s}}{2\pi} \, P_{a \to bc} \left(z = \frac{x}{x'} \right).$$

Then we have defined a *conditional probability*: if parton b is present at scale Q^2, what is the probability that it will turn out to have come from a branching $a \to b\,c$ at some infinitesimally *smaller* scale? (Recall that the original Eq. (4) was defined for increasing virtuality.) As in the case of FSR, this expression has to be modified by a Sudakov factor to preserve total probability, and this factor is again the exponent of the real-emission expression with a negative sign, integrated over Q^2 from an upper starting scale Q^2_{max} down to the Q^2 of the hypothetical branching.

The approach is now clear. First a hard scattering is selected, making use of the Q^2-evolved parton densities. Then, with the hard process as the upper maximum scale, a succession of ISR branchings are reconstructed at lower and lower Q^2 scales, going "backwards in time" towards the early low-virtuality initiators of the cascades. Again some cutoff needs to be introduced when the nonperturbative regime is reached.

Unfortunately the story does not end there. For FSR we discussed the need to take into account coherence effects and the possibility of using different variables. Such issues exist for the treatment of ISR as well, but there are also additional ones. For example, the evolution need not be strictly ordered in Q^2, and non-ordered chains in some cases can be important. Another issue is that there can be so many partons evolving inside a hadron that they become close packed, which leads to recombinations.

5 Combining Matrix Elements and Parton Showers

As we have seen, both matrix elements (ME) and parton showers (PS) have advantages and disadvantages. ME allow a systematic expansion in powers of α_s, and thereby offer a controlled approach towards higher precision. Calculations can be done with several partons in the final state, so long as only Born-level results are required, and it is possible to tailor the phase-space cuts for these partons precisely to the experimental needs. On the other hand, loop calculations are much more difficult and the mathematically correct cancellation between real- and virtual-emission graphs in the soft/collinear regions is not physically sensible. Therefore ME cannot be used to explore the internal structure of a jet, and are difficult to match to hadronisation models, which are supposed to take over in the very soft/collinear region.

PS are clearly an approximate description and do not come with a precise prediction for well-separated jets. You cannot control the probabilistic evolution of a shower too much, and therefore the efficiency for obtaining events in a specific region of phase space can be quite low. On the other hand, PS are universal and so for any new model you only need to provide the basic hard process, which PS will turn into reasonably realistic multiparton topologies. The use of Sudakov factors ensures a physically sensible behaviour in the soft/collinear regions and it is also here that the PS formalism is supposed to be most reliable. It is therefore possible to obtain a good picture of the internal structure of jets, and to provide a good match to hadronisation models.

In a nutshell: ME are good for well-separated jets, PS for the structure inside jets. Clearly the two complement each other and a marriage is highly desirable. This is less trivial to do without double counting or gaps in the phase space coverage and several alternative approaches have been developed. In the following we will discuss three main options: merging, vetoed parton showers and NLO matching, roughly ordered in increasing complexity. Which of these to use may well depend on the task at hand.

5.1 Merging

The aspiration of merging is to cover the whole phase space with a smooth transition from ME to PS. The typical case would be a process where the lowest-order (LO) ME is known, as well as the next-to-leading-order (NLO) real-emission one, say of an additional gluon. The shower should then reproduce

$$W^{\mathrm{ME}} = \frac{1}{\sigma(\mathrm{LO})} \frac{\mathrm{d}\sigma(\mathrm{LO} + g)}{\mathrm{d}(\mathrm{phasespace})} \tag{5}$$

starting from an LO topology. If the shower populates phase space according to W^{PS} this implies that a correction factor $W^{\mathrm{ME}}/W^{\mathrm{PS}}$ needs to be applied.

At first glance this does not appear to make sense: if all we do is get back W^{ME}, then what did we gain? However, the trick is to recall that the PS formula comes in two parts: the real-emission answer and a Sudakov factor that ensures total conservation of probability. What we have called W^{PS} above should only be the real-emission part of the story. It is also this one that we know *will* agree with W^{ME} in the soft and collinear regions. Actually, with some moderate amount of effort it is often possible to ensure that $W^{\mathrm{ME}}/W^{\mathrm{PS}}$ is of order unity over the whole phase space, and to adjust the showers in the hard region so that the ratio always is below unity, i.e., so that standard Monte Carlo rejection techniques can be used. What the Sudakov factor then does is introduce some ordering variable Q^2, so that the whole phase space is covered starting from "hard" emissions and moving to "softer" ones. This results in a distribution over phase space of the form

$$W^{\mathrm{PS}}_{\mathrm{actual}}(Q^2) = W^{\mathrm{ME}}(Q^2) \exp\left(-\int_{Q^2}^{Q^2_{\mathrm{max}}} W^{\mathrm{ME}}(Q'^2) \, \mathrm{d}Q'^2\right).$$

Here, we have used the PS choice of the evolution variable to provide an exponentiated version of the ME answer. As such it agrees with the ME answer in the hard region, where the Sudakov factor is close to unity, and with the PS in the soft/collinear regions, where $W^{\mathrm{ME}} \approx W^{\mathrm{PS}}$.

In PYTHIA this approach is used for essentially all resonance decays in the Standard Model and minimal supersymmetric extensions thereof [21]: $\gamma^*/Z^0 \to q\bar{q}$, $t \to bW^+$, $W^+ \to u\bar{d}$, $H \to b\bar{b}$, $\chi^0 \to \tilde{q}\bar{q}$, $\tilde{q} \to q\tilde{g}$, ... It is also used in ISR to describe, e.g., $q\bar{q} \to \gamma^*/Z^0/W^{\pm}$.

Merging is also used for several processes in HERWIG, such as $\gamma^*/Z^0 \to q\bar{q}$, $t \to bW^+$ and $q\bar{q} \to \gamma^*/Z^0/W^{\pm}$. A special problem here is that the angular-ordered algorithms, both for FSR and for ISR, leave some "dead zones" of hard emissions that are kinematically forbidden for the shower to populate. It is therefore necessary to start directly from higher-order matrix elements in these regions. A consistent treatment still allows a smooth joining across the boundary.

5.2 Vetoed Parton Showers

The objective of using vetoed parton showers is again to combine the real-emission behaviour of ME with the emission-ordering-variable-dependent Sudakov factors of PS. While the merging approach only works for combining the LO and NLO expressions, however, the vetoed parton showers offer a generic approach for combining several different orders.

To understand how the algorithm works, consider a lowest-order process such as $q\bar{q} \to W^{\pm}$. For each higher-order one additional jet would be added to the final state, so long as only real-emission graphs are considered: in first order, e.g., $q\bar{q} \to W^{\pm}g$, in second order, e.g., $q\bar{q} \to W^{\pm}gg$, and so on. Call these (differential) cross sections σ_0, σ_1, σ_2, ... It should then come as no surprise that each σ_i, $i \geq 1$ contains soft and collinear divergences. We therefore need to impose some set of ME phase-space cuts, e.g., on invariant masses of parton pairs, or on parton energies and angular separation between them. When these cuts are varied, so that, e.g., the mass or energy thresholds are lowered towards zero, all of these σ_i, $i \geq 1$, increase without bounds.

The reason is that in the ME approach without virtual corrections there is no "detailed balance," wherein the addition of cross section to σ_{i+1} is compensated by a depletion of σ_i. That is, if you have an event with i jets at some resolution scale, and a lowering of the minimal jet energy reveals the presence of one additional jet, then you should reclassify the event from being i-jet to being $i + 1$-jet. Add one, subtract one, with no net change in $\sum_i \sigma_i$. To solve this, Sudakovs of showers are used to ensure this detailed balance. Of course, in a complete description the cancellation between real and virtual corrections is not completely exact but leaves a finite net contribution, which is not predicted in this approach.

A few alternative algorithms exist along these lines. All share the three first steps as follows:

(1) Pick a hard process within the ME-cuts-allowed phase-space region, in proportions provided by the ME integrated over the respective allowed

region, $\sigma_0 : \sigma_1 : \sigma_2 : \ldots$ Use for this purpose an α_{s0} larger than the α_s values that will be used below.

(2) Reconstruct an imagined shower history that describes how the event could have evolved from the lowest-order process to the actual final state. This provides an ordering of emissions by whatever shower-evolution variable is intended.

(3) The "best-bet" choice of α_s scale in showers is known to be the squared transverse momentum of the respective branching. Therefore a factor $W_\alpha = \prod_{\text{branchings}}(\alpha_s(p_{\perp i}^2)/\alpha_{s0})$ provides the probability that the event should be retained.

Now the different algorithms diverge. In the CKKW and L approaches the subsequent steps are:

(4) Evaluate Sudakov factors for all the "propagator" lines in the shower history reconstructed in step 2, i.e., for intermediate partons that split into further partons, and also for the evolution of the final partons down to the ME cuts without any further emissions. This provides an acceptance weight $W_{\text{Sud}} = \prod_{\text{"propagators"}} \text{Sudakov}(Q_{\text{beg}}^2, Q_{\text{end}}^2)$ where Q_{beg}^2 is the large scale where a parton is produced by a branching and Q_{end}^2 either is the scale at which the parton branches or the ME cuts, depending on the particular case.

(4a) In the CKKW approach [22] the Sudakovs are evaluated by analytical formulae, which is fast.

(4b) In the L approach [23] trial showers are used to evaluate Sudakovs, which is slower but allows a more precise modelling of kinematics and phase space than offered by the analytic expression.

(5) Now the matrix-element configuration can be evolved further, to provide additional jets below the ME cuts used. In order to avoid doublecounting of emissions, any branchings that might occur above the ME cuts must be vetoed.

The MLM approach [24] is rather different. Here the steps instead are:

(4′) Allow a complete parton shower to develop from the selected parton configuration.

(5′) Cluster these partons back into a set of jets, e.g., using a cone-jet algorithm, with the same jet-separation criteria as used when the original parton configuration was picked.

(6′) Try to match each jet to its nearest original parton.

(7′) Accept the event only if the number of clustered jets agrees with the number of original partons, and if each original parton is sensibly matched to its jet. This would not be the case, e.g., if one parton gave rise to two jets, or two partons to one jet, or an original b quark migrated outside of the clustered jet. The point of the MLM approach is that the probability of *not* generating any additional fatal jet activity during the shower evolution is provided by the Sudakovs used in step 4′.

The different approaches have been compared both on an experimental and a theoretical level, to understand the differences and possible shortcomings [25, 26].

5.3 NLO Matching

Matching to next-to-leading order in some respects is the most ambitious approach: it aims to get not only real but also virtual contributions correctly included, so that cross sections are accurate to NLO, and that NLO results are obtained for all observables when formally expanded in powers of α_s. Thus hard emissions should again be generated according to ME, while soft and collinear ones should fall within the PS regime. There are two main approaches on the market: MC@NLO and POWHEG.

For MC@NLO [27] the scheme works as follows in simplified terms:

(1) Calculate the NLO ME corrections to an n-body process, including $n+1$-body real corrections and n-body virtual ones.

(2) Calculate analytically how a first branching in a shower starting from an n-body topology would populate $n + 1$-body phase space, excluding the Sudakov factor.

(3) Subtract the shower expression from the $n + 1$ ME one to obtain the "true" $n + 1$ events, and consider the rest as belonging to the n-body class. The PS and ME expressions agree in the soft and collinear limits, so the singularities in these regions cancel, leaving finite cross sections both for the n- and $n + 1$-body event classes.

(4) Now add showers to both kinds of events.

Several processes have been considered, such as Z^0, $b\bar{b}$, $t\bar{t}$ and W^+W^- production. A technical problem is that, although ME and PS converge in the collinear region, it is not guaranteed that ME is everywhere above PS. This is solved by having a small fraction of events with negative weights.

The POWHEG approach [28, 29] is very closely related to the merging approach presented earlier, but is more differential in phase space:

$$d\sigma = \bar{B}(v)d\Phi_v \left[\frac{R(v,r)}{B(v)} \exp \left(- \int_{p_\perp} \frac{R(v,r')}{B(v)} d\Phi_r' \right) d\Phi_r \right] \qquad (6)$$

where

$$\bar{B}(v) = B(v) + V(v) + \int d\Phi_r [R(v,r) - C(v,r)],$$

and $v, d\Phi_v$ are the Born-level n-body variables and differential phase space, $r, d\Phi_r$ are extra $n + 1$-body variables and differential phase space, $B(v)$ the Born-level cross section, $V(v)$ the virtual corrections, $R(v,r)$ the real-emission cross section, and $C(v,r)$ the counterterms for collinear factorization of parton densities. The basic idea is to pick the real emission with the largest transverse momentum according to complete MEs, including the Sudakov factor derived from an exponentiation of the real-emission expression, with the NLO normalization as a prefactor. (Note that the expression inside the square bracket of Eq. (6) integrates to unity for any v if real emissions are allowed down to the soft/collinear singularities. With some effective cutoff a few events will not have any emissions at all above this cut.) Thereafter normal showers can be used to do the subsequent evolution downwards from the $p_\perp$ scale picked by the above equations.

The MC@NLO and POWHEG methods are formally equivalent to NLO, but not beyond, so differences are useful for exploring higher orders. The latter approach may be more appealing, since it eliminates the negative-weights problem and agrees with the concept of $p_\perp$ as the natural hardness scale, also, e.g., for showers. However, as of now, MC@NLO has been worked out for more processes.

Note that the real-emission $n+1$-body part is only handled to LO accuracy, and higher-order jet topologies not at all. The NLO methods thus are useful for precision measurements of the total cross section of a process such as Z^0 or top production, but for studies of multiparton topologies the vetoed showers are more appropriate. Each tool to its task.

6 Multiparton Interactions

The cross section for $2 \to 2$ QCD parton processes is dominated by t-channel gluon exchange, as we already mentioned, and thus diverges like $dp_\perp^2/p_\perp^4$ for $p_\perp \to 0$. The cross section σ_{int} can be calculated by introducing a lower cut $p_{\perp\text{min}}$ and integrating the interaction cross section above this, properly convoluted with parton densities. At LHC energies this $\sigma_{\text{int}}(p_{\perp\text{min}})$ reaches around 100 mb for $p_{\perp\text{min}} = 5$ GeV, and 1000 mb at around 2 GeV. Since each interaction gives two jets to lowest order, the jet cross section is twice as large. This should be compared with an expected *total* cross section of the order of 100 mb. In addition, at least a third of the total cross section is related to elastic scattering $pp \to pp$ and low-mass diffractive states $pp \to pX$ that could not contain jets.

So can it really make sense that $\sigma_{\text{int}}(p_{\perp\text{min}}) > \sigma_{\text{tot}}$? Yes, it can! The point is that each incoming hadron is a bunch of partons. You can have several (more or less) independent parton–parton interactions when these two bunches pass through each other. Then an event with n interactions above $p_{\perp\text{min}}$ counts once for the total cross section but once *for each interaction* when the interaction rate is calculated. That is,

$$\sigma_{\text{tot}} = \sum_{n=0}^{\infty} \sigma_n \qquad \text{while} \qquad \sigma_{\text{int}} = \sum_{n=0}^{\infty} n\,\sigma_n,$$

where σ_n is the cross section for events with n interactions. Thus $\sigma_{\text{int}} > \sigma_{\text{tot}}$ is equivalent to $\langle n \rangle > 1$, i.e., each event on the average contains more than one interaction. Furthermore, if interactions do occur independently when two hadron pass by each other, then one would expect the number of interactions, n, to have a Poissonian distribution of the form: $\mathcal{P}_n = \langle n \rangle^n \exp(-\langle n \rangle)/n!$, so that several interactions could occur occasionally also when $\sigma_{\text{int}}(p_{\perp\text{min}}) < \sigma_{\text{tot}}$, e.g., for a larger $p_{\perp\text{min}}$ cut. However, energy–momentum conservation ensures that interactions are never truly independent, and also other effects enter (see below). Nonetheless the Poissonian ansatz is still a useful starting point.

However, multiparton interactions (MPI) can only be half the solution. The divergence for $p_{\perp\text{min}} \to 0$ would seem to imply an infinite average number of interactions. But what one should realize is that, in order to calculate the $d\hat{\sigma}/d\hat{t}$ matrix elements within standard perturbation theory, it has to be assumed that free quark and gluon states exist at negative and positive infinity. This does not take into account the confinement of colour into hadrons of finite size. So obviously perturbation theory has to break down at a $p_{\perp\text{min}}$ given by:

$$p_{\perp\text{min}} \simeq \frac{\hbar}{r_p} \approx \frac{0.2 \text{ GeV} \cdot \text{fm}}{0.7 \text{ fm}} \approx 0.3 \text{ GeV} \simeq \Lambda_{\text{QCD}}.$$

The nature of the breakdown is also easy to understand. A small-$p_{\perp}$ gluon, to be exchanged between the two incoming hadrons, has a large transverse wavelength and thus almost the same phase across the extent of each hadron. The contributions from all the colour charges in a hadron thus add coherently, and that means that they add to zero since the hadron is a colour singlet.

What is the typical scale of such colour screening effects, i.e., at what $p_{\perp}$ has the interaction rate dropped to ~half of what it would have been if the quarks and gluons of a proton had all been free to interact fully independently? That ought to be related to the typical separation distance between a given colour and its opposite anticolour. When a proton contains many partons this characteristic screening distance can well be much smaller than the proton radius. Empirically we need to introduce a $p_{\perp\text{min}}$ scale of the order of 2 GeV to describe Tevatron data, i.e., of the order of 0.1-fm separation. However, this number should not be taken too seriously without a detailed model of the space–time structure of a hadron.

The 2 GeV number is very indirect and does not really explain exactly how the dampening occurs. One can use a simple recipe, with a step-function cut at this scale, or a physically more reasonable dampening by a factor $p_\perp^4/(p_{\perp 0}^2 + p_\perp)^2$, plus a corresponding shift of the α_s argument,

$$\frac{d\hat{\sigma}}{dp_\perp^2} \propto \frac{\alpha_s^2(p_\perp^2)}{p_\perp^4} \to \frac{\alpha_s^2(p_{\perp 0}^2 + p_\perp^2)}{(p_{\perp 0}^2 + p_\perp^2)^2}, \tag{7}$$

with $p_{\perp 0}$ a dampening scale that also lands at around 2 GeV. This translates into a typical number of 2–3 interactions per event at the Tevatron and 4–5 at LHC. For events with jets or other hard processes the average number is likely to be higher.

6.1 Multiparton-Interactions Models

A good description of Tevatron data has been obtained with a simple model [30] based on the following principles:

(1) Only address MPI for inelastic nondiffractive events, with cross section σ_{nd}, which form the bulk of what is triggered as minimum bias.

(2) Dampen the perturbative jet cross section using the smooth turnoff of Eq. (7), to cover the whole $p_\perp$ range down to $p_\perp = 0$.

(3) Hadrons are extended, and therefore partons are distributed in (transverse) coordinates. To allow a flexible parametrization and yet have an easy-to-work-with expression, a double Gaussian

$$\rho_{\mathrm{matter}}(\mathbf{r}) = N_1 \exp\left(-r^2/r_1^2\right) + N_2 \exp\left(-r^2/r_2^2\right)$$

is used, where N_2/N_1 and r_2/r_1 are tunable parameters.

(4) The matter overlap during a collision as calculated by

$$\mathcal{O}(b) = \int d^3\mathbf{x}\, dt\, \rho_{1,\mathrm{matter}}^{\mathrm{boosted}}(\mathbf{x}, t)\rho_{2,\mathrm{matter}}^{\mathrm{boosted}}(\mathbf{x}, t)$$

directly determines the average activity in events at different impact parameter b: $\langle n(b)\rangle \propto \mathcal{O}(b)$, where central collisions tend to have more activity and peripheral collisions less activity.

(5) An event has to contain at least one interaction to be an event at all. This provides a natural dampening of the cross section at large-impact parameters. Normalizations have to be picked such that the b-integrated probability for having at least one interaction gives σ_{nd}, while the b-integrated rate of all interactions gives the (dampened) jet cross section. Further, $p_{\perp 0}$ has to be selected sufficiently small that $\sigma_{\mathrm{int}} > \sigma_{\mathrm{nd}}$.

(6) To first approximation the number of interactions at a given impact parameter obeys a Poissonian distribution, with the 0-interaction rate removed. Since central collisions have a larger mean and peripheral ones a smaller, the end result is a distribution broader than a Poissonian.

(7) The interactions are generated in an ordered sequence of decreasing $p_\perp$ values: $p_{\perp 1} > p_{\perp 2} > p_{\perp 3} > \ldots$. This is possible using the standard Sudakov kind of trick:

$$\frac{\mathrm{d}\mathcal{P}}{\mathrm{d}p_{\perp i}} = \frac{1}{\sigma_{\mathrm{nd}}} \frac{\mathrm{d}\sigma}{\mathrm{d}p_\perp} \exp\left[-\int_{p_\perp}^{p_{\perp(i-1)}} \frac{1}{\sigma_{\mathrm{nd}}} \frac{\mathrm{d}\sigma}{\mathrm{d}p'_\perp} \mathrm{d}p'_\perp\right],$$

with a starting $p_{\perp 0} = E_{\mathrm{cm}}/2$.

(8) The ordering of emissions allows parton densities to be rescaled in x after each interaction, so that energy–momentum is not violated. Thereby the tail towards large multiplicities is reduced.

(9) For technical reasons the model was simplified after the first interaction, so that there only gg or $q\bar{q}$ outgoing pairs were allowed, and no showers were added to these further $2 \to 2$ interactions.

Already this simple PYTHIA-based model is able to "explain" a large set of experimental data.

More recently a number of improvements have been included [17, 31, 32]:

(1) The introduction of junction fragmentation, wherein the confinement field between the three quarks in a baryon is described as a Y-shaped topology. This allows the handling of topologies where several valence quarks are kicked out, thus allowing arbitrary flavours and showering in all interactions in an event.

(2) Parton densities are not only rescaled for energy–momentum conservation, but also to take into account the number of remaining valence quarks, or that sea quarks have to occur in $q\bar{q}$ pairs.

(3) The introduction of $p_\perp$-ordered showers allows the selection of new ISR and FSR branchings and new interactions to be interleaved in one common sequence of falling $p_\perp$ values. Thereby the competition between especially ISR and MPI, which both remove energy from the incoming beams, is modelled more realistically.

(4) Rescattering is optionally allowed, wherein one parton may undergo successive scatterings.

The traditional HERWIG soft underlying event (SUE) approach to this issue has its origin in the UA5 Monte Carlo. In it a number of clusters are distributed almost independently in rapidity and transverse momentum, but shifted so that energy–momentum is conserved, and the clusters then decay isotropically. The multiplicity distribution of clusters and their y and $p_\perp$ spectra are tuned to give the observed inclusive hadron spectra. No jets are produced in this approach.

The JIMMY program [33] started as an add-on to HERWIG, but is nowadays an integrated part [34]. It replaces the SUE model with an MPI-based one more similar to the PYTHIA ones above, e.g., with an impact-parameter-based picture for the multiparton-interactions rate. Many technical differences exist, e.g., JIMMY interactions are not picked to be $p_\perp$-ordered and thus energy–momentum issues are handled differently.

The DPMJET/DTUJET/PHOJET family of programs [35–37] come from the "historical" tradition of soft physics, wherein multiple $p_\perp \approx 0$ "pomeron" exchanges fill a role somewhat similar to the hard MPI above. Jet physics was originally not included, but later both hard and soft interactions have been allowed. A feature of this framework is that it naturally includes diffractive events.

6.2   Multiparton-Interactions Studies

How do we know that MPI exist? The key problem is that it is not possible to identify jets coming from $p_\perp \approx 2$ GeV partons. Therefore we either have to use indirect signals for the presence of interactions at this scale or we have to content ourselves with studying the small fraction of events where two interactions occur at visibly large $p_\perp$ values.

An example of the former is the total charged multiplicity distribution in high-energy $pp/p\bar{p}$ collision. This distribution is very broad, measured in terms of the width over the average, $\sigma(n_{\mathrm{ch}})/\langle n_{\mathrm{ch}} \rangle$, and is found to be getting broader with increasing energy. By contrast, recall that for a Poissonian distribution this quantity scales like $1/\sqrt{n_{\mathrm{ch}}}$ and thus should get narrower. Simple models with at most one interaction and with a fragmentation framework in agreement with LEP data cannot explain this. They predict distributions that are far too narrow and have the wrong energy behaviour. If MPI are included the additional variability in the number of interactions per event offers the missing piece. The variable impact parameter improves the description further.

Another related example is forward–backward correlations. Consider the charged multiplicity n_f and n_b in a forward and a backward rapidity bin, each of width one unit, separated by a central rapidity gap of size Δy. It is not unnatural that n_f and n_b are somewhat correlated in two-jet events, and for small Δy one may also be sensitive to the tails of jets. But the correlation coefficient, although falling with Δy, is still appreciable even out to $\Delta y = 5$, and here again traditional one-interaction models fail to describe the data. In an MPI scenario each interaction provides additional particle production over a large rapidity range, and this additional number-of-MPI variability leads to good agreement with data.

Direct evidence comes from the study of four-jet events. These can be caused by two separate interactions, but also by a single one where higher orders (call it ME or PS) have allowed two additional branchings in a basic two-jet topology. Fortunately the kinematics should be different. Assume the four jets are ordered in $p_\perp$, $p_{\perp 1} > p_{\perp 2} > p_{\perp 3} > p_{\perp 4}$. If coming from two separate interactions the jets should pair up into two separately balancing sets, $|\mathbf{p}_{\perp 1} + \mathbf{p}_{\perp 2}| \approx 0$ and $|\mathbf{p}_{\perp 3} + \mathbf{p}_{\perp 4}| \approx 0$. If an azimuthal angle φ is introduced between the two jet axes this also should be flat if the interactions are uncorrelated. By contast the higher-order graph offers no reason why the jets should occur in balanced pairs, and the φ distribution ought to be peaked at small values, corresponding to the familiar collinear singularity. The first to observe an MPI signal this way was the AFS collaboration at ISR (*pp* at 62 GeV), but with large uncertainties. A more convincing study was made by CDF, who obtained a clear signal in a sample with three jets plus a photon. In fact the deduced rate was almost a factor of three higher than naive expectations, but quite in agreement with the impact-parameter-dependent picture in which correlations of this kind are enhanced. Recently also D0 have shown a comparable signal with even higher statistics.

A topic that has been quite extensively studied in CDF is that of the jet pedestal [38], i.e., the increased activity seen in events with a jet, even away from the jet itself, and away from the recoiling jet that should be there. Some effects come from the showering activity, i.e., the presence of additional softer jets, but much of it rather finds its explanation in MPI, as a kind of "trigger bias" effect, as follows:

(1) Central collisions tend to produce many interactions, peripheral ones few.

(2) If an event has n interactions there are n chances that one of them is hard.

Combine the two and one concludes that events with hard jets are biased towards central collisions and many additional interactions. The rise of the pedestal with triggger-jet energy saturates once $\sigma_{\text{int}}(p_{\perp\min} = p_{\perp\text{jet}}) \ll \sigma_{\text{nd}}$, however, because by then events are already maximally biased towards small impact parameter. And this is indeed what is observed in the data: a rapid rise of the pedestal up to $p_{\perp\text{jet}} \approx 10$ GeV, and then a slower increase that is mainly explained by showering contributions.

In more detailed studies of this kind of pedestal effects there are also some indications of a jet substructure in the pedestal, i.e., that indeed the pedestal is associated with the production of additional (soft) jet pairs.

In spite of many qualitative successes, and even some quantitative ones, one should not be led to believe that all is understood. Possibly the most troublesome issue is how colours are connected between all the outgoing partons that come from several different interactions. A first, already difficult, question is how colours are correlated between all the partons that are taken out from an incoming hadron. These colours are then mixed up by the respective scattering, which in principle, is calculated approximately. But, finally, all the outgoing partons will radiate further and overlap with one another on the way out, and how much that may mess up colours is an open question.

A sensitive quantity is $\langle p_\perp \rangle (n_{\text{ch}})$, i.e., how the average transverse momentum of charged particles varies as a function of their multiplicity. If interactions are uncorrelated in colour this curve tends to be flat: each further interaction adds about as much $p_\perp$ as $n_{\rm ch}$. If colours somehow would rearrange themselves, so that the confinement colour fields would not have to criss-cross the event, then the multiplicity would not rise as fast for each further interaction, and so a positive slope would result. It is interesting to note that the CDF Monte Carlo tunes tend to come up with values that are about 90% on the way to being maximally rearranged, which is far more than one would have guessed. Obviously further modelling and tests are necessary here.

Another issue is whether the $p_{\perp 0}$ regularization scale should be energy dependent. In olden days there was no need for this, but it became necessary when HERA data showed that parton densities rise faster at small x values than had commonly been assumed. This means that the partons become more close-packed and that the colour screening increases faster with increasing collision energy. Therefore an energy-dependent $p_{\perp 0}$ is not unreasonable, but also cannot be predicted. If one assumes that $p_{\perp 0} \propto E_{\rm cm}^p$, with some power p, then the debate has centered on the range $p = 0.16 - 0.26$, with the current best tunes [39, 40] leaning towards the higher end of this range. Recall that a larger p implies a larger $p_{\perp 0}$ at LHC energies, and thus a smaller multiplicity, but we must still allow for some range of uncertainty.

7   Hadronisation

The physics mechanisms discussed so far are mainly being played out on the partonic level, while experimentalists observe hadrons. In between exists the very important hadronisation phase, where all the outgoing partons end up confined inside hadrons of a typical 1 GeV mass scale. This phase cannot (so far) be described from first principles, but has to involve some modelling. The main approaches in use today are string fragmentation [41] and cluster fragmentation [42].

Hadronisation models start from some ideologically motivated principles, but then have to add "cookbook recipes" with free parameters to arrive at a complete picture of all the nitty gritty details. This should come as no surprise, given that there are hundreds of known hadron species to take into account, each with its mass, width, wavefunction, couplings, decay patterns and other properties that could influence the structure of the observable hadronic state, and with many of those properties being poorly or not at all known. In that sense, it is sometimes more surprising that the models can work as well as they do rather than that they fail to describe everything.

While non-perturbative QCD is not solved, lattice QCD studies lend support to a linear confinement picture (in the absence of dynamical quarks), i.e., the energy stored in the colour dipole field between a charge and an anticharge increases linearly with the separation between the charges, if the short-distance Coulomb term is neglected. This is quite different from the behaviour in QED, and is related to the presence of a three-gluon vertex in QCD. The details are not yet well understood, however.

The assumption of linear confinement provides the starting point for the string model, most easily illustrated for the production of a back-to-back $q\bar{q}$ jet pair. As the partons move apart, the physical picture is that of a colour flux tube (or maybe colour vortex line) being stretched between the q and the $\bar{q}$. The transverse dimensions of the tube are of typical hadronic sizes, roughly 1 fm. If the tube is assumed to be uniform along its length, this automatically leads to a confinement picture with a linearly rising potential. In order to obtain a Lorentz covariant and causal description of the energy flow due to this linear confinement, the most straightforward way is to use the dynamics of the massless relativistic string with no transverse degrees of freedom. The mathematical one-dimensional string can be thought of as parameterising the position of the axis of a cylindrically symmetric flux tube. From hadron spectroscopy, the string constant, i.e., the amount of energy per unit length, is deduced to be $\kappa \approx 1$ GeV/fm.

As the q and $\bar{q}$ move apart, the potential energy stored in the string increases, and the string may break by the production of a new $q'\bar{q}'$ pair, so that the system splits into two colour singlet systems $q\bar{q}'$ and $q'\bar{q}$. If the invariant mass of either of these string pieces is large enough, further breaks may occur. In the Lund string model, the string break-up process is assumed to proceed until only on-mass-shell hadrons remain, each hadron corresponding to a small piece of string.

In order to generate the quark–antiquark pairs $q'\bar{q}'$, which lead to string break-ups, the Lund model invokes the idea of quantum mechanical tunnelling. This leads to a flavour-independent Gaussian spectrum for the transverse momentum of $q'\bar{q}'$ pairs. Tunnelling also implies a suppression of heavy quark production, $u : d : s : c \approx 1 : 1 : 0.3 : 10^{-11}$. Charm and heavier quarks hence are not expected to be produced in the soft fragmentation.

A tunnelling mechanism can also be used to explain the production of baryons. This is still a poorly understood area. In the simplest possible approach, a diquark in a colour antitriplet state is treated just like an ordinary antiquark, such that a string can break either by quark–antiquark or antidiquark–diquark pair production. A more complex scenario is the "popcorn" model, where diquarks as such do not exist, but rather quark–antiquark pairs are produced one after the other.

In general, the different string breaks are causally disconnected. This means that it is possible to describe the breaks in any convenient order, e.g., from the quark end inwards. Results, at least not too close to the string endpoints, should be the same if the process is described from the q end or from the $\bar{q}$ one. This "left–right" symmetry constrains the allowed shape of fragmentation functions $f(z)$, where z is the fraction of $E + p_L$ that the next particle will take out of whatever remains. Here p_L is the longitudinal momentum along the direction of the respective endpoint, opposite for the q and the $\bar{q}$. Two free parameters remain, which have to be determined from data.

If several partons are moving apart from a common origin, the details of the string drawing become more complicated. For a $q\bar{q}g$ event, a string is stretched from the q end via the g to the $\bar{q}$ end, i.e., the gluon is a kink on the string, carrying energy and momentum. As a consequence, the gluon has two string pieces attached, and the ratio of gluon/quark string forces is 2, a number that can be compared with the ratio of colour charge Casimir operators, $N_C/C_F = 2/(1 - 1/N_C^2) = 9/4$. In this, as in other respects, the string model can be viewed as a variant of QCD, where the number of colours N_C is not 3 but infinite. Fragmentation along this kinked string proceeds along the same lines, as sketched for a single straight string piece. Therefore no new fragmentation parameters have to be introduced.

The concept of cluster fragmentation offers the great promise of a simple, local and universal description of hadronisation. At the end of the shower evolution all gluons are split into $q\bar{q}$ pairs, and $q\bar{q}'$ colour singlets can be formed from them by keeping track of the colour flow in the event. Typically the q and $\bar{q}'$ of such a singlet were formed in adjacent shower branches, and therefore tend to have a rather small mass. These so-called clusters are assumed to be the basic units from which the hadrons are produced. A cluster is ideally characterized only by its total mass and total flavour content, i.e., unlike a string it does not possess an internal structure. If the shower-evolution cutoff is chosen such that most clusters have a mass of a few GeV, the cluster mass spectrum may be thought of as a superposition of fairly broad (i.e., short-lived) resonances. Phase–space aspects may then be expected to dominate the decay properties. This applies both for the selection of decay channels, and for the kinematics of the decay. Thus a decay is assumed to be isotropic in the rest frame of the cluster. This gives a compact description with few parameters. This approach has been successful in explaining the particle composition in terms of very few parameters. The momentum distributions and correlations are a bit more tricky, and in practice some string ideas are needed, e.g., to break large-mass clusters into smaller ones along a string direction.

One general conclusion is that neither of the two models is well constrained from first principles. In the string model many parameters are needed for the flavour composition setup, while the energy–momentum (and space–time) picture is very economical. In the cluster model it is the other way around.

8   Summary and Outlook

In these lectures we have followed the flow of generators roughly "inwards out," i.e., from short-distance processes to long-distance ones. At the core lies the hard process, described by matrix elements. It is surrounded by initial- and final-state showers that should be properly matched to the hard process. Multiple parton–parton interactions can occur, and the colour flow is tied up with the structure of beam remnants. At longer timescales the partons turn into hadrons, many of which are unstable and decay further. This basic pattern is likely to remain in the future, but many aspects will change.

One such aspect, that stands a bit apart, is that of languages. The traditional event generators, like PYTHIA and HERWIG, have been developed in Fortran — up until the end of the LEP era this was the main language in high-energy physics. But now the experimental community has switched to C++ for heavy computing. The older generators are still being used, hidden under C++ wrappers, but this can only be a temporary solution, for several reasons. One is that younger experimentalists often need to look into the code of generators and tailor some parts to their specific needs, and if then the code is in an unknown language this will not work. Another is that theory students who apply for non-academic positions are much better off if their résumés say "object-oriented programming guru" rather than "Fortran fan."

A conversion program thus has begun on many fronts. SHERPA, as the youngest of the general-purpose generators, was conceived from the onset as a C++ package and thus is some steps ahead of the other programs in this respect. HERWIG++ [43] is a complete reimplementation of HERWIG, as is PYTHIA 8 [44] of PYTHIA 6. Both conversions have taken longer than originally hoped, but by now the new programs are fully operational and starting to be used by the LHC collaborations. THEPEG [45] is a generic toolkit for event generators, used in particular by HERWIG++.

The authors of these event generators have joined in MCnet, which currently is funded by the European Union as a Marie Curie Research Training Network. In addition a few other projects are pursued, notably the RIVET package that implements various experimental analyses from the literature, and the PROFESSOR framework [40] that takes RIVET input as the starting point for semiautomatic tuning of event generators. MCnet arranges summer schools each year [2–4], alone or in collaboration with the CTEQ school. There are also funds to allow graduate students in theory and experiment to come and work with a generator author on a specific project for a few months. If you are interested, have a look at `http://www.montecarlonet.org/`

To summarize these lectures, there are many aspects where we have seen progress in recent years and can hope for more:

- Faster, better and more user-friendly general-purpose matrix-element generators with an improved sampling of phase space

- New ready-made libraries of physics processes, in particular with full NLO corrections included

- More precise parton showers

- Better matching between matrix elements and parton showers

- Improved models for minimum-bias physics and underlying events

- Some upgrades of hadronisation models and decay descriptions

In general one would say that generators are getting better all the time, but at the same time the experimental demands are also increasing, so it is a close race. However, given that typical hadronic final states at LHC will contain hundreds of particles and quite complex patterns buried in that, it is difficult to see that there are any alternatives to the Monte Carlo generator approach.

References

[1] T. Sjöstrand, http://www.thep.lu.se/~torbjorn/ (click on "Talks").

[2] MCnet (2007), Summer School, Durham:
http://conference.ippp.dur.ac.uk/conferenceTimeTable.py?
confId=3&detailLevel=contribution&viewMode=parallel.

[3] CTEQ-MCnet (2008), Summer School, Debrecen:
http://conference.ippp.dur.ac.uk/conferenceOtherViews.py?
view=ippp&confId=156.

[4] MCnet (2009), Summer School, Lund:
http://conference.ippp.dur.ac.uk/conferenceOtherViews.py?
view=ippp&confId=264#2009-07-01.

[5] G. Corcella *et al.*, JHEP **0101** (2001) 010.

[6] T. Sjöstrand *et al.*, JHEP **0605** (2006) 026.

[7] T. Gleisberg *et al.*, JHEP **0902** (2009) 007.

[8] E. Boos *et al.*, in Proceedings of the Workshop on Physics at TeV Colliders, Les Houches, France, 2001, [hep-ph/0109068].

[9] J. Alwall *et al.*, Computer Physics Comm. **176** (2007) 300.

[10] D. Bourilkov *et al.* [hep-ph/0605240];
http://projects.hepforge.org/lhapdf/.

[11] P. Skands *et al.*, JHEP **0407** (2004) 036.

[12] B. Allanach *et al.*, Computer Physics Comm. **180** (2009) 8.

[13] M. Dobbs and J. B. Hansen, Computer Physics Comm. **134** (2001 41).

[14] C. Amsler *et al.* (Particle Data Group), Phys. Lett. **B667** (2008) 1.

[15] G. Gustafson and U. Pettersson, Nucl. Phys. **B306** (1988) 746.

[16] L. Lönnblad, Computer Physics Commun. **71** (1992) 15.

[17] T. Sjöstrand and P. Skands, Eur. Phys. J. **C39** (2005) 129.

[18] S. Schumann and F. Krauss, JHEP **0803** (2008) 038.

[19] J. Winter and F. Krauss, JHEP **0807** (2008) 040.

[20] S. Gieseke *et al.*, JHEP **0312** (2003) 045.

[21] E. Norrbin and T. Sjöstrand, Nucl. Phys. **B603** (2001) 297.

[22] S. Catani *et al.*, JHEP **0111** (2001) 063.

[23] L. Lönnblad, JHEP **0205** (2002) 046.

[24] M. L. Mangano *et al.*, JHEP **0701** (2007) 013.

[25] J. Alwall *et al.*, Eur. Phys. J. **C53** (2008) 473.

[26] N. Lavesson and L. Lönnblad, JHEP **0804** (2008) 085.

[27] S. Frixione and B. R. Webber, JHEP **0206** (2002) 029.

[28] P. Nason, JHEP **0411** (2004) 040.

[29] S. Frixione *et al.*, JHEP **0711** (2007) 070.

[30] T. Sjöstrand and M. van Zijl, Phys. Rev. **D36** (1987) 2019.

[31] T. Sjöstrand and P. Skands, JHEP **0403** (2004) 053.

[32] R. Corke and T. Sjöstrand, JHEP **1001** (2010) 035.

[33] J. M. Butterworth *et al.*, Z. Phys. **C72** (1996) 637.

[34] M. Bähr *et al.*, JHEP **0807** (2008) 076.

[35] P. Aurenche *et al.*, Computer Physics Commun. **83**, 107 (1994).

[36] R. Engel and J. Ranft, Phys. Rev. **D54** (1996) 4244.

[37] S. Roesler *et al.* [hep-ph/0012252].

[38] Field, R D (1999–2009) presentations partly on behalf of the CDF Collaboration:
     `http://www.phys.ufl.edu/~rfield/cdf/rdf_talks.html`.

[39] P. Skands, arXiv:0905.3418 [hep-ph].

[40] A. Buckley *et al.*, arXiv:0907.2973 [hep-ph].

[41] B. Andersson *et al.*, Phys. Rep. **97** (1983) 31.

[42] B. R. Webber, Nucl. Phys. **B238** (1984) 492.

[43] M. Bähr *et al.*, Eur. Phys. J. **C58** (2008) 639.

[44] T. Sjöstrand *et al.*, Computer Physics Comm. **178** (2008) 852.

[45] L. Lönnblad, Nucl. Instrum. Meth. **A559** (2008) 246.

Topics in Statistical Data Analysis for HEP

Glen Cowan

Royal Holloway, University of London, UK

1 Introduction

When a high-energy physics experiment enters the phase of data collection and analysis, the daily tasks of its postgraduate students are often centred not around the particle physics theories one is trying to test but rather on statistical methods. These methods are the tools needed to compare data with theory and quantify the extent to which one stands in agreement with the other. Of course one must understand the physical basis of the models being tested and so the theoretical emphasis in postgraduate education is no doubt well founded. But with the increasing cost of HEP experiments it has become important to exploit as much of the information as possible in the hard-won data, and to quantify as accurately as possible the inferences one draws when confronting the data with model predictions.

Despite efforts to make the lectures self contained, some familiarity with basic ideas of statistical data analysis is assumed. Introductions to the subject can be found, for example, in the reviews of the Particle Data Group [1] or in the texts [2–6].

In these lectures we will discuss two topics that are becoming increasingly important: Bayesian statistics and multivariate methods. In Section 2 we will review briefly the concept of probability and see how this is used differently in the frequentist and Bayesian approaches. In Section 2.2 we will discuss a simple example, the fitting of a straight line to a set of measurements, in both the frequentist and Bayesian approaches and compare different aspects of the two. This will include in Section 2.2.3 a brief description of Markov chain Monte Carlo (MCMC), one of the most important tools in Bayesian computation. We generalize the treatment in Section 2.3 to include systematic errors.

In Section 3 we take up the general problem of how to distinguish between two classes of events, say, signal and background, on the basis of a set of characteristics measured for each event. We first describe how to quantify the performance of a classification method in the framework of a statistical test. Although the Neyman–Pearson lemma indicates that this problem has an optimal solution using the likelihood ratio, this usually cannot be used in practice and one is forced to seek other methods. In Section 3.1 we look at a specific example of such a method, the boosted decision tree. Using this example we describe several issues common to many classification methods, such as overtraining. Finally, some conclusions are mentioned in Section 4.

2   Bayesian Statistical Methods for High-Energy Physics

In this section we look at the basic ideas of Bayesian statistics and explore how these can be applied in particle physics. We will contrast these with the corresponding notions in frequentist statistics, and to make the treatment largely self contained, the main ideas of the frequentist approach will be summarized as well.

2.1   The role of probability in data analysis

We begin by defining probability with the axioms written down by Kolmogorov [7] using the language of set theory. Consider a set S containing subsets $A, B, \ldots$. We define the probability P as a real-valued function with the following properties:

1. For every subset A in S, $P(A) \geq 0$;

2. For disjoint subsets (i.e., $A \cap B = \emptyset$), $P(A \cup B) = P(A) + P(B)$;

3. $P(S) = 1$.

In addition, we define the conditional probability $P(A|B)$ (read P of A given B) as

$$P(A|B) = \frac{P(A \cap B)}{P(B)}. \tag{1}$$

From this definition and using the fact that $A \cap B$ and $B \cap A$ are the same, we obtain *Bayes' theorem*,

$$P(A|B) = \frac{P(B|A)P(A)}{P(B)}. \tag{2}$$

From the three axioms of probability and the definition of conditional probability, we can derive the *law of total probability*,

$$P(B) = \sum_i P(B|A_i)P(A_i), \tag{3}$$

for any subset B and for disjoint A_i with $\cup_i A_i = S$. This can be combined with Bayes' theorem (2) to give

$$P(A|B) = \frac{P(B|A)P(A)}{\sum_i P(B|A_i)P(A_i)}, \tag{4}$$

where the subset A could, for example, be one of the A_i.

The most commonly used interpretation of the subsets of the sample space are outcomes of a repeatable experiment. The probability $P(A)$ is assigned a value equal to the limiting frequency of occurrence of A. This interpretation forms the basis of *frequentist statistics*.

The subsets of the sample space can also be interpreted as *hypotheses*, i.e., statements that are either true or false, such as "The mass of the W boson lies between 80.3 and 80.5 GeV." In the frequency interpretation, such statements are either always or never true, i.e., the corresponding probabilities would be 0 or 1. Using *subjective probability*, however, $P(A)$ is interpreted as the degree of belief that the hypothesis A is true.

Subjective probability is used in *Bayesian* (as opposed to frequentist) statistics. Bayes' theorem can be written

$$P(\text{theory}|\text{data}) \propto P(\text{data}|\text{theory})P(\text{theory}), \tag{5}$$

where "theory" represents some hypothesis and "data" are the outcome of the experiment. Here $P(\text{theory})$ is the *prior* probability for the theory, which reflects the experimenter's degree of belief before carrying out the measurement, and $P(\text{data}|\text{theory})$ is the probability to have gotten the data actually obtained, given the theory, which is also called the *likelihood*.

Bayesian statistics provides no fundamental rule for obtaining the prior probability; this is necessarily subjective and may depend on previous measurements, theoretical prejudices, etc. Once this has been specified, however, Eq. (5) tells how the probability for the theory must be modified in the light of the new data to give the *posterior* probability, $P(\text{theory}|\text{data})$. As Eq. (5) is stated as a proportionality, the probability must be normalized by summing (or integrating) over all possible hypotheses.

The difficult and subjective nature of encoding personal knowledge into priors has led to what is called *objective Bayesian statistics*, where prior probabilities are based not on an actual degree of belief but rather derived from formal rules. These give, for example, priors which are invariant under a transformation of parameters or which result in a maximum gain in information for a given set of measurements. For an extensive review see, for example, Ref. [8].

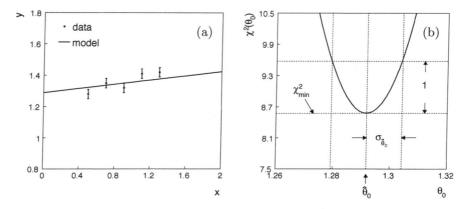

Figure 1. *(a) Illustration of fitting a straight line to data (see text). (b) The χ^2 as a function of the parameter θ_0, illustrating the method to determine the estimator $\hat{\theta}_0$ and its standard deviation $\sigma_{\hat{\theta}_0}$.*

2.2   An Example: Fitting a Straight Line

In Section 2.2 we look at the example of a simple fit in both the frequentist and Bayesian frameworks. Suppose we have independent data values y_i, $i = 1, ..., n$, that are each made at a given value x_i of a control variable x. Suppose we model the y_i as following a Gaussian distribution with given standard deviations σ_i and mean values μ_i given by a function that we evaluate at the corresponding x_i,

$$\mu(x; \theta_0, \theta_1) = \theta_0 + \theta_1 x. \tag{6}$$

We would like to determine values of the parameters θ_0 and θ_1 such that the model best describes the data. The ingredients of the analysis are illustrated 'n Fig. 1(a).

Now suppose the real goal of the analysis is only to estimate the parameter The slope parameter θ_1 must also be included in the model to obtain a description of the data, but we are not interested in its value as such. 'er to θ_0 as the parameter of interest, and θ_1 as a *nuisance parameter*. ollowing sections we treat this problem using both the frequentist and approaches.

Frequentist Approach

tes that the measurements are Gaussian distributed, i.e., the sity function (pdf) for the ith measurement y_i is

$$f(y_i; \boldsymbol{\theta}) = \frac{1}{\sqrt{2\pi}\sigma_i} e^{-(y_i - \mu(x_i; \boldsymbol{\theta}))^2 / 2\sigma_i^2}, \tag{7}$$

where $\boldsymbol{\theta} = (\theta_0, \theta_1)$.

The *likelihood function* is the joint pdf for all of the y_i, evaluated with the y_i obtained and regarded as a function of the parameters. Since we are assuming that the measurements are independent, the likelihood function is in this case given by the product

$$L(\boldsymbol{\theta}) = \prod_{i=1}^{n} f(y_i; \boldsymbol{\theta}) = \prod_{i=1}^{n} \frac{1}{\sqrt{2\pi}\sigma_i} e^{-(y_i - \mu(x_i; \boldsymbol{\theta}))^2/2\sigma_i^2}. \tag{8}$$

In the frequentist approach we construct estimators $\hat{\boldsymbol{\theta}}$ for the parameters $\boldsymbol{\theta}$, usually by finding the values that maximize the likelihood function (we will use a hat to indicate an estimator for a parameter). In this case one can see from (8) that this is equivalent to minimizing the quantity

$$\chi^2(\boldsymbol{\theta}) = \sum_{i=1}^{n} \frac{(y_i - \mu(x_i; \boldsymbol{\theta}))^2}{\sigma_i^2} = -2\ln L(\boldsymbol{\theta}) + C, \tag{9}$$

where C represents terms that do not depend on the parameters. Thus for the case of independent Gaussian measurements, the maximum likelihood (ML) estimators for the parameters coincide with those of the method of least squares (LS).

Suppose first that the slope parameter θ_1 is known exactly, and so it is not adjusted to maximize the likelihood (or minimize the χ^2) but rather held fixed. The quantity χ^2 versus the single adjustable parameter θ_0 would be as shown in Fig. 1(b), where the minimum indicates the value of the estimator $\hat{\theta}_0$.

Methods for obtaining the standard deviations of estimators — the statistical errors of our measured values — are described in many references such as [1–6]. Here in the case of a single fitted parameter the rule boils down to moving the parameter away from the estimate until χ^2 increases by one unit (i.e., $\ln L$ decreases from its maximum by $1/2$) as indicated in the figure.

It may be, however, that we do not know the value of the slope parameter θ_1, and so even though we do not care about its value in the final result, we are required to treat it as an adjustable parameter in the fit. Minimizing $\chi^2(\boldsymbol{\theta})$ results in the estimators $\hat{\boldsymbol{\theta}} = (\hat{\theta}_0, \hat{\theta}_1)$, as indicated schematically in Fig. 2(a). Now the recipe to obtain the statistical errors, however, is not simply a matter of moving the parameter away from its estimated value until the χ^2 goes up by one unit. Here the standard deviations must be found from the tangent lines (or in higher-dimensional problems, the tangent hyperplanes) to the contour defined by $\chi^2(\boldsymbol{\theta}) = \chi_{\min}^2 + 1$, as shown in the figure.

The tilt of the contour in Fig. 2(a) reflects the correlation between the estimators $\hat{\theta}_0$ and $\hat{\theta}_1$. A useful estimate for the inverse of the matrix of covariances $V_{ij} = \text{cov}[V_i, V_j]$ can be found from the second derivative of the log-likelihood evaluated at its maximum,

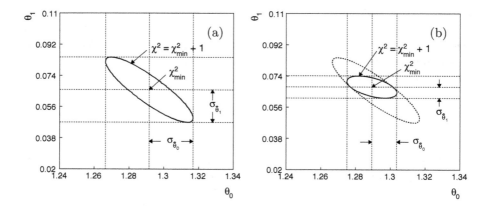

Figure 2. *Contour of $\chi^2(\boldsymbol{\theta}) = \chi^2_{\min} + 1$ centred about the estimates $(\hat{\theta}_0, \hat{\theta}_1)$ (a) with no prior measurement of θ_1 and (b) when a prior measurement of θ_1 is included.*

$$\widehat{V}_{ij}^{-1} = -\left.\frac{\partial^2 \ln L}{\partial \theta_i \partial \theta_j}\right|_{\boldsymbol{\theta}=\hat{\boldsymbol{\theta}}}. \tag{10}$$

More information on how to extract the full covariance matrix from the contour can be found, for example, in Refs. [1–6]. The point to note here is that the correlation between the estimators for the parameter of interest and the nuisance parameter has the result of inflating the standard deviations of both. That is, if θ_1 were known exactly, then the distance one would have to move θ_0 away from its estimated value to make the χ^2 increase by one unit would be less, as one can see from the figure. So although we can improve the ability of a model to describe the data by including additional nuisance parameters, this comes at the price of increasing the statistical errors. This is an important theme which we will encounter often in data analysis.

Now consider the case where we have a prior measurement of θ_1. For example, we could have a measurement t_1 which we model as following a Gaussian distribution centred about θ_1 and having a given standard deviation σ_{t_1}. If this measurement is independent of the other y_i values, then the full likelihood function is obtained simply by multiplying the original one by a Gaussian, and so when we find the new χ^2 from $-2 \ln L$ there is an additional term, namely,

$$\chi^2(\boldsymbol{\theta}) = \sum_{i=1}^{n} \frac{(y_i - \mu(x_i; \boldsymbol{\theta}))^2}{\sigma_i^2} + \frac{(\theta_1 - t_1)^2}{\sigma_{t_1}^2}. \tag{11}$$

As shown in Fig. 2(b), the new (solid) contour of $\chi^2 = \chi^2_{\min} + 1$ is compressed relative to the old (dashed) one in the θ_1 direction, and this compression has the effect of decreasing the error in θ_0 as well. The lesson is: by better constraining nuisance parameters, one improves the statistical accuracy of the parameters of interest.

2.2.2 The Bayesian Approach

To treat the example above in the Bayesian framework, we write Bayes' theorem (2) as

$$p(\boldsymbol{\theta}|\mathbf{y}) = \frac{L(\mathbf{y}|\boldsymbol{\theta})\pi(\boldsymbol{\theta})}{\int L(\mathbf{y}|\boldsymbol{\theta})\pi(\boldsymbol{\theta})\,d\boldsymbol{\theta}}. \tag{12}$$

Here $\boldsymbol{\theta} = (\theta_0, \theta_1)$ symbolizes the hypothesis whose probability we want to determine. The likelihood $L(\mathbf{y}|\boldsymbol{\theta})$ is the probability to obtain the data $\mathbf{y} = (y_1, \ldots, y_n)$ given the hypothesis, and the prior probability $\pi(\boldsymbol{\theta}|\mathbf{y})$ represents our degree of belief about the parameters before seeing the outcome of the experiment. The posterior probability $p(\boldsymbol{\theta})$ encapsulates all of our knowledge about $\boldsymbol{\theta}$ when the data $\mathbf{y}$ is combined with our prior beliefs. The denominator in (12) serves to normalize the posterior pdf to unit area.

The likelihood $L(\mathbf{y}|\boldsymbol{\theta})$ is the same as the $L(\boldsymbol{\theta})$ that we used in the frequentist approach above. The slightly different notation here simply emphasizes its role as the conditional probability for the data given the parameter.

To proceed we need to write down a prior probability density $\pi(\theta_0, \theta_1)$. This phase of a Bayesian analysis, sometimes called the *elicitation of expert opinion*, is in many ways the most problematic, as there are no universally accepted rules to follow. Here we will explore some of the important issues that come up.

In general, prior knowledge about one parameter might affect knowledge about the other, and if so this must be built into $\pi(\theta_0, \theta_1)$. Often, however, one may regard the prior knowledge about the parameters as independent, in which case the density factorizes as

$$\pi(\theta_0, \theta_1) = \pi_0(\theta_0)\pi_1(\theta_1). \tag{13}$$

For purposes of the present example we will assume that this holds.

For the parameter of interest θ_0, it may be that we have essentially no prior information, so the density $\pi_0(\theta_0)$ should be very broad. Often one takes the limiting case of a broad distribution simply to be a constant, i.e.,

$$\pi_0(\theta_0) = \text{const.} \tag{14}$$

Now one apparent problem with Eq. (14) is that it is not normalizable to unit area, and so does not appear to be a valid probability density. It is said to be an *improper prior*. The prior always appears in Bayes' theorem multiplied by the likelihood, however, and as long as this falls off quickly enough as a function of the parameters, then the resulting posterior probability density can be normalized to unit area.

A further problem with uniform priors is that if the prior pdf is flat in $\boldsymbol{\theta}$, then it is not flat for a nonlinear function of $\boldsymbol{\theta}$, and so a different parameterisation of the problem would lead in general to a non-equivalent posterior pdf.

For the special case of a constant prior, one can see from Bayes' theorem (12) that the posterior is proportional to the likelihood, and therefore the mode (peak position) of the posterior is equal to the ML estimator. The posterior mode, however, will change in general upon a transformation of parameter. A summary statistic other than the mode may be used as the Bayesian estimator, such as the median, which is invariant under a monotonic parameter transformation. But this will not in general coincide with the ML estimator.

For the prior $\pi_1(\theta_1)$, let us assume that our prior knowledge about this parameter includes the earlier measurement t_1, which we modelled as a Gaussian distributed variable centred about θ_1 with standard deviation σ_{t_1}. If we had taken, even prior to that measurement, a constant prior for θ_1, then the "intermediate-state" prior that we have before looking at the y_i is simply this flat prior times the Gaussian likelihood, i.e., a Gaussian prior in θ_1:

$$\pi_1(\theta_1) = \frac{1}{\sqrt{2\pi}\sigma_{t_1}} e^{-(\theta_1 - t_1)^2 / 2\sigma_{t_1}^2}. \tag{15}$$

Putting all of these ingredients into Bayes' theorem gives

$$p(\theta_0, \theta_1 | \mathbf{y}) \propto \prod_{i=1}^{n} \frac{1}{\sqrt{2\pi}\sigma_i} e^{-(y_i - \mu(x_i; \theta_0, \theta_1))^2 / 2\sigma_i^2} \pi_0 \frac{1}{\sqrt{2\pi}\sigma_{t_1}} e^{-(\theta_1 - t_1)^2 / 2\sigma_{t_1}^2}, \tag{16}$$

where π_0 represents the constant prior in θ_0 and the equation has been written as a proportionality with the understanding that the final posterior pdf should be normalized to unit area.

What Bayes' theorem gives us is the full joint pdf $p(\theta_0, \theta_1 | \mathbf{y})$ for both the parameter of interest θ_0 as well as the nuisance parameter θ_1. To find the pdf for the parameter of interest only, we simply integrate (marginalize) the joint pdf, i.e.,

$$p(\theta_0 | \mathbf{y}) = \int p(\theta_0, \theta_1 | \mathbf{y}) \, d\theta_1. \tag{17}$$

In this example, it turns out that we can do the integral in closed form. We find a Gaussian posterior,

$$p(\theta_0 | \mathbf{y}) = \frac{1}{\sqrt{2\pi}\sigma_{\theta_0}} e^{-(\theta_0 - \hat{\theta}_0)^2 / 2\sigma_{\theta_0}^2}, \tag{18}$$

where $\hat{\theta}_0$ is in fact the same as the ML (or LS) estimator found above with the frequentist approach, and σ_{θ_0} is the same as the standard deviation of that estimator $\sigma_{\hat{\theta}_0}$.

So we find something that looks just like the frequentist answer, although here the interpretation of the result is different. The posterior pdf $p(\theta_0|\mathbf{y})$ gives our degree of belief about the location of the parameter in the light of the data. We will see below how the Bayesian approach can, however, lead to results that differ both in interpretation as well as in numerical value from what would be obtained in a frequentist calculation. First, however, we need to pause for a short digression on Bayesian computation.

2.2.3 Bayesian Computation and MCMC

In most real Bayesian calculations, the marginalization integrals cannot be carried out in closed form, and if the number of nuisance parameters is too large then they can also be difficult to compute with standard Monte Carlo methods. However, *Markov chain Monte Carlo* (MCMC) has become the most important tool for computing integrals of this type and has revolutionized Bayesian computation. In-depth treatments of MCMC can be found, for example, in the texts by Robert and Casella [9], Liu [10], and the review by Neal [11].

The basic idea behind using MCMC to marginalize the joint pdf $p(\theta_0, \theta_1|\mathbf{y})$ is to sample points $\boldsymbol{\theta} = (\theta_0, \theta_0)$ according to the posterior pdf but then only to look at the distribution of the component of interest, θ_0. A simple and widely applicable MCMC method is the Metropolis–Hastings algorithm, which allows one to generate multidimensional points $\boldsymbol{\theta}$ distributed according to a target pdf that is proportional to a given function $p(\boldsymbol{\theta})$, which here will represent our posterior pdf. It is not necessary to have $p(\boldsymbol{\theta})$ normalized to unit area, which is useful in Bayesian statistics, as posterior probability densities are often determined only up to an unknown normalization constant, as is the case in our example.

To generate points that follow $p(\boldsymbol{\theta})$, one first needs a proposal pdf $q(\boldsymbol{\theta}; \boldsymbol{\theta}_0)$, which can be (almost) any pdf from which independent random values $\boldsymbol{\theta}$ can be generated, and which contains as a parameter another point in the same space $\boldsymbol{\theta}_0$. For example, a multivariate Gaussian centred about $\boldsymbol{\theta}_0$ can be used. Beginning at an arbitrary starting point $\boldsymbol{\theta}_0$, the Hastings algorithm iterates the following steps:

1. Generate a value $\boldsymbol{\theta}$ using the proposal density $q(\boldsymbol{\theta}; \boldsymbol{\theta}_0)$;

2. Form the Hastings test ratio, $\alpha = \min\left[1, \frac{p(\boldsymbol{\theta})q(\boldsymbol{\theta}_0; \boldsymbol{\theta})}{p(\boldsymbol{\theta}_0)q(\boldsymbol{\theta}; \boldsymbol{\theta}_0)}\right]$;

3. Generate a value u uniformly distributed in $[0, 1]$;

4. If $u \leq \alpha$, take $\boldsymbol{\theta}_1 = \boldsymbol{\theta}$. Otherwise, repeat the old point, i.e., $\boldsymbol{\theta}_1 = \boldsymbol{\theta}_0$.

If one takes the proposal density to be symmetric in $\boldsymbol{\theta}$ and $\boldsymbol{\theta}_0$, then this is the *Metropolis*–Hastings algorithm, and the test ratio becomes

$$\alpha = \min[1, p(\boldsymbol{\theta})/p(\boldsymbol{\theta}_0)].$$

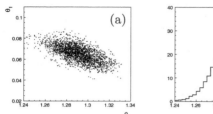 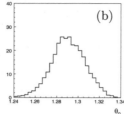 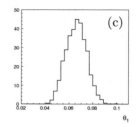

Figure 3. *MCMC marginalization of the posterior pdf $p(\theta_0, \theta_1 | \mathbf{y})$: (a) scatter plot of points in (θ_0, θ_1) plane and the marginal distribution of (b) the parameter of interest θ_0 and (c) the nuisance parameter θ_1.*

That is, if the proposed $\boldsymbol{\theta}$ is at a value of probability higher than $\boldsymbol{\theta}_0$, the step is taken. If the proposed step is rejected, the new point is taken to be the same as the old one.

Methods for assessing and optimizing the performance of the algorithm are discussed, for example, in Refs. [9–11]. One can, for example, examine the autocorrelation as a function of the lag k, i.e., the correlation of a sampled point with one k steps removed. This should decrease as quickly as possible for increasing k. Generally one chooses the proposal density so as to optimize some quality measure such as the autocorrelation. For certain problems it has been shown that one achieves optimal performance when the acceptance fraction, that is, the fraction of points with $u \leq \alpha$, is around 40%. This can be adjusted by varying the width of the proposal density. For example, one can use for the proposal pdf a multivariate Gaussian with the same covariance matrix as that of the target pdf, but scaled by a constant.

For our example above, MCMC was used to generate points according to the posterior pdf $p(\theta_0, \theta_1)$ by using a Gaussian proposal density. The result is shown in Fig. 3.

From the (θ_0, θ_1) points in the scatter plot in Fig. 3(a) we simply look at the distribution of the parameter of interest, θ_0 [Fig. 3(b)]. The standard deviation of this distribution is what we would report as the statistical error in our measurement of θ_0. The distribution of the nuisance parameter θ_1 from Fig. 3(c) is not directly needed, although it may be of interest in some other context where that parameter is deemed interesting.

In fact one can go beyond simply summarizing the width of the distributions with a statistic such as the standard deviation. The full form of the posterior distribution of θ_0 contains useful information about where the parameter's true value is likely to be. In this example the distributions will in fact turn out to be Gaussian, but in a more complex analysis there could be non-Gaussian tails and this information can be relevant in drawing conclusions from the result.

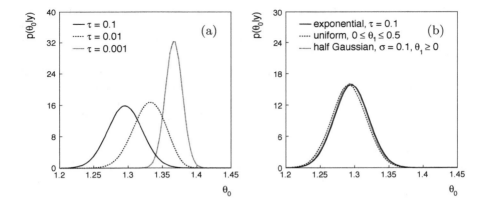

Figure 4. *Posterior probability densities for the parameter θ_0 obtained using (a) an exponential prior for θ_0 of different widths and (b) several different functional forms for the prior.*

2.2.4 Sensitivity Analysis

The posterior distribution of θ_0 obtained above encapsulates all of the analyst's knowledge about the parameter in the light of the data, given that the prior beliefs were reflected by the density $\pi(\theta_0, \theta_1)$. A different analyst with different prior beliefs would in general obtain a different posterior pdf. We would like the result of a Bayesian analysis to be of value to the broader scientific community, not only to those who share the prior beliefs of the analyst. And therefore it is important in a Bayesian analysis to show by how much the posterior probabilities would change upon some reasonable variation in the prior. This is sometimes called the *sensitivity analysis* and is an important part of any Bayesian calculation.

In the example above, we can imagine a situation where there was no prior measurement t_1 of the parameter θ_1, but rather a theorist had told us that, based on considerations of symmetry, consistency, aesthetics, etc., θ_1 was "almost certainly" positive, and had a magnitude "probably less than 0.1 or so." When pressed to be precise, the theorist sketches a curve roughly resembling an exponential with a mean of 0.1. So we can express this prior as

$$\pi_1(\theta_1) = \frac{1}{\tau}e^{-\theta_1/\tau} \quad (\theta_1 \geq 0), \tag{19}$$

with $\tau \approx 0.1$. We can substitute this prior into Bayes' theorem (16) to obtain the joint pdf for θ_0 and θ_1, and then marginalize to find the pdf for θ_0. Doing this numerically with MCMC results in the posterior distributions shown in Fig. 4(a).

Now the theorist who proposed this prior for θ_1 may feel reluctant to be pinned down, and so it is important to recall (and to reassure the theorist about) the "if-then" nature of a Bayesian analysis. One does not have to be absolutely certain about the prior in Eq. (19). Rather, Bayes' theorem simply says that *if* one were to have these prior beliefs, *then* we obtain certain posterior beliefs in the light of the data.

One simple way to vary the prior here is to try different values of the mean τ, as shown in Fig. 4(a). We see here the same basic feature as shown already in the frequentist analysis, namely, that when one increases the precision with which the nuisance parameter, θ_1, is determined, then the knowledge about the parameter of interest, θ_0, is improved.

Alternatively (or in addition) we may try different functional forms for the prior, as shown in Fig. 4(b). In this case using a uniform distribution for $\pi_1(\theta_1)$ with $0 \le \theta_1 \le 0.5$ or Gaussian with $\sigma = 0.1$ truncated for $\theta_1 < 0$ both give results similar to the exponential with a mean of 0.1. So one concludes that the result is relatively insensitive to the detailed nature of the tails of $\pi_1(\theta_1)$.

2.3 A Fit with Systematic Errors

We can now generalize the example of Section 2.2 to explore some further aspects of a Bayesian analysis. Let us suppose that we are given a set of n measurements as above, but now in addition to the statistical errors we also are given systematic errors. That is, we are given $y_i \pm \sigma_i^{\text{stat}} \pm \sigma_i^{\text{sys}}$ for $i = 1, \ldots, n$ where the measurements as before are each carried out for a specified value of a control variable x.

More generally, instead of having $y_i \pm \sigma_i^{\text{stat}} \pm \sigma_i^{\text{sys}}$ it may be that the set of measurements comes with an $n \times n$ covariance matrix V^{stat} corresponding to the statistical errors and another matrix V^{sys} for the systematic ones. Here the square roots of the diagonal elements give the errors for each measurement, and the off-diagonal elements provide information on how they are correlated.

As before we assume some functional form $\mu(x; \boldsymbol{\theta})$ for the expectation values of the y_i. This could be the linear model of Eq. (6) or something more general, but in any case it depends on a vector of unknown parameters $\boldsymbol{\theta}$. In this example, however, we will allow that the model is not perfect, but rather could have a systematic bias. That is, we write that the true expectation value of the ith measurement can be written

$$E[y_i] = \mu(x_i; \boldsymbol{\theta}) + b_i, \tag{20}$$

where b_i represents the bias. The b_i can be viewed as the systematic errors of the model, present even when the parameters $\boldsymbol{\theta}$ are adjusted to give the best description of the data. We do not know the values of the b_i. If we did, we would account for them in the model and they would no longer be biases. We do not in fact know that their values are nonzero, but we are allowing for the possibility that they could be. The reported systematic errors are intended as a quantitative measure of how large we expect the biases to be.

As before, the goal is to make inferences about the parameters $\boldsymbol{\theta}$; some of these may be of direct interest and others may be nuisance parameters. In Section 2.3.1 we will try to do this using the frequentist approach, and in Section 2.3.2 we will use the Bayesian method.

2.3.1 A Frequentist Fit with Systematic Errors

If we adopt the frequentist approach, we need to write down a likelihood function such as Eq. (8), but here we know in advance that the model $\mu(x; \boldsymbol{\theta})$ is not expected to be fully accurate. Furthermore, it is not clear how to insert the systematic errors. Often, perhaps without a clear justification, one simply adds the statistical and systematic errors in quadrature, or in the case where one has the covariance matrices V^{stat} and V^{sys}, they are summed to give a sort of "full" covariance matrix:

$$V_{ij} = V_{ij}^{\mathrm{stat}} + V_{ij}^{\mathrm{sys}}. \tag{21}$$

One might then use this in a multivariate Gaussian likelihood function, or equivalently it could be used to construct the χ^2,

$$\chi^2(\boldsymbol{\theta}) = (\mathbf{y} - \boldsymbol{\mu}(\boldsymbol{\theta}))^T V^{-1} (\mathbf{y} - \boldsymbol{\mu}(\boldsymbol{\theta})), \tag{22}$$

which is then minimized to find the LS estimators for $\boldsymbol{\theta}$. In Eq. (22) the vector $\mathbf{y} = (y_1, \ldots, y_n)$ should be understood as a column vector, $\boldsymbol{\mu}(\boldsymbol{\theta}) = (\mu(x_1; \boldsymbol{\theta}), \ldots, \mu(x_n; \boldsymbol{\theta}))$ is the corresponding vector of model values, and the superscript T represents the transpose (row) vector. Minimizing this χ^2 gives the generalized LS estimators $\hat{\boldsymbol{\theta}}$, and the usual procedures can be applied to find their covariances, which now in some sense include the systematics.

But in what sense is there any formal justification for adding the covariance matrices in Eq. (21)? Next we will treat this problem in the Bayesian framework and see that there is indeed some reason behind this recipe, but with limitations, and further we will see how to get around these limitations.

2.3.2 The Equivalent Bayesian Fit

In the corresponding Bayesian analysis, one treats the statistical errors as given by V^{stat} as reflecting the distribution of the data $\mathbf{y}$ in the likelihood. The systematic errors, through V^{sys}, reflect the width of the prior probabilities for the bias parameters b_i. That is, we take

$$L(\mathbf{y}|\boldsymbol{\theta}, \mathbf{b}) \propto \exp\left[-\frac{1}{2}(\mathbf{y} - \boldsymbol{\mu}(\boldsymbol{\theta}) - \mathbf{b})^T V_{\mathrm{stat}}^{-1}(\mathbf{y} - \boldsymbol{\mu}(\boldsymbol{\theta}) - \mathbf{b})\right], \tag{23}$$

$$\pi_b(\mathbf{b}) \propto \exp\left[-\frac{1}{2}\mathbf{b}^T V_{\mathrm{sys}}^{-1}\mathbf{b}\right], \qquad \pi_\theta(\boldsymbol{\theta}) = \mathrm{const.}, \tag{24}$$

$$p(\boldsymbol{\theta}, \mathbf{b}|\mathbf{y}) \propto L(\mathbf{y}|\boldsymbol{\theta}, \mathbf{b})\pi_\theta(\boldsymbol{\theta})\pi_b(\mathbf{b}), \tag{25}$$

where in (25), Bayes' theorem is used to obtain the joint probability for the parameters of interest, $\boldsymbol{\theta}$, and also the biases $\mathbf{b}$. To obtain the probability for $\boldsymbol{\theta}$ we integrate (marginalize) over $\mathbf{b}$,

$$p(\boldsymbol{\theta}|\mathbf{y}) = \int p(\boldsymbol{\theta}, \mathbf{b}|\mathbf{y}) \, d\mathbf{b}. \tag{26}$$

One finds that the mode of $p(\boldsymbol{\theta}|\mathbf{y})$ is at the same position as the least-squares estimates, and its covariance will be the same as obtained from the frequentist analysis where the full covariance matrix was given by the sum $V = V^{\text{stat}} + V^{\text{sys}}$. So this can be taken in effect as the formal justification for the addition in quadrature of statistical and systematic errors in a least-squares fit.

2.3.3   The Error on the Error

If one stays with the prior probabilities used above, the Bayesian and least-squares approaches deliver essentially the same result. An advantage of the Bayesian framework, however, is that it allows one to refine the assessment of the systematic uncertainties as expressed through the prior probabilities.

For example, the least-squares fit including systematic errors is equivalent to the assumption of a Gaussian prior for the biases. A more realistic prior would take into account the experimenter's own uncertainty in assigning the systematic error, i.e., the "error on the error." Suppose, for example, that the ith measurement is characterized by a reported systematic uncertainty σ_i^{sys} and an unreported factor s_i, such that the prior for the bias b_i is

$$\pi_b(b_i) = \int \frac{1}{\sqrt{2\pi} s_i \sigma_i^{\text{sys}}} \exp\left[-\frac{1}{2} \frac{b_i^2}{(s_i \sigma_i^{\text{sys}})^2}\right] \pi_s(s_i) \, ds_i. \tag{27}$$

Here the "error on the error" is encapsulated in the prior for the factor s, $\pi_s(s)$. For this we can take whatever function is deemed appropriate. For some types of systematic error it could be close to the ideal case of a delta function centred about unity. Many reported systematics are, however, at best rough guesses, and one could easily imagine a function $\pi_s(s)$ with a mean of unity but a standard deviation of, say, 0.5 or more. Here we show examples using a gamma distribution for $\pi_s(s)$, which results in substantially longer tails for the prior $\pi_b(b)$ than those of the Gaussian. This can be seen in Fig. 5, which shows $\ln \pi_b(b)$ for different values of the standard deviation of $\pi_s(s)$, σ_s. Related studies using an inverse gamma distribution can be found in Refs. [12, 13], which have the advantage that the posterior pdf can be written down in closed form.

Using a prior for the biases with tails longer than those of a Gaussian results in a reduced sensitivity to outliers, which arise when an experimenter overlooks an important source of systematic uncertainty in the estimated error of a measurement. As a simple test of this, consider the sample data shown in Fig. 6(a). Suppose these represent four independent measurements of the same quantity, here a parameter called μ, and the goal is to combine the

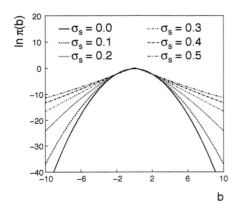

Figure 5. *The log of the prior pdf for a bias parameter b for different values of the standard deviation of* $\pi_s(s)$.

measurements to provide a single estimate of μ. That is, we are effectively fitting a horizontal line to the set of measured y values, where the control variable x is just a label for the measurements.

In this example, suppose that each measurement y_i, $i = 1, \ldots 4$, is modelled as Gaussian distributed about μ, having a standard deviation $\sigma_{\text{stat}} = 0.1$, and furthermore each measurement has a systematic uncertainty $\sigma_{\text{sys}} = 0.1$, which here is taken to refer to the standard deviation of the Gaussian component of the prior $\pi_b(b_i)$. This is then folded together with $\pi_s(s_i)$ to get the full prior for b_i using Eq. (27), and the joint prior for the vector of bias parameters is simply the product of the corresponding terms, as the systematic errors here are treated as being independent. These ingredients are then assembled according to the recipe of Eqs. (23)–(26) to produce the posterior pdf for μ, $p(\mu|\mathbf{y})$.

Results of the exercise are shown in Fig. 6. In Fig. 6(a), the four measurements y_i are reasonably consistent with one another. Figure 6(c) shows the corresponding posterior $p(\mu|\mathbf{y})$ for two values of σ_s, which reflect differing degrees of non-Gaussian tails in the prior for the bias parameters, $\pi_b(b_i)$. For $\sigma_s = 0$, the prior for the bias is exactly Gaussian, whereas for $\sigma_s = 0.5$, the non-Gaussian tails are considerably longer, as can be seen from the corresponding curves in Fig. 5. The posterior pdfs for both cases are almost identical, as can be seen in Fig. 6(c). Determining the mean and standard deviation of the posterior for each gives $\hat{\mu} = 1.000 \pm 0.71$ for the case of $\sigma_s = 0$, and $\hat{\mu} = 1.000 \pm 0.72$ for $\sigma_s = 0.5$. So assuming a 50% "error on the error" here one only inflates the error of the averaged result by a small amount.

Now consider the case where one of the measured values is substantially different from the other three, as shown in Fig. 6(b). Here using the same priors for the bias parameters results in the posteriors shown in Fig. 6(d). The posterior means and standard deviations are $\hat{\mu} = 1.125 \pm 0.71$ for the case of $\sigma_s = 0$, and $\hat{\mu} = 1.093 \pm 0.089$ for $\sigma_s = 0.5$.

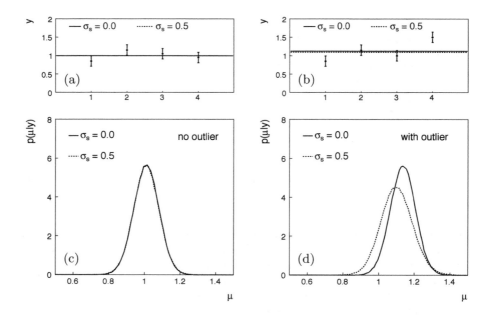

Figure 6. *(a) Data values which are relatively consistent and (b) a data set with an outlier; the horizontal lines indicate the posterior mean for two different values of the parameter σ_s. (c) and (d) show the posterior distributions corresponding to (a) and (b), respectively. (The dashed and solid curves in (a) and (c) overlap.)*

When we assume a purely Gaussian prior for the bias ($\sigma_s = 0.0$), the presence of the outlier has in fact no effect on the width of the posterior. This is rather counter-intuitive and results from our assumption of a Gaussian likelihood for the data and a Gaussian prior for the bias parameters. The posterior mean is, however, pulled substantially higher than the three other measurements, which are clustered around 1.0. If the priors $\pi_b(b_i)$ have longer tails, as occurs when we take $\sigma_s = 0.5$, then the posterior is broader, and furthermore it is pulled less far by the outlier, as can be seen in Fig. 6(d).

The fact is that the width of the posterior distribution, which effectively tells us the uncertainty on the parameter of interest μ, becomes coupled to the internal consistency of the data. In contrast, in the (frequentist) least-squares method, or in the Bayesian approach using a Gaussian prior for the bias parameters, the final uncertainty on the parameter of interest is unaffected by the presence of outliers. And in many cases of practical interest, it would be in fact appropriate to conclude that the presence of outliers should indeed increase one's uncertainty about the final parameter estimates. The example shown here can be generalized to cover a wide variety of model uncertainties by including prior probabilities for an enlarged set of model parameters.

2.4 Summary of Bayesian methods

In these lectures we have seen how Bayesian methods can be used in parameter estimation, and this has also given us the opportunity to discuss some aspects of Bayesian computation, including the important tool of Markov chain Monte Carlo. Although Bayesian and frequentist methods may often deliver results that agree numerically, there is an important difference in their interpretation. Furthermore, Bayesian methods allow one to incorporate prior information that may be based not on other measurements but rather on theoretical arguments or purely subjective considerations. And as these considerations may not find universal agreement, it is important to investigate how the results of a Bayesian analysis would change for a reasonable variation of the prior probabilities.

It is important to keep in mind that in the Bayesian approach, all information about the parameters is encapsulated in the posterior probabilities. So if the analyst also wants to set upper limits or determine intervals that cover the parameter with a specified probability, then this is a straightforward matter of finding the parameter limits such that the integrated posterior pdf has the desired probability content. A discussion of Bayesian methods to the important problem of setting upper limits on a Poisson parameter is covered in Ref. [1] and references therein; we will not have time in these lectures to go into that question here.

We will also unfortunately not have time to explore Bayesian model selection. This allows one to quantify the degree to which the the data prefer one model over the other using a quantity called the Bayes factor. These have not yet been widely used in particle physics but should be kept in mind as providing important complementary information to the corresponding outputs of frequentist hypothesis testing such as p-values. A brief description of Bayes factors can be found in Ref. [1] and a more in-depth treatment is given in Ref. [14].

3 Topics in Multivariate Analysis

In the second part of these lectures we will take a look at the important topic of multivariate analysis. In-depth information on this topic can be found in the textbooks [15–18]. In a particle physics context, multivariate methods are often used when selecting events of a certain type using some potentially large number of measurable characteristics for each event. The basic framework we will use to examine these methods is that of a frequentist hypothesis test.

The fundamental unit of data in a particle physics experiment is the "event," which in most cases corresponds to a single particle collision. In some cases it could be instead a decay, and the picture does not change much if we look, say, at individual particles or tracks. But to be concrete let us suppose that we want to search for events from proton–proton collisions at the LHC that correspond to some interesting "signal" process, such as super-symmetry.

When running at full intensity, the LHC should produce close to a billion events per second. After a quick sifting, the data from around 200 per second are recorded for further study, resulting in more than a billion events per year. But only a tiny fraction of these are of potential interest. If one of the speculative theories such as supersymmetry turns out to be realized in Nature, then this will result in a subset of events having characteristic features, and the SUSY events will simply be mixed in randomly with a much larger number of Standard Model events. The relevant distinguishing features depend on what new physics Nature chooses to reveal, but one might see, for example, high p_T jets, leptons, missing energy.

Unfortunately, background processes (e.g., Standard Model events) can often mimic these features and one will not be able to say with certainty that a given event shows a clear evidence for something new such as supersymmetry. For example, even Standard Model events can contain neutrinos which also escape undetected. The typical amount and pattern of missing energy in these events differs on average, however, from what a SUSY event would give, and so a statistical analysis can be applied to test whether something besides Standard Model events is present.

In a typical analysis there is a class of event we are interested in finding (signal), and these, if they exist at all, are mixed in with the rest of the events (background). The data for each event are some collection of numbers $\mathbf{x} = (x_1, \ldots, x_n)$ representing particle energies, momenta, etc. We will refer to these as the *input variables* of the problem. And the probabilities are joint densities for $\mathbf{x}$ given the signal (s) or background (b) hypotheses: $f(\mathbf{x}|s)$ and $f(\mathbf{x}|b)$.

To illustrate the general problem, consider the scatter plots shown in Fig. 7. These show the distribution of two variables, x_1 and x_2, which represent two out of a potentially large number of quantities measured for each event. The circles could represent the sought-after signal events, and the triangles the background. In each of the three figures there is a decision boundary representing a possible way of classifying the events.

Figure 7(a) represents what is commonly called the "cut-based" approach. One selects signal events by requiring $x_1 < c_1$ and $x_2 < c_2$ for some suitably chosen cut values c_1 and c_2. If x_1 and x_2 represent quantities for which one has some intuitive understanding, then this can help guide one's choice of the cut values.

Another possible decision boundary is made with a diagonal cut as shown in Fig. 7(b). One can show that for certain problems a linear boundary has optimal properties, but in the example here, because of the curved nature of the distributions, neither the cut-based nor the linear solution is as good as the nonlinear boundary shown in Fig. 7(c).

The decision boundary is a surface in the n-dimensional space of input variables, which can be represented by an equation of the form $y(\mathbf{x}) = y_{cut}$,

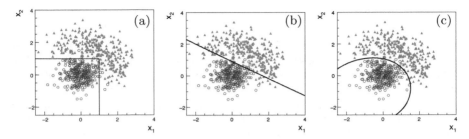

Figure 7. *Scatter plots of two variables corresponding to two hypotheses: signal and background. Event selection could be based, e.g., on (a) cuts, (b) a linear boundary, (c) a nonlinear boundary.*

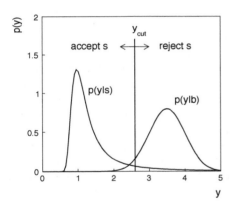

Figure 8. *Distributions of the scalar test statistic $y(\mathbf{x})$ under the signal and background hypotheses.*

where y_{cut} is some constant. We accept events as corresponding to the signal hypothesis if they are on one side of the boundary, e.g., $y(\mathbf{x}) \leq y_{\text{cut}}$ could represent the acceptance region and $y(\mathbf{x}) > y_{\text{cut}}$ could be the rejection region.

Equivalently we can use the function $y(\mathbf{x})$ as a scalar *test statistic*. Once its functional form is specified, we can determine the pdfs of $y(\mathbf{x})$ under both the signal and background hypotheses, $p(y|s)$ and $p(y|b)$. The decision boundary is now effectively a single cut on the scalar variable y, as illustrated in Fig. 8.

To quantify how good the event selection is, we can define the *efficiency* with which one selects events of a given type as the probability that an event will fall in the acceptance region. That is, the signal and background efficiencies, respectively, are

$$\varepsilon_{\text{s}} \;=\; P(\text{accept event}|s) = \int_A f(\mathbf{x}|s)\, d\mathbf{x} = \int_{-\infty}^{y_{\text{cut}}} p(y|s)\, dy, \qquad (28)$$

$$\varepsilon_{\text{b}} \;=\; P(\text{accept event}|b) = \int_A f(\mathbf{x}|b)\, d\mathbf{x} = \int_{-\infty}^{y_{\text{cut}}} p(y|b)\, dy. \qquad (29)$$

where the region of integration A represents the acceptance region.

Dividing the space of input variables into two regions where one accepts or rejects the signal hypothesis is essentially the language of a frequentist statistical test. If we regard background as the "null hypothesis," then the background efficiency is the same as what in a statistical context would be called the significance level of the test, or the rate of "type-I error." Viewing the signal process as the alternative, the signal efficiency is then what a statistician would call the power of the test; it is the probability to reject the background hypothesis if in fact the signal hypothesis is true. Equivalently, this is one minus the rate of "type-II error."

The use of a statistical test to distinguish between two classes of events (signal and background) comes up in different ways. Sometimes both event classes are known to exist, and the goal is to select one class (signal) for further study. For example, proton–proton collisions leading to the production of top quarks is a well-established process. By selecting these events one can carry out precise measurements of the top quark's properties such as its mass. In other cases, the signal process could represent an extension to the Standard Model, say, supersymmetry, whose existence is not yet established, and the goal of the analysis is to see if one can do this. Rejecting the Standard Model with a sufficiently high significance level amounts to discovering something new, and of course one hopes that the newly revealed phenomena will provide important insights into how Nature behaves.

What the physicist would like to have is a test with maximal power with respect to a broad class of alternative hypotheses. For specific signal and background hypotheses, it turns out that there is a well-defined optimal solution to our problem. The *Neyman–Pearson* lemma states that one obtains the maximum power relative for the signal hypothesis for a given significance level (background efficiency) by defining the acceptance region such that, for **x** inside the region, the *likelihood ratio*, i.e., the ratio of pdfs for signal and background,

$$\lambda(\mathbf{x}) = \frac{f(\mathbf{x}|s)}{f(\mathbf{x}|b)}, \tag{30}$$

is greater than or equal to a given constant, and it is less than this constant everywhere outside the acceptance region. This is equivalent to the statement that the ratio (30) represents the test statistic with which one obtains the highest signal efficiency for a given background efficiency, or equivalently, for a given signal purity.

In principle the signal and background theories should allow us to work out the required functions $f(\mathbf{x}|s)$ and $f(\mathbf{x}|b)$, but in practice the calculations are too difficult and we do not have explicit formulae for these. What we have instead of $f(\mathbf{x}|s)$ and $f(\mathbf{x}|b)$ are complicated Monte Carlo programs, that is, we can sample **x** to produce simulated signal and background events. Because of the multivariate nature of the data, where **x** may contain at least several or perhaps even hundreds of components, it is a nontrivial problem to construct a test with a power approaching that of the likelihood ratio.

In the usual case where the likelihood ratio (30) cannot be used explicitly, there exists a variety of other multivariate classifiers that effectively separate different types of events. Methods often used in HEP include *neural networks* or *Fisher discriminants*. Recently, further classification methods from machine learning have been applied in HEP analyses; these include *probability density estimation (PDE)* techniques, *kernel-based PDE (KDE or Parzen window)*, *support vector machines*, and *decision trees*. Techniques such as "boosting" and "bagging" can be applied to combine a number of classifiers into a stronger one with greater stability with respect to fluctuations in the training data. Descriptions of these methods can be found, for example, in the textbooks [15–18] and in proceedings of the PHYSTAT conference series [19]. Software for HEP includes the TMVA [20] and StatPatternRecognition [21] packages, although support for the latter has unfortunately been discontinued.

As we will not have the time to examine all of the methods mentioned above, in the following section we look at a specific example of a classifier to illustrate some of the main ideas of a multivariate analysis: the boosted decision tree.

3.1 Boosted Decision Trees

Boosted decision trees (BDTs) exploit relatively recent developments in machine learning and have gained significant popularity in HEP. First in Section 3.1.1 we describe the basic idea of a decision tree, and then in Section 3.1.2 we will say how the the technique of "boosting" can be used to improve its performance.

3.1.1 Decision Trees

A decision tree is defined by a collection of successive cuts on the set of input variables. To determine the appropriate cuts, one begins with a sample of N training events which are known to be either signal or background, e.g., from Monte Carlo. The set of n input variables measured for each event constitutes a vector $\mathbf{x} = (x_1, \ldots x_n)$. Thus we have N instances of $\mathbf{x}$, $\mathbf{x}_1, \ldots \mathbf{x}_N$, as well as the corresponding N true class labels $y_1, \ldots, y_N$. It is convenient to assign numerical values to the labels so that, e.g., $y = 1$ corresponds to signal and $y = -1$ for background.

In addition we will assume that each event can be assigned a weight, w_i, with $i = 1, \ldots, N$. For any subset of the events and for a set of weights, the signal fraction (purity) is taken to be

$$p = \frac{\sum_{i \text{ins}} w_i}{\sum_{i \text{ins}} w_i + \sum_{i \text{inb}} w_i}, \tag{31}$$

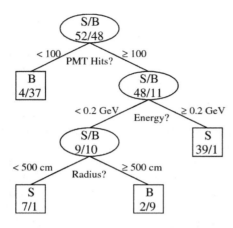

Figure 9. *Illustration of a decision tree used by the MiniBooNE experiment [22] (see text).*

where s and b refer to the signal and background event types, respectively. The weights are not strictly speaking necessary for a decision tree, but will be used in connection with boosting in Section 3.1.2. For a decision tree without boosting we can simply take all the weights to be equal.

To quantify the degree of separation achieved by a classifier for a selected subset of the events one can use, for example, the *Gini coefficient* [23], which historically has been used as a measure of dispersion in economics and is defined as

$$G = p(1 - p). \tag{32}$$

The Gini coefficient is zero if the selected sample is either pure signal or background. Another measure is simply the misclassification rate,

$$\varepsilon = 1 - \max(p, 1 - p). \tag{33}$$

This quantity is defined such that either high or low p corresponds to good separation of the two event classes and hence gives low ε.

The idea behind a decision tree is illustrated in Fig. 9, from an analysis by the MiniBooNE neutrino oscillation experiment at Fermilab [22].

One starts with the entire sample of training events in the root node, shown in the figure with 52 signal and 48 background events. Out of all of the possible input variables in the vector $\mathbf{x}$, one finds the component that provides the best separation between signal and background by use of a single cut. This requires a definition of what constitutes "best separation," and there are a number of reasonable choices. For example, for a cut that splits a set of events a into two subsets b and c, one can define the degree of separation through the weighted change in the Gini coefficients,

$$\Delta = W_a G_a - W_b G_b - W_c G_c, \tag{34}$$

where

$$W_a = \sum_{i \text{ in } a} w_i, \tag{35}$$

and similarly for W_b and W_c. Alternatively one may use a quantity similar to (34) but with the misclassification rate (33), for example, instead of the Gini coefficient. More possibilities can be found in Ref. [20].

For whatever chosen measure of degree of separation, Δ, one finds the cut on the variable amongst the components of $\mathbf{x}$ that maximizes it. In the example of the MiniBooNE experiment shown in Fig. 9, this happened to be a cut on the number of PMT hits with a value of 100. This splits the training sample into the two daughter nodes shown in the figure, one of which is enhanced in signal and the other in background events.

The algorithm requires a stopping rule based, for example, on the number of events in a node or the misclassification rate. If, for example, the number of events or the misclassification rate in a given node falls below a certain threshold, then this is defined as a terminal node or "leaf." It is classified as a signal or background leaf based on its predominant event type. In Fig. 9, for example, the node after the cut on PMT hits with 4 signal and 37 background events is classified as a terminal background node.

For nodes that have not yet reached the stopping criterion, one iterates the procedure and finds, as before, the variable that provides the best separation with a single cut. In Fig. 9 this is an energy cut of $0.2\,\text{GeV}$. The steps are continued until all nodes reach the stopping criterion.

The resulting set of cuts effectively divides the $\mathbf{x}$ space into two regions: signal and background. To provide a numerical output for the classifier we can define

$$f(\mathbf{x}) = \begin{cases} 1 & \mathbf{x} \text{ in signal region,} \\ -1 & \mathbf{x} \text{ in background region.} \end{cases} \tag{36}$$

Equation (36) defines a decision tree classifier. In this form, these tend to be very sensitive to statistical fluctuations in the training data. One can easily see why this is, for example, if two of the components of $\mathbf{x}$ have similar discriminating power between signal and background. For a given training sample, one variable may be found to give the best degree of separation and is chosen to make the cut, and this affects the entire further structure of the tree. In a different statistically independent sample of training events, the other variable may be found to be better, and the resulting tree could look very different. Boosting is a technique that can decrease the sensitivity of a classifier to such fluctuations, and we describe this in the following section.

3.1.2 Boosting

Boosting is a general method of creating a set of classifiers which can be combined to give a new classifier that is more stable and has a smaller misclassification rate than any individual one. It is often applied to decision trees, precisely because they suffer from sensitivity to statistical fluctuations in the training sample, but the technique can be applied to any classifier.

Let us suppose as above that we have a sample of N training events, i.e., N instances of the data vector, $\mathbf{x}_1, \ldots, \mathbf{x}_N$, and N true class labels $y_1, \ldots, y_N$, with $y = 1$ for signal and $y = -1$ for background. Also as above assume we have N weights $w_1^{(1)}, \ldots, w_N^{(1)}$, where the superscript (1) refers to the fact that this is the first training set. We initially set the weights equal and normalized such that

$$\sum_{i=1}^{N} w_i^{(1)} = 1. \tag{37}$$

The idea behind boosting is to create from the initial sample, a series of further training samples which differ from the initial one in that the weights will be changed according to a specific rule. A number of boosting algorithms have been developed, and these differ primarily in the rule used to update the weights. We will describe the AdaBoost algorithm of Freund and Schapire [24], as it was one of the first such algorithms and its properties have been well studied.

One begins with the initial training sample and from it derives a classifier. We have in mind here a decision tree, but it could be any type of classifier for where the training employs the event weights. The resulting function $f_1(\mathbf{x})$ will have a certain misclassification rate ε_1. In general for the kth classifier (i.e., based on the kth training sample), we can write the error rate as

$$\varepsilon_k = \sum_{i=1}^{N} w_i^{(k)} I(y_i f_k(\mathbf{x}_i) \leq 0), \tag{38}$$

where $I(X) = 1$ if the Boolean expression X is true, and is zero otherwise. We then assign a score to the classifier based on its error rate. For the AdaBoost algorithm this is

$$\alpha_k = \ln \frac{1 - \varepsilon_k}{\varepsilon_k}, \tag{39}$$

which is positive as long as the error rate is lower than 50%, i.e., the classifier does better than random guessing.

Having carried out these steps for the initial training sample, we define the second training sample by updating the weights. More generally, the weights for step $k + 1$ are found from those for step k by

$$w_i^{(k+1)} = w_i^{(k)} \frac{e^{-\alpha_k f_k(\mathbf{x}_i) y_i / 2}}{Z_k}, \tag{40}$$

where the factor Z_k is chosen so that the sum of the updated weights is equal to unity. Note that if an event is incorrectly classified, then the true class label y_i and the value $f_k(\mathbf{x}_i)$ have opposite signs, and thus the new weights are greater than the old ones. Correctly classified events have their weights decreased. This means that the updated training set will pay more attention in the next iteration to those events that were not correctly classified, the idea being that it should try harder to get it right the next time around.

After K iterations of this procedure one has classifiers $f_1(\mathbf{x}), \ldots, f_K(\mathbf{x})$, each with a certain error rate and score based on Eqs. (38) and (39). In the case of decision trees, the set of new trees is called a *forest*. From these one defines an averaged classifier as

$$y(\mathbf{x}) = \sum_{k=1}^{K} \alpha_k f_k(\mathbf{x}). \tag{41}$$

Equation (41) defines a boosted decision tree (or more generally, a boosted version of whatever classifier was used).

One of the important questions to be addressed is how many boosting iterations to use. One can show that for a sufficiently large number of iterations, a boosted decision tree will eventually classify all of the events in the training sample correctly. Similar behaviour is found with any classification method where one can control to an arbitrary degree the flexibility of the decision boundary. The user can arrange the boundary twists and turns so as to get all of the events on the right side.

In the case of a neural network, for example, one can increase the number of hidden layers, or the number of nodes in the hidden layers; for a support vector machine, one can adjust the width of the kernel function and the regularization parameter to increase the flexibility of the boundary. An example is shown in Fig. 10(a), where an extremely flexible classifier has managed to enclose all of the signal events and exclude all of the background.

Of course if we were now to take the decision boundary shown in Fig. 10(a) and apply it to a statistically independent data sample, there is no reason to believe that the contortions that led to such good performance on the training sample will still work. This can be seen in Fig. 10(b), which shows the same boundary with a new data sample. In this case the classifier is said to be *overtrained*. Its error rate calculated from the same set of events used to train the classifier underestimates the rate on a statistically independent sample.

To deal with overtraining, one estimates the misclassification rate not only with the training data sample but also with a statistically independent test sample. We can then plot these rates as a function of the parameters that regulate the flexibility of the decision boundary, e.g., the number of boosting iterations used to form the BDT. For a small number of iterations, one will

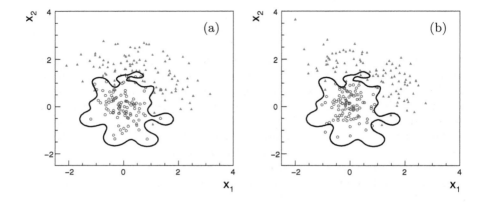

Figure 10. *Scatter plot of events of two types and the decision boundary determined by a particularly flexible classifier. Plot (a) shows the events used to train the classifier, and (b) shows an independent sample of test data.*

find in general that the error rates for both samples drop. The error rate based on the training sample will continue to drop, eventually reaching zero. But at some point the error rate from the test sample will cease to decrease and in general will increase. One chooses the architecture of the classifier (number of boosting iterations, number of nodes or layers in a neural network, etc.) to minimize the error rate on the test sample.

As the test sample is used to choose between a number of competing architectures based on the minimum observed error rate, this in fact gives a biased estimate of the true error rate. In principle one should use a third validation sample to obtain an unbiased estimate of the error rate. In many cases the bias is small and this last step is omitted, but one should be aware of its potential existence.

In some applications, the training data are relatively inexpensive; one simply generates more events with Monte Carlo. But often event generation can take a prohibitively long time and one may be reluctant to use only a fraction of the events for training and the other half for testing. In such cases, procedures such as *cross validation* (see, e.g., Refs. [15,16]) can be used where the available events are partitioned in a number of different ways into training and test samples and the results averaged.

Boosted decision trees have become increasingly popular in particle physics in recent years. One of their advantages is that they are relatively insensitive to the number of input variables used in the data vector **x**. Components that provide little or no separation between signal and background are rarely chosen as for the cut that provides separation, i.e., to split the tree, and thus they are effectively ignored. Decision trees have no difficulty in dealing with different types of data; these can be real, integer, or they can simply be labels for which there is no natural ordering (categorical data). Furthermore, boosted decision trees are surprisingly insensitive to overtraining. That is, although the error rate on the test sample will not decrease to zero as one increases the number of boosting iterations (as is the case for the training sample), it tends not to increase. Further discussion of this point can be found in Ref. [25].

3.2 Summary of Multivariate Methods

The boosted decision tree is an example of a relatively modern development in Machine Learning that has attracted substantial attention in HEP. Support Vector Machines (SVMs) represent another such development and will no doubt also find further application in particle physics; further discussion on SVMs can be found in Refs. [15, 16] and references therein. Linear classifiers and neural networks will no doubt continue to play an important role, as will probability density estimation methods used to approximate the likelihood ratio.

Multivariate methods have the advantage of exploiting as much information as possible out of all of the quantities measured for each event. In an environment of competition between experiments, this can be a natural motivation to use them. Some caution should be exercised, however, before placing too much faith in the performance of a complicated classifier, to say nothing of a combination of complicated classifiers. These may have decision boundaries that indeed exploit nonlinear features of the training data, often based on Monte Carlo. But if these features have never been verified experimentally, then they may or may not be present in the real data. There is thus the risk of, say, underestimating the rate of background events present in a region where one looks for signal, which could lead to a spurious discovery. Simpler classifiers are not immune to such dangers either, but in such cases the problems may be easier to control and mitigate.

One should therefore keep in mind the following quote, often heard in the multivariate analysis community:

> *Keep it simple. As simple as possible. Not any simpler.*
> — A. Einstein

To this we can add the more modern variant,

> *If you believe in something you don't understand, you suffer . . . ,*
> —Stevie Wonder

Having made the requisite warnings, however, it seems clear that multivariate methods will play an important role in the discoveries we hope to make at the LHC. One can easily imagine, for example, that 5-sigma evidence for New Physics from a highly performant, and complicated, classifier would be regarded by the community with some scepticism. But if this is backed up by, say, 4-sigma significance from a simpler, more transparent analysis, then the conclusion would be more easily accepted, and the team that pursues both approaches may well win the race.

4 Summary and Conclusions

In these lectures we have looked at two topics in statistics, Bayesian methods and multivariate analysis, which will play an important role in particle physics in the coming years. Bayesian methods provide important tools for analysing systematic uncertainties, where prior information may be available that does not necessarily stem solely from other measurements, but rather from theoretical arguments or other indirect means. The Bayesian framework allows one to investigate how the posterior probabilities change upon variation of the prior probabilities. Through this type of sensitivity analysis, a Bayesian result becomes valuable to the broader scientific community.

As experiments become more expensive and the competition more intense, one will always be looking for ways to exploit as much information as possible from the data. Multivariate methods provide a means to achieve this, and advanced tools such as boosted decision trees have in recent years become widely used. And while their use will no doubt increase as the LHC experiments mature, one should keep in mind that a simple analysis also has its advantages. As one studies the advanced multivariate techniques, however, their properties become more apparent and the community will surely find ways of using them so as to maximize the benefits without excessive risk.

Acknowledgements

I wish to convey my thanks to the students and organizers of the 65th SUSSP in St. Andrews for a highly stimulating environment. The friendly atmosphere and lively discussions created a truly enjoyable and productive school.

References

[1] C. Amsler et al. (Particle Data Group), Phys. Letts. **B667** (2008) 1; available at pdg.lbl.gov.

[2] G. D. Cowan, *Statistical Data Analysis*, Oxford University Press, 1998.

[3] L. Lyons, *Statistics for Nuclear and Particle Physicists*, Cambridge University Press, 1986.

[4] R.J. Barlow, *Statistics: A Guide to the Use of Statistical Methods in the Physical Sciences*, Wiley, 1989.

[5] F. James, *Statistical Methods in Experimental Physics*, 2nd ed., World Scientific, 2006.

[6] S. Brandt, *Data Analysis*, 3rd ed., Springer, 1999.

[7] A. N. Kolmogorov, *Grundbegriffe der Wahrscheinlichkeitsrechnung*, Springer, 1933; *Foundations of the Theory of Probability*, 2nd ed., Chelsea, 1956.

[8] R. E. Kass and L. Wasserman, The Selection of Prior Distributions by Formal Rules, J. Am. Stat. Assoc., Vol. 91, No. 435 1343–1370 (1996).

[9] C. P. Robert and G. Casella, *Monte Carlo Statistical Methods*, 2nd ed., Springer, 2004.

[10] J. S. Liu, *Monte Carlo Strategies in Scientific Computing*, Springer, 2001.

[11] R. M. Neal, Probabilistic Inference Using Markov Chain Monte Carlo Methods, Technical Report CRG-TR-93-1, Dept. of Computer Science, University of Toronto, available from **www.cs.toronto.edu/~radford/res-mcmc.html**.

[12] G. D'Agostini, Sceptical Combination of Experimental Results; General Considerations and Application to ε'/ε, hep-ex/9910036.

[13] V. Dose and W. von der Linden, Outlier Tolerant Parameter Estimation, in XVIII Workshop on Maximum Entropy and Bayesian Methods, Kluwer, 1999.

[14] R. E. Kass and A. E. Raftery, Bayes Factors, J. Am. Stat. Assoc., Vol. 90, No. 430 773–795 (1995).

[15] C. M. Bishop, *Pattern Recognition and Machine Learning*, Springer, 2006.

[16] T. Hastie, R. Tibshirani and J. Friedman, *The Elements of Statistical Learning*, 2nd ed., Springer, 2009.

[17] R. Duda, P. Hart and D. Stork, *Pattern Classification*, 2nd ed., Wiley, 2001.

[18] A. Webb, *Statistical Pattern Recognition*, 2nd ed., Wiley, 2002.

[19] Links to the Proceedings of the PHYSTAT conference series (Durham 2002, Stanford 2003, Oxford 2005, and Geneva 2007) can be found at **phystat.org**.

[20] A. Höcker et al., TMVA Users Guide, **physics/0703039** (2007); software available from **tmva.sf.net**.

[21] I. Narsky, StatPatternRecognition: A C++ Package for Statistical Analysis of High Energy Physics Data, **physics/0507143** (2005); software available from **sourceforge.net/projects/statpatrec**.

[22] B. Roe et al., Boosted Decision Trees as an Alternative to Artificial Neural Networks for Particle Identification, NIM A543 577–584 (2005); H. J. Yang, B. Roe and J. Zhu, Studies of Boosted Decision Trees for MiniBooNE Particle Identification, NIM A555 370–385 (2005).

[23] C. W. Gini, Variabilità e Mutabilità, Studi Economicogiuridici Università di Cagliari, III, 2a, Bologna, 1–156 (1912).

[24] Y. Freund and R. E. Schapire, A Decision-Theoretic Generalization of On-line Learning and an Application to Boosting, Journal of Computer and System Sciences, 55(1):119–139, August 1997.

[25] Y. Freund and R. E. Schapire, A Short Introduction to Boosting, Journal of Japanese Society for Artificial Intelligence, 14(5):771–780, September, 1999.

Grid Computing

Philippe Charpentier

CERN, Geneva, Switzerland

1 What Is the Grid?

1.1 Introduction

Each of the four LHC experiments typically expect to record data from their detectors at a rate of 200 to 300 events per second (1 MB in size), which for a typical running time of 5×10^6 seconds gives 1.5 PB of data every year.

Depending on the detector complexity, the reconstruction time varies between 1 and 20 seconds per event. This implies the usage of between 1000 and 5000 processing cores permanently during data taking. For re-processing of the data, it is expected this should take place in a much shorter time, and therefore require four to five times more processing power.

Simulation of the detector response to various types of physics events is essential for assessing efficiencies, identifying backgrounds and so on. For this purpose, the simulation time may take up to 15 minutes on a powerful processor, and therefore several 10,000s processing cores are needed permanently.

Finally, reconstructed or simulated data must be analysed by physics groups in order to produce final results. Usually this task requires much less time, but there are several hundreds of physicists who are willing to look at many millions of events. Therefore it is expected that the computing resources required for analysis will be of the same order as those for data reconstruction.

In an ideal world, these tasks could be fulfilled by a huge Computing Centre housing several hundreds of thousands of processing cores, tens of petabytes of data storage, all interconnected with ultrafast and reliable networks. There are, however, risks associated to such a paradigm (if the site fails, all activities are stopped) and political aspects (due to the centralisation of resources).

The alternative was proposed in the late 1990s in the form of the so-called Grid Computing, also known as distributed computing. The name Grid is an analogy to the power Grid that interconnects multiple provider resources to a large number of users. The Computing network plays the role of the electric network and the Grid Computing Centres play the role of the electric plants.

Grid Computing Centres house at the same time large processing capabilities, large storage capacities (possibly with archival facilities) and high bandwidth network connectivity.

Users should see the Grid as a huge Worldwide Computing Centre, and not have to care about where data are located nor where their tasks are being executed. Their unique concern is that they want to be able to get their results reliably available as soon as possible!

1.2　The MONARC Model

Just on the turn of the century, a hierarchical model was developed (MONARC [1]) that identifies several levels of computing facilities (called Tiers), each of which fulfills a specific role in the whole Grid.

The site at which data are being produced (CERN in the case of the LHC) plays a particular role, in that this is where data are initially produced and stored. It is also a federative location for the Collaborations that run the LHC detectors. This central role is that of the Tier0.

A small number of large computing centres, worldwide distributed, supplement the Tier0 for real data archival (to minimise destruction risks), providing a high level of resilience as well as a high quality of service. These sites (Tier1s) can be shared by several LHC experiments or dedicated to only one of them. They can also serve as computing facilities for other HEP experiments and even other domains of science.

Smaller computing centres, distributed worldwide, provide computing resources also to the LHC experiments. Their sizes in terms of computing power as well as in storage capacity are smaller, and the quality of service doesn't need to be as high as in Tier1s. The sites or federation of such sites are called Tier2s.

Finally the MONARC model recognises the existence of multiple clusters of computers at the level of an institute department (Tier3) and even at the level of individuals' desktops (Tier4). In this model, the data flow according to the well-established hierarchy of sites: reconstruction at Tier0 and Tier1, large-scale analysis at Tier2s and individual end-user analysis at Tier3 and Tier4.

1.3　The Grid Sites Hierarchy

The Grid Distributed Computing architecture uses the same terminology (Tiers) as MONARC. However only the first three levels are considered part of the Grid infrastructure officially.

The LHC Computing Grid is an implementation of the Grid paradigm over a variety of Grid infrastructures (EGEE/EGI in Europe and Asia, OSG in the USA, NorduGrid in the Nordic European countries).

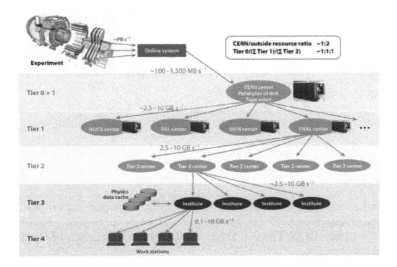

Figure 1. *A schematic of the* MONARC *model [2].*

In the LCG implementation, there is a less clear hierarchical dependency between sites, although the Tier1 in a specific country is usually providing support to the Tier2s in the same country. The usage of Tiers, however, is driven by each experiment's computing model and not rigidly by a MONARC model.

1.4 The Computing Models

Each of the LHC experiments has designed and described its computing model in the form of a technical design report in 2005 [3–7]. The computing models describe the functionality expected from the Grid and how the functions map onto various Tiers.

The main functions required by all experiments are the following:

- Real data collection and recording in real time

- Real data reconstruction in quasi-real time

- Real data re-reconstruction after months or years

- Monte Carlo simulation

- Data analysis by physicists, either on large data sets, or for final extraction of physics results

	Tier0	Tier1	Tier2
	Central data recording	Data mass storage	
ALICE			
ATLAS	First event reconstruction Calibration Analysis	Event re-reconstruction	AOD storage Simulation Analysis
CMS			
LHCb	Event reconstruction Analysis		Simulation

Figure 2. *Roles of the Grid Tiers in each LHC experiment's computing model.*

The mapping of these functions to Tiers differs from one experiment to the next, depending on specific characteristics of the experiment. Figure 2 shows the roles of the LCG Tiers in the computing model of each of the four major LHC experiments.

1.5 The Grid Abstraction

Broadly speaking the Grid can be considered as a large worldwide computing centre. Its functionality can be broken down as a set of services, that can be grouped in three large systems:

- Storage and data transfer (Data Management)

- Job submission (Workload Management)

- Global infrastructure (Computing Centres and user support, networks, security, etc.)

1.6 Grid Security

Providers of Grid services and Grid resources need to be able to trace who has done what and when (traceability), and on the collaboration side, one needs to control who is allowed to performed certain operations in order to protect data, for example.

To this purpose, each Grid user has to obtain his passport in the form of a "certificate," authorised by a certification authority (e.g., an authority in his country or at CERN). When using the Grid, the user creates a short-term copy of his/her certificate, called a proxy. This adds a level of security in case the certificate gets stolen: the proxy can only be used for a short time (hours or days).

In addition, the same user may have to perform different types of operation, depending on his/her current activity. Authorisation given as a simple user may not be sufficient to perform Data Management operations on behalf of the whole collaboration, for example. This is why when creating a proxy, one can specify which "role" this proxy will allow the user to take for performing an action. Roles can be granted to Grid users by the manager of the group of users, called a Virtual Organisation (VO).

1.7 Grid Infrastructures

Grid computing relies on underlying infrastructures, usually supported by countries or groups of countries. The Grid infrastructure can be used by various Grid initiatives that support certain areas of science. For the LHC experiments and in general high Energy Physics, the Worldwide LHC Computing Grid (WLCG) was created in 2001. The WLCG is mainly based on three Grid infrastructures:

- In Europe and Asia, the EGEE infrastructure (Enabling Grids for E-sciencE), now terminated and replaced by EGI (European Grid Initiative) that coordinates National Grid Initiatives (NGIs)

- In the USA, the OSG (Open Science Grid)

- In Scandinavia, NDGF, a.k.a. NorduGrid

2 Grid Data Management

2.1 Data Management Services

Data Management is responsible for the storage, transfer and access to data in the Distributed Computing environment. Data Management is based on a set of services:

- Storage Elements (SE): They are responsible for the actual storage of data, on disk servers and (for Tier0 and Tier1s) on mass-storage (a.k.a. tape storage). The SE provides all services for getting access to data, managing files and storage space.

- File Catalogs (FC): They contain all information on the files owned by a VO on the Grid. For each file, some metadata are recorded (size, creation data, checksum...) and in order to know where to find the file, a list of all instances of the file (a.k.a. replicas) is kept up to date.

- File transfer services: They are responsible for optimising the data transfers between two SEs

- Data access libraries: Those libraries allow transparent access from the applications to files stored in different technologies.

2.2   Storage Elements

Grid sites have the freedom to chose between a set of technologies for implementing their local storage solution. In WLCG, four basic technologies are in use:

- Castor is an integrated disk and mass-storage (a.k.a. tape) system developed at CERN. It is used currently at CERN, RAL and AGSC (Taiwan).

- dCache is a disk cache management that can use multiple mass storage systems (MSS). It is used at most other Tier1s and some Tier2s (in which case there is no MSS backend).

- DPM (Disk Pool Manager) is a disk only storage developed at CERN and used at most Tier2s.

- StoRM is a disk cache manager based on a parallel file system (GPFS), developed at CNAF. It can have a MSS backend. It is used at most Italian sites (with MSS at the Italian Tier1 CNAF).

2.3   Storage Resources Manager (SRM)

With these four brands of SEs, it was felt there was a need for an abstraction interface to them so that users deal with a single set of requests and commands. This is provided by the Storage Resource Manager (SRM). The SRM specifications were written down well before the WLCG started and got interested in it. However, in order to accommodate the use cases of the LHC experiments, these specifications were amended in 2006, in particular in order to support the partitioning of the storage space, each subspace having its characteristics in terms of access and latency policy. A pictorial view of the SRM standard interface is given in Fig. 3.

SRM "spaces" may have three access and latency policies:

- T1D0: Mass storage copy, but no permanent disk copy

- T1D1: Mass storage copy with a permanent disk copy

- T0D1: Permanent disk copy without mass storage

- This set of policies can be extended, for example, with T2 or D2 policies. Scratch files would correspond to T0D0 policy (volatile files).

SRM enables us to use access control lists for various types of action. The set of actions is an extension of those available on disk file systems (for example, for controlling the ability to recall back files from tape to disk).

An SRM file can be accessed through its so-called storage URL (SURL) which has a standard format:

```
srm://<endpoint>:<port>/<base-path>/<path>
```

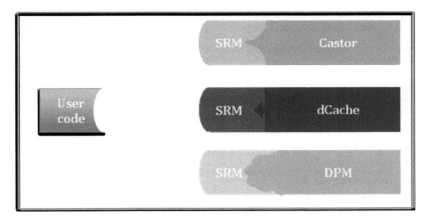

Figure 3. *Pictorial view of the standard SRM interface to storage.*

where `<base-path>` can depend on the VO, in order to partition the namespace of files. `<path>` is to the choice of the user, while `<endpoint>` is the IP name of the SRM server.

In order to access a file (e.g., for transferring it or for opening it with an application), the user fist has to get a transport URL (tURL) that depends on the protocol he wants to use for this operation:

- Data transfers: gsiftp protocol

- POSIX-like access: file, rfio, dcap, gsidcap, xrootd depending on the protocol supported by the implementation

SRM allows any file system directory and file manipulation (srm-mkdir, srm-rm, srm-ls, etc.).

To hide from users the internal complexity of the SRM interface, some higher-level tools are available (gfal, lcg_utils).

2.4   File Catalogs

Files on the Grid may have multiple copies (a.k.a. replicas) distributed on several SEs, each of which has its own SURL. However, users need a unique way to refer to a file, as they do not know which replica they will use (it depends, for example, where the job that needs it, will run).

For this purpose files are given a Logical File Name (LFN) that is human readable and allows implementation of a hierarchical namespace (like on a normal file system). For internal convenience, files are also assigned a unique identifier (GUID) that is immutable, and used in particular as a file identifier when internally cross-referencing files. The GUID is unique by construction, and assigned at the creation of the file.

```
[lxplus227] ~ > lfc-ls -l /grid/lhcb/user/p/phicharp/TestDVAssociators.root
-rwxw-sr-x   1 19503   2703   430 Feb 16 16:36 /grid/lhcb/user/p/phicharp/TestDVAssociators.root

[lxplus227] ~ > lcg-lr lfn:/grid/lhcb/user/p/phicharp/TestDVAssociators.root
srm://srm-lhcb.cern.ch/castor/cern.ch/grid/lhcb/user/p/phicharp/TestDVAssociators.root

[lxplus227] ~ > lcg-gt srm://srm-lhcb.cern.ch/castor/cern.ch/grid/lhcb/user/p/phicharp/
TestDVAssociators.root rfio
rfio://castorlhcb.cern.ch:9002/?svcClass=lhcbuser&castorVersion=2&path=/castor/cern.ch/grid/
lhcb/user/p/phicharp/TestDVAssociators.root

[lxplus227] ~ > lcg-gt srm://srm-lhcb.cern.ch/castor/cern.ch/grid/lhcb/user/p/phicharp/
TestDVAssociators.root gsiftp
gsiftp://lxfsrj0501.cern.ch:2811://castor/cern.ch/grid/lhcb/user/p/phicharp/TestDVAssociators.root

[lxplus227] ~ > lcg-cp srm://srm-lhcb.cern.ch/castor/cern.ch/grid/lhcb/user/p/phicharp/
TestDVAssociators.root file:test.root
```

Figure 4. *Examples of lcg_utils commands.*

File Catalogues enable the mapping between LFN, GUID and a list of replicas. They also contain some metadata such as the file size, the creation date, the file checksum, etc.

In the LHC experiments, three types of catalogs are used:

- The LCG File Catalog (LFC): Used by ATLAS and LHCb

- The AliEn File Catalog: Developed and used in ALICE

- The trivial File Catalog: Used by CMS

2.5   High-Level Tools

These integrate the File Catalogs, the SRM and also possibly the transfer protocol. A set of basic libraries is provided by gfal (grid file access layer), which are then used in command tools (grouped under the name lcg_utils). Figure 4 shows a few examples of lcg_utils commands and their output.

When transferring large datasets, one needs to control the data throughput as well as the network bandwidth in order to optimise the data transfer rate. A File Transfer Service (FTS) has been developed in order to achieve such transfers. It is primarily used for transfer between large sites (Tier1s) over a dedicated optical network (LHCOPN). FTS manages transfers over channels (from A to B). It takes care of transfer retries and checking. The network bandwidth can be controlled by site managers for sharing the network between different VOs.

2.6   VO Data Management Frameworks

In order to implement their computing models, the VOs have to elaborate from the basic functionality provided by the middleware. In particular users need to be able to find out which files they should analyse, corresponding to a certain number of VO-specific criteria: data-taking conditions, processing level, type of data (RAW, ESD, AOD...). No generic solution was ever designed to achieve this and therefore each VO has developed its own including AliEn for ALICE, DDM/DQ2 for ATLAS and DIRAC for LHCb.

3 Workload Management

3.1 The Great Picture

Ideally what we would like to build for running jobs on the Grid is a fully distributed huge Computing Centre! The sequence of operations is rather straightforward:

- Prepare and test locally your software

- Distribute it to all the Grid sites

- Submit jobs to the Grid

- Monitor their progress

- When completed, retrieve their results

It is irrelevant to the user where the job ran, but just whether it was successful!

We shall see in the following that the real life scenario is slightly more complicated!

3.2 Workload Management Components

On each and every site, jobs are scheduled and monitored by a local batch system. There are various brands of batch system (LSF, PBS, BQS...) and the choice is up to the sites. Batch systems allow applying shares between sets of users and provide monitoring and accounting of resources. They are responsible for the most efficient use of the local resources.

In order to abstract this large variety of batch systems, sites are providing a set of entry points called Computing Elements (CEs). One or more CEs are available at each Grid site, and their number depends on the amount of resources the site has, since CEs are only able to handle a limited number of jobs at once. CEs publish the state of their underlying resources (number of slots, number of running and waiting jobs...).

Users do not want to know about which particular batch system is used at a site or where their jobs are going to run. That is why they submit their jobs to a central service that then takes care of dispatching them to a specific CE onto the Grid, based on certain criteria like availability of input data, platform requirements, CPU resources and free resources availability.

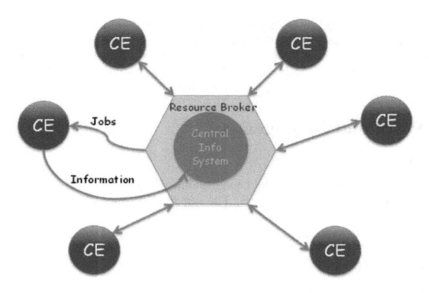

Figure 5. *Central resource brokering and information system.*

The parameters that are used for matching resources with a job that was centrally submitted have to be expressed in a uniform language.

3.3   Resource Brokering

Ideally the central job dispatching service should be a kind of "Big Brother" that knows everything about all sites on the Grid: their available resources (qualitative and quantitative) and which files are accessible there.

This knowledge should be permanently updated in order to reflect the exact state of the Grid and therefore allow the best decision for scheduling jobs. For this, Resource Brokers (RBs) need a central information system that is permanently updated by the resources themselves (see Fig. 5), and a file catalogue for locating file replicas.

There are, however, practical limitations with central resource brokering: information cannot be updated instantaneously by the CEs, therefore the state of the system as "seen" by the RB may have been altered by the time it makes its decision. Furthermore a single RB cannot handle and keep track of the very large number of jobs that need to run on the Grid (several hundred thousand). In order to cope with this scaling issues, one needs to use multiple instances of RBs, which even worsens the knowledge of the state of the system, as they act independently and may decide for example to all submit a bunch of jobs to the same site, on the basis that it had enough free slots, but only for one instance.

The choice of where to send the job is also not simple: What is the criterion that should be used before submitting a job? Ideally, one would like a job to go to the place that will minimise its return time. This is however a very difficult parameter to estimate.

Therefore real life use of RBs is somewhat more pragmatic. Each VO uses several RB instances, each of which may handle up to 10,000 jobs. The information system (a.k.a. BDII) is updated at regular intervals (typically 2–3 minutes) and therefore this refresh rate sets a limitation on the quality of the decision for brokering. A retry mechanism may be used in the BK in case a job was erroneously sent to a CE that was not in a position to accept it. Therefore the interaction between the RB and the CE is quite complex.

In addition to resources being available, one should ensure that the services made available on Worker Nodes (WNs) at the site are adequate for running the job: software availability, operating system and CPU constraints.

When the job finally arrives on a WN, the Grid user identity (a.k.a. Distinguished Name, DN) is translated into a regular local UNIX identifier that is chosen out of a local pool of user accounts. This mapping is defined by the site and may vary from site to site, although most sites create a one-to-one mapping between DNs and identifiers, to be able to trace easily the owners of jobs.

3.4 An Alternative to Resource Brokering

In view of all the difficulties described earlier with central resource brokering, the LHC VOs have invented an alternative, called "pilot jobs." The basic idea is to send to the WNs a small "job" whose task will be to reserve a slot in the batch system. Once this Pilot Job (PJ) is running, it can discover exactly the resources available on the WN. It can then ask centrally for some task to be performed that matches those capabilities.

In this paradigm, the actual jobs that know which task to perform (a.k.a. payload jobs) are sitting in a central task queue until a pilot pulls them onto a WN. Therefore it implements a "pull" paradigm as opposed to a "push" paradigm implemented by the RB. The late binding of the job to the WN allows applying specific policies on the jobs that are in the central task queue, such as applying priorities. A PJ can also, after it has executed a payload job, ask for another one if its remaining CPU resources allow. This decreases the load on the batch system as well as on the RBs and CEs that are still used for submitting the PJ to the sites.

The PJ mechanism was pioneered in 2003 by ALICE and LHCb and is now widely used by all LHC experiments.

An example of such a framework is given in Fig. 6, which shows a block diagram of the DIRAC framework developed and used by LHCb.

Pilot Jobs enable us to easily federate different flavours of Grids, as it is sufficient to submit pilot jobs to the system, without the need to define complex requirements that may depend on the Grid flavour. Usually the VOs

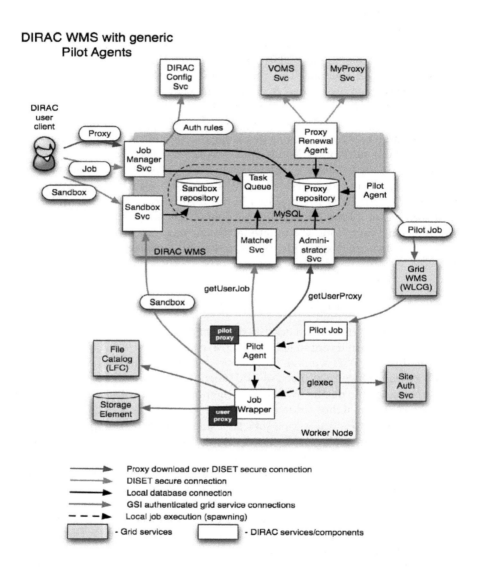

Figure 6. *Block diagram of the DIRAC Workload Management System (WMS).*

have a "pilot factory" that submits pilots, based on the set of jobs that are in their central task queue.

Similar systems are AliEn (ALICE), PANDA (ATLAS) and CMSGlideins (CMS).

4 The Grid Experience

The basic infrastructure for the WLCG Grid is there: EGEE (now replaced by EGI) in Europe, OSG in the USA and NorduGrid in the European Nordic countries.

The basic middleware exists although with limited functionality and performance compared to the initial hopes:

- Storage-ware (mainly developed in the HEP community): Castor, dCache, DPM, StoRM, SRM

- Data Management middleware: gfal, lcg_utils, FTS

- Workload Management middleware: glite-WMS, LCG-CE now being replaced by the CREAM-CE

Very good coordination structures exist (EGEE/EGI, OSG), coordinated for the LHC experiments by the WLCG. The relationships between all these components are, however, quite complex and make changes very difficult and heavy. It is quite difficult to adapt to the ever-evolving possibilities of computing (cloud computing, virtual machines, multi-core CPUs, etc.).

Large VOs such as the LHC experiments had to develop complex frameworks in order to integrate the basic middleware. Pilot Job paradigm is part of these frameworks for coping with some limitations of the basic middleware.

In addition, based on their Grid frameworks, the LHC experiments have developed equally complex Production systems and user analysis frameworks.

4.1 Production Systems

Traditionally the name "production" is used for designating the computing activities within an HEP collaboration that are centrally managed and dedicated to the whole or a large set of users in the collaboration. Usually these activities are complex, CPU-intensive and consist of applying the same transformation to a large number of input datasets. Some examples are: real data reconstruction, large Monte Carlo detector simulations and generic analysis on behalf of large physics working groups.

Production systems allow defining the task (i.e., its workflow and all the associated parameters), defining the input datasets, generating the jobs, submitting and monitoring them and finally collecting and publishing the results.

Figure 7. *Screenshot of an LHCb production request (using a web browser).*

Each VO has its own dedicated and tailored production system(s). We shall again use as an example the LHCb production system that is one of the components of DIRAC.

When a group of users needs a production activity to be performed, they first define the needs (transformation, datasets). This production request is then validated by application experts as well as by a body in the collaboration that evaluates the real need of this production. After these validation steps, the production is launched, jobs are generated and are monitored until the full request was completed.

Figure 7 shows a screenshot of the LHCb production request web page, on which one can follow in the rightmost column the progress of the different requests while jobs are running and completing, publishing their output datasets.

Production activities are well organised and agreed by the whole collaboration or a delegating body. They enable strict application of all policies defined by the collaboration's computing model (such as priorities, data placement, etc.). The Production teams also take care of archiving or deleting obsolete datasets.

4.2 User Analysis

It is much more difficult to control the activity from individual physicists in the collaborations. Therefore this activity is much less organised than the production activities.

Analysis applications are not so well tested as production applications, and in particular no thorough checks can be made by a team of experts before jobs are submitted. Ideally users would like a rapid development cycle, running on limited datasets until they are satisfied with their application, try it further

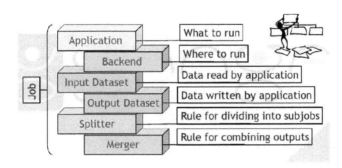

Figure 8. *Ganga functional chart.*

with a slightly larger though not complete data sample and finally run it on the full data sample when they are satisfied with the output. The results of an analysis may also be of different kinds: histograms, Ntuples, reduced datasets. Therefore the output may vary from a few Mbytes to several GBytes.

When they need to run on large data samples, physicists may have a big advantage when using the Grid, as it provides a unique set of aggregated resources. The frameworks that have been developed by the collaborations allow them to benefit from many good features of the Grid like easy file replication and registration (catalogues). However, for testing and running on smaller samples, it is still advantageous to run the applications locally. Local analysis is also very useful for the so-called "end analysis," for example producing histograms or diagrams for physics publications.

ATLAS and LHCb have jointly developed a tool for user analysis called "ganga" that allows users to prepare their applications, test them locally, change datasets easily and eventually submit their tasks to the Grid on large data samples in a seamless manner. Ganga is based on a set of functional building blocks that can easily be specialised by means of plugin techniques. Therefore ganga can easily be used by various communities, even outside HEP. Figure 8 shows a functional diagram of the various components of ganga.

Other experiments have similar user interfaces to the Grid (AliEn for AL-ICE and CRAB for CMS). They are mainly intended at hiding the complexity of the Grid usage to users, allowing them to concentrate on their application.

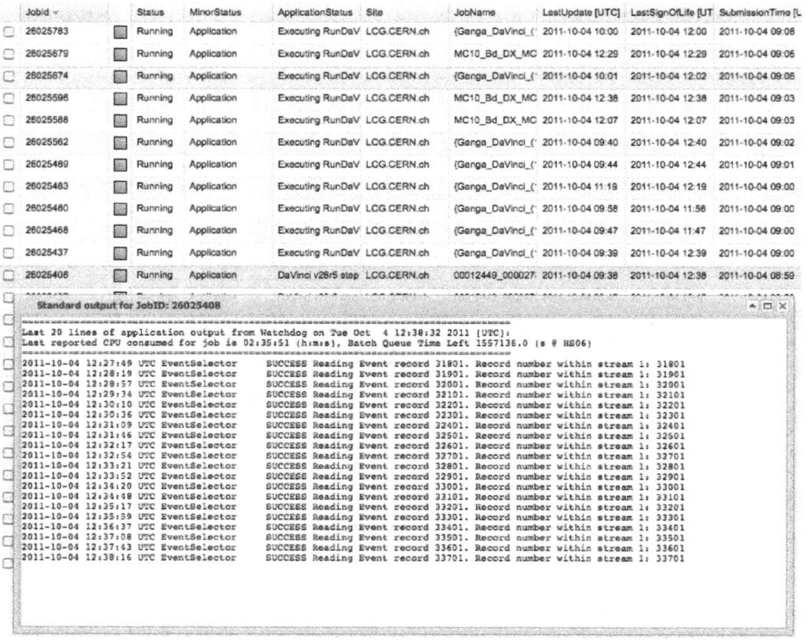

Figure 9. *Example of detailed job monitoring web page (DIRAC system).*

4.3 Monitoring

Owing to the high complexity of the Grid, it is very important to perform a careful and permanent monitoring of its functioning as well as an accounting of the resources consumed by users. Also at the level of the experiment-specific framework, it is very important to be able to follow the progress of jobs or data transfers, access information in case of failure and take curative actions. Figure 9 shows an example of detailed job monitoring.

In order to reach the current level of reliability, the Grid infrastructures and the experiments have undergone now for years a series of "challenges" that were trying to reproduce as closely as possible the load expected when the LHC would be running. Some of these challenges were topical (workload management or data management/transfers), others were more generic, involving all levels of computing and mimicking the experiments' Computing Models.

Despite the Grid being now run as a production service, sites and experiments are still on a learning curve. Not all operational issues are yet understood, nor are procedures clearly defined when things go wrong, for example when a service is out of operation for a very long time. Many metrics have been put in place for assessing site performance. In particular, specific tests are being run by the central operations team of the Grid and by each

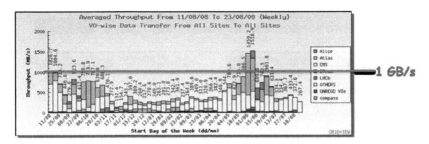

Figure 10. *Overall transfer throughput in the WLCG Grid from August 2008 to August 2009.*

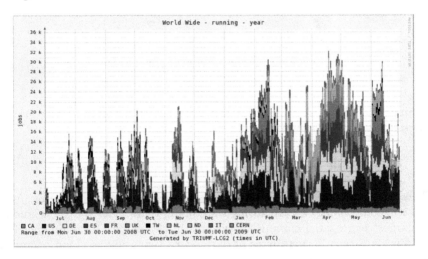

Figure 11. *Number of concurrently running jobs for ATLAS over a period of one year (June 2008 to June 2009).*

experiment aiming at identifying failures before they damage the quality of service. For example, when an SE misbehaves, it is important that jobs do not try and access datasets on that SE to the risk of deteriorating even more the quality of service. Figures 10 through 12 show monitoring plots extracted during Summer 2009.

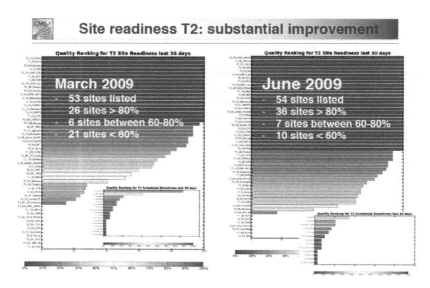

Figure 12. *Tier2 site readiness for CMS showing the great improvement between March and June 2009.*

5 Outlook

There cannot be a conclusion to such an overview of Grid Computing, as the field will continuously evolve during the coming years. The WLCG is amongst the largest aggregated computing systems in the world. Its main specificity is being data-centric (petabytes of data every year), unlike cloud computing which is more CPU-centric.

Running Grid services distributed all over the world requires a huge effort, from the funding agencies providing the hardware and manpower infrastructure, from the coordinating bodies (EGEE/EGI, OSG, WLCG), from the user communities. It requires permanent flexibility in order to accommodate changes. Although now with the first data coming from the LHC it is important to concentrate on the stability of the services, it is also primordial to think of an evolution of the computing (usage of Virtual machines, emergence of multi-core and many-core CPUs) that will completely change again the picture of massive computing in the coming decades.

References

[1] M. Aderholz *et al.*, "The MONARC Final report," CERN/LCB/2000-001, March 2000.

[2] I. Bird, "Computing for the Large Hadron Collider," Ann. Rev. Nucl. Part. Sci. **61** (2011) 99.

[3] "ALICE Computing Technical Design Report," ALICE TDR 012, June 2005.

[4] "ATLAS Computing Technical Design Report," ATLAS TDR–017, CERN-LHCC-2005-022, June 2005.

[5] "CMS Computing Technical Design Report," CMS TDR 7, CERN-LHCC-2005-023, June 2005.

[6] "LHCb Computing Technical Design Report,"LHCb-TDR-11, CERN-LHCC-2005-019, May 2005.

[7] "LHC Computing Grid Technical Design Report," LCG-TDR-001, CERN-LHCC-2005-024, June 2005.

Index